RESIDENTIAL CHANGE
AND DEMOGRAPHIC CHALLENGE

Residential Change and Demographic Challenge
The Inner City of East Central Europe in the 21st Century

Edited by

ANNEGRET HAASE
ANNETT STEINFÜHRER
SIGRUN KABISCH
KATRIN GROSSMANN
RAY HALL

Routledge
Taylor & Francis Group

LONDON AND NEW YORK

First published 2011 by Ashgate Publishing

2 Park Square, Milton Park, Abingdon, Oxon OX14 4RN

711 Third Avenue, New York, NY 10017, USA

Routledge is an imprint of the Taylor & Francis Group, an informa business

First issued in paperback 2016

British Library Cataloguing in Publication Data
Residential change and demographic challenge : the inner
    city of East Central Europe in the 21st century.
    1. Europe, Eastern--Population. 2. Europe, Eastern--
    Population--Case studies. 3. Inner cities--Europe,
    Eastern. 4. City dwellers--Europe, Eastern. 5. City
    dwellers--Housing--Europe, Eastern. 6. Europe, Eastern--
    Social conditions--21st century.
    I. Haase, A.
    304.6'2'0943-dc22

Library of Congress Cataloging-in-Publication Data
Residential change and demographic challenge : the inner city of East Central Europe in the
21st century / by Annegret Haase ... [et al.].
    p. cm.
    Includes bibliographical references and index.
    ISBN 978-0-7546-7934-9 (hbk.)
    1. Cities and towns--Europe, Eastern. 2. Housing--Europe, Eastern. 3. Europe,
Eastern--Population. I. Haase, A.
    HT169.E85R47 2010
    307.3'3616094371--dc22

                                                                      2010036462

ISBN 978-0-7546-7934-9 (hbk)
ISBN 978-1-138-25443-5 (pbk)

Contents

PART I: CONCEPTUAL BACKGROUND, CONTEXT CONDITIONS AND METHODOLOGICAL CONSIDERATIONS

PART III: SUMMARY, CONCLUSION AND OUTLOOK

List of Figures

List of Tables

List of Contributors

Adam Bierzyński, researcher and doctoral fellow at Institute of Geography and Spatial Organisation of the Polish Academy of Sciences, Warsaw

Maja Grabkowska, doctoral fellow at University of Gdańsk, Department of Economic Geography

Katrin Grossmann, PhD, researcher at Helmholtz Centre for Environmental Research – UFZ, Leipzig

Annegret Haase, PhD, researcher at Helmholtz Centre for Environmental Research – UFZ, Leipzig

Ray Hall, PhD, senior research fellow at Queen Mary, University of London, Department of Geography

Sigrun Kabisch, PhD, senior researcher at Helmholtz Centre for Environmental Research – UFZ, Leipzig, Head of the Department of Urban and Environmental Sociology

Petr Klusáček, PhD, researcher at Institute of Geonics, Academy of Sciences of the Czech Republic v.v.i., Brno

Andreas Maas, staff member at the Federal Labour Office, Statistical Service Southeast, Nürnberg

Jana Mair, doctoral fellow at Institute of Ethnology, Academy of Sciences of the Czech Republic v.v.i., Prague

Stanislav Martinát, PhD fellow and researcher at Institute of Geonics, Academy of Sciences of the Czech Republic v.v.i., Ostrava

Philip E. Ogden, PhD, Professor of Geography and Senior Vice-Principal of Queen Mary, University of London

Jana Pospíšilová, PhD, senior researcher at Institute of Ethnology, Academy of Sciences of the Czech Republic v.v.i., head of branch Brno

Iwona Sagan, PhD, professor at University of Gdańsk and Chair of the Department of Economic Geography

Annett Steinführer, PhD, researcher at the Institute of Rural Studies, Johann Heinrich von Thünen Institute (vTI), Braunschweig

Zdeněk Uherek, PhD, Head of the Institute of Ethnology, Academy of Sciences of the Czech Republic v.v.i., Prague

Antonin Vaishar, PhD, senior researcher at Institute of Geonics, Academy of Sciences of the Czech Republic v.v.i., Brno and senior lecturer at Mendel University of Agriculture and Forestry, Brno

Grzegorz Węcławowicz, PhD and Professor of Geography, Institute of Geography and Spatial Organisation of the Polish Academy of Sciences, Warsaw, Head of the Department of Urban and Population Geography

Jana Zapletalová, PhD, senior researcher at Institute of Geonics, Academy of Sciences of the Czech Republic v.v.i., Brno

Foreword
Living in the European City: Demographic Change and Residential Patterns

Philip E. Ogden
Queen Mary, University of London

The way in which people live and work in the European city is fast evolving. The last two decades in particular have seen wide-ranging and fundamental changes in the way that European cities function and this volume aims to show the extent to which common trends in some key respects underlie the diversity of size, economy and social structure. Perhaps the single most important message in the chapters presented here, which focus on demographic change and residential patterns, is that there is always a variety of processes, sometimes apparently contradictory, at work at a single moment. The contributions aim to untangle some of these dimensions, not least in the manner in which they have manifested themselves in different ways and to different timescales in the cities of East Central Europe and in their inner cities in particular. In focusing on post-socialist cities, the volume draws of course also on the experience of Western European cities and represents almost a decade of collaboration between researchers in Western and Eastern Europe on these themes, authors being drawn from Germany, Poland, the Czech Republic and the UK. In this brief foreword I should like to draw attention to four main themes.

First, fundamental to the ideas that lie behind the volume is the recent rapid evolution of demographic behaviour which has profound effects on the way people live, on their experience of family, friends and residence in the city. We have elsewhere recently described (Buzar et al. 2005: 413–36) these trends as the 'quiet demography of urban transformation' echoing to an extent the ideas underlying the *Secret Life of Cities* by Helen Jarvis and her colleagues (Jarvis et al. 2001). Focusing on demographic changes is not without controversy – particularly if it is construed as drawing attention away from issues of class and economic disadvantage – but the argument here is that analysis of post-socialist cities has thus far underplayed the effects of fundamental shifts in family formation. The 'second demographic transition' draws attention to the effects of declining birth rates on family size, to the postponement of marriage and the increase in cohabitation, a rise of divorce and couple separation and the consequent increase in single-parent families, and to increased longevity. One of the key aggregate effects of these changes is to increase the number of households (even when overall population might be in decline) and to increase the diversity of household types (for example, people living alone both at younger and older ages). There is also an increase in the number of transitions between different household types made by the individual during the life course. These changes have profound effects in turn on housing demand and on residential mobility and these relationships are a key focus of enquiry in the chapters in this volume.

Second, the volume focuses explicitly on second-order rather than capital cities and on their inner-city areas. The focus on the inner cities arises from the question: are there signs of reurbanization in the middle-ranking cities of East Central Europe in the ways that have been posited in their western counterparts? This is particularly interesting in the case-study countries of the Czech Republic and Poland because the common perception of these cities is of inner decline rather than growth. The inner cities are of particular interest because of the extent to which the renovation of the historic cores has the potential to attract some of the 'new' households referred to above. The extensive literature on gentrification has already drawn attention to the way in which class-based residential choices can transform neighbourhoods and the question posed in this volume is whether wider 'reurbanization' processes are also at work. A concentration on the inner city should not of course distract attention from the fate of the suburbs, as the pattern of demographic behaviour and its effects on household formation is also bringing about profound changes in traditional conceptions of suburban populations. Certainly, the extent to which recent trends might point to an increased attraction of the 'urbanity' represented by the heart of the city interests a number of the authors in this volume.

Third, the volume of course focuses on the particular circumstances of post-socialist cities. As I write this foreword, the British press is full of stories marking the 20th anniversary of the fall of the Berlin Wall and these chapters contribute to our understanding of the impact of political change on cities and their populations. Twenty years is a brief period in the life of the cities studied, all of which have long histories; yet the impact of changing legal, institutional and political systems is already evident and, by any standards, these two decades have been remarkable. The study of household demography is much influenced by complex changes in social attitudes, economic possibilities and the regulatory framework of the law. In all these respects, East Central Europe over the last two decades is a fascinating laboratory. A key part of the debate in the volume is the extent to which there is something distinctive about the post-socialist urban transition or whether such an approach risks masking wider processes common to Europe as a whole. The empirical material here, derived from work in Poland and the Czech Republic, points strongly to a shared European experience. Although they are still, and will be further, influenced by the legacies of state socialism and post-socialism, the case studies in this volume underline the similarities of trends in behaviour to those documented more generally in Europe over recent decades. The cities discussed here are not reducible to their 'post-socialist' fate, but rather patterns of demographic and social change point to a common European heritage and to an increasing convergence.

Finally, the volume draws attention to the need to understand not just the statistical data describing changes in demography and residence but also the impact on the way people experience the city, the way they live and interact with others within and beyond their immediate household. Alongside increased individual choice is a 'pluralization of life styles, both over the life course of the individual and in the cross-sectional aggregate stock of living arrangements at one moment in time' (Kuijsten 1996: 140).

With the decline in traditional family structures comes also an interest in wider social networks of friends and also the extent to which some groups may have a renewed interest in living in the central city. The focus in these chapters on residential change and the role of households recalls an earlier appeal for more research in a field whose relative neglect is an unintended result of the way empirical research has been drawn to the 'public' and dominant sphere of the formal economy and of state-based politics, and turned away from the 'private' and subordinate sphere of household and residential life (Mullins 1995). Though these words were written in respect of Australian cities in the 1990s, they seem to be also very apt as a preface to some of the concerns in the chapters in the present volume.

Preface and Acknowledgements

This volume is the outcome of long-term research on the development of cities in Poland and the Czech Republic. The central question of the research is how demographic change interrelates with urban processes and the conditions of transformation and persistency in the course of post-socialist transition. The impetus for this work was the conviction that processes like the downsizing of households, increasing diversity of living arrangements, ageing and the migration of heterogeneous social groups impact decisively on the shape and use of urban and housing space as well as on housing mobility. Urban challenges in Poland and the Czech Republic have generally been analyzed in the context of widely-known concepts such as gentrification, segregation and polarization. Having been uneasy with the simple application of these terms and their underlying conceptualizations, we deliberately decided to step further back and to focus on the analytical category of residential change in order to better understand what happens in the cities, how urban and demographic transformations are intertwined, and how they shape social, residential and spatial processes. Putting residential change on the agenda also enabled us to look at both the residents (in terms of residential population) and their place of residence (in terms of home and residential environment). Based on this systematization the discussion considers the extent to which there is gentrification, segregation or polarization in Polish and Czech large cities.

This volume deals with Polish and Czech large cities in two specific ways. First, it is concerned with second-order cities, that is non-capital cities, cities which in comparison with capital cities are often neglected by international research. Second, it examines the inner parts of these cities with respect to their residential function. Throughout, the focus is on less 'spectacular' processes and neighbourhoods; in other words, on areas that are unlikely to become centres of luxury renovation, large-scale gentrification or social exclusion within the near future.

The volume supports the idea of urban research as a multidisciplinary endeavour. Urban geographers, sociologists and anthropologists have worked together each bringing their particular approaches, methods and understanding of the interlinkages of urban and demographic issues. The application of the concept of residential change formed the unifying approach for the different disciplinary ways of thinking, theories and methods. The consortium included not only different disciplines but also researchers from four countries (Poland, the Czech Republic, Germany and the UK) and indeed, from different generations, which is especially important with respect to the examination of post-socialist transition and long-term legacies of state socialism. We are confident that the multicultural character of the consortium has enriched the quality of the research presented here.

Acknowledgements

The editors wish to thank all contributing authors for their work on this volume, Jana Kaiser for proof-reading, Anne Wessner, Daniela Siedschlag, Andreas Schneider and Anna Kunath for technical support and layout preparation, all interview partners and local stakeholders in Łódź, Gdańsk, Brno and Ostrava as well as all our colleagues and friends who uncomplainingly supported our work and enabled it to be brought to a satisfactory conclusion. The volume draws on the ideas and findings of the international research project conDENSE (*Social and spatial consequences of demographic change for East Central European cities*, http://www.condense-project.org/) which ran from 2006 to 2009 and was funded by the German Volkswagen foundation, to whom the editors also want to express their grateful thanks.

We would especially like to thank Sabine Linke who contributed greatly to the production of this volume but sadly died before its publication. We dedicate the book to her.

Annegret Haase,
Annett Steinführer,
Sigrun Kabisch,
Katrin Grossmann,
Ray Hall
Leipzig, Braunschweig and London

PART I
Conceptual Background, Context Conditions and Methodological Considerations

Chapter 1

Introduction: Idea, Premises and Background of this Volume

Annegret Haase, Annett Steinführer, Sigrun Kabisch,
Katrin Grossmann, Ray Hall

Idea and Background of the Research

During the last 20 years, there has been increasing interest in demographic change by both academic researchers and public policy makers, not only within Europe but also worldwide. Issues of concern range from population growth to decline, below-replacement fertility and future negative demographic growth scenarios. Within this context, the countries of East Central Europe[1] developed as centres of demographic change in the 1990s, becoming demographically much more like Western and Southern Europe: birth rates fell to very low levels, and one-child families became much more prevalent. Today, birth rates remain very low, contributing to the rapid ageing of East Central Europe's population (Vojtěchovská 2000, Kotowska 2001). In addition, migration patterns have changed dramatically since 1989. The exodus from some parts of East Central Europe, in particular from eastern Germany, during the first half of the 1990s has developed into a complex pattern of movements, ranging from labour-related seasonal out-migration to newly emerging re-migration (Raymer and Willekens 2008). Whereas some metropolitan regions and their suburban fringes are booming and growing, other areas – especially old industrialized urban regions and rural peripheries – are locked in a trap of economic and population decline.

Since 1989, cities in East Central Europe have been the foci of demographic and social changes, and the consequences of these changes are most obvious there. It is in the cities of East Central Europe that childlessness and one-person households

---

1   A distinct and for all research efforts standardized delimitation of East Central Europe does not exist. In general, a broader understanding can be distinguished from a narrow one that is oriented partly towards historical and partly towards current affiliations. In the narrow sense, East Central Europe includes Poland, the Czech Republic, Slovakia and Hungary. In a broader sense, and depending on the research questions, bordering countries such as Slovenia, the Baltic States, the Ukraine, Belarus, Romania, Croatia and Serbia could be included. In the following – because of the numerous historical parallels to the development in Western Europe and the recent integration in the European Union – the narrow interpretation of East Central Europe is used.

are most wide-spread. It is here that in- and out-migration play a key role in the changing socio-spatial patterns, for example, in the form of rejuvenation by the influx of 'new' households or ageing due to the outflow of younger age groups to the suburbs. And – apart from some rural areas – it is mainly large cities where rapidly increasing numbers of older people are concentrated (Kurek 2008).

There is a great deal of literature dealing with demographic change in East Central Europe (for example, Rychtaříková 1999, 2000, Kotowska 1999b, 2001, Kučera et al. 2000, Rabušic 2001, Sobotka et al. 2003, Surkyn and Lesthaeghe 2004, Okólski 2006), but a closer examination of the urban dimension of demographic change reveals three large gaps in current research. *First,* there is little connection between debates on demographic change and urban development: demographic change is well documented on the national scale but systematic analyses of its consequences for cities and towns are rare. *Second,* demographic research usually does not consider the household level that is decisive for urban environments. It is households and not individuals who are actors on the urban 'stage' because, for example, they can decide to relocate or to remain at the current place of residence. *Finally,* urban and demographic change in East Central Europe is, almost always, conceptualized with reference to the post-socialist condition. Comparisons with Western European developments have rarely been undertaken. The countries and cities of Eastern Europe are generally subsumed under the heading 'post-socialist' or 'in-transition countries', referring both to the background of their current development and also to their perceived specifics. This usually leads to a statement of their non-conformity with 'general' or 'Western' European patterns. Consequently, other explanatory factors are not considered and East Central Europe is excluded from consideration within the context of overarching European trends (Steinführer and Haase 2007).

This volume aims to overcome these deficits by bringing together urban and demographic strands of research and by focusing on households as important urban actors. It goes beyond the approach of explaining all changes in East Central Europe as the result of the post-socialist transition. Starting from these premises, we focus on *residential change* in East Central European inner cities. We conceptualize residential change as the process that occurs through the interplay of people, places, local contexts and society in terms of housing and daily life. Within this process, residential patterns are in constant flux. We take residential patterns to mean small-scale clusters of interrelated social and physical characteristics. Residential patterns are thus the result of the dimensions of residential change – the residents and their residences – as well as intermediary structures. The residents comprise both individuals and households (that is the residing population) of an urban area and they exhibit specific socio-demographic and socio-economic characteristics, residential orientations (in terms of preferences and perceptions), as well as residential behaviours (such as mobility or adaptation in place). Residences comprise physical aspects, such as the building and housing stock, infrastructure, amenities and environmental conditions, as well as social aspects, which include personal networks in the neighbourhoods. Residents and residences are brought together – or kept apart – by intermediary structures, such as local housing markets or existing patterns of residential segregation, as well as by broader factors, such as urban governance and local labour markets. Residential change evolves through the mediated interactions between residents and their places of residence. It is embedded into the overall societal framework (see Chapter 2 for more details).

Residential change can be induced by general societal shifts, by the cumulative effects of individual agency (individuals who create new structures by using the given space for action), as well as by change in the local context, for instance the city development strategy of local authorities or the influence of private developers. Residential change can also occur in a variety of forms. Processes such as gentrification, reurbanization, studentification or suburbanization would be specific forms of residential change. Our conceptualization relates closely to the interconnection of residential change with demographic change as an overarching societal process (Chapters 2 and 3 provide more details).

Starting from the assumption that demographic processes within a specific socio-economic context always have a spatial dimension, this publication discusses demographic changes and their impact on residential structures as a challenge for cities, their housing markets, neighbourhoods, actors and fortunes by taking the cities of Łódź and Gdańsk in Poland and Brno and Ostrava in the Czech Republic as examples. They all represent cities that rank second in the respective national urban hierarchy, irrespective of their population size (which ranges from 300–750,000; see Chapter 6 for more details). We focus on the *inner parts of these cities,* because residential change there is under-researched. Drawing on knowledge from western research of the last decades (for example, Helbrecht 1996, Bromley et al. 2005, Buzar et al. 2007), one might expect that the interdependencies between residential and demographic change would be particularly prominent in these inner-city areas. Moreover, it is presumed that East Central European inner cities have undergone fundamental changes after the decades of decline and dilapidation they experienced during the period of state socialism. Although our focus is on the inner city, the volume also discusses the evidence for simultaneous processes of population growth and losses, as well as of changing residential and household structures on different scales (ranging from the inner city to single neighbourhoods, buildings or even flats) over time. The city as a whole serves as a reference framework. The analyses mainly refer to the two decades between 1988 and 2008. Thus, the volume provides a framework for linking theoretical and empirical urban and demographic research and embedding the findings into a broader European framework.

Embedding and Contextualizing the Research

The analysis of inner-city residential change in East Central European cities relates to a number of research contexts and strands of discussion (Figure 1.1).

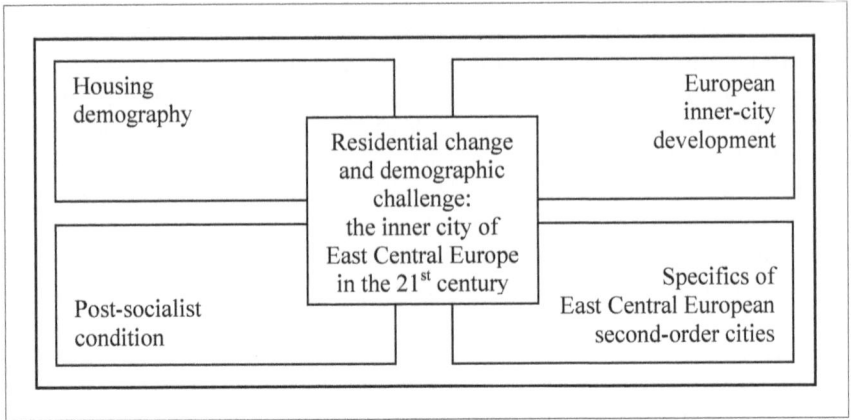

Figure 1.1 Embedding and contextualizing the volume's themes
Source: Authors' work.

The volume deals with (1) the interconnections of demographic and urban development ('housing demography'), by setting the focus on (2) European inner cities (in Poland and the Czech Republic) that share the (3) specifics of the post-socialist background and (4), as second-order cities, fulfil a specific function in their respective national urban hierarchies. Chapters 3–6 will expand on these contexts in more detail.

The major part of the debate on international inner-city resurgence relates to social and economic aspects. Over the last few decades, processes such as gentrification, reurbanization, revitalization, regeneration and renewal have been discussed (Davidson and Lees 2005, Bromley et al. 2005, Haase et al. 2010). The challenge of '*housing demography*', that is the way demographic behaviour and, in particular, household formation and change interact both with housing and the appropriation and use of urban space, has remained surprisingly under-researched (Myers 1990b). But, as argued above, cities are major arenas of the far-reaching demographic changes of the second half of the 20th century for which demographers have developed the concept of the second demographic transition (SDT). The concept also underlines the importance of household-driven changes, an area in which research is particularly lacking. The role of households in the stabilization of urban neighbourhoods and the reshaping of residential perceptions, wants and needs has often been neglected. Although analyses of the 'quiet demography of urban transformation' (Buzar et al. 2005) and the interrelations between changing living arrangements and housing (Mulder and Dieleman 2002) have recently gained some attention, there is still a lack of empirical research to show the complexity of these processes and their impact on urban development. For a long time, a 'mysterious separation' (Myers 1990a: 5) between housing and population studies existed, with a paucity of attempts to foster a more open exchange of ideas (Ogden and Hall 2004). Our aim is to provide evidence that this separation can be overcome. We argue that cities, and especially inner cities, are crucial departure points for household-driven urban change. The proportion of one-person households has risen very rapidly to high levels in inner cities and the numbers of other, so-called 'new' household types, such as cohabiting couples, same-sex unions and flat sharers, have also increased. This diversification of living and housing arrangements transforms the affected neighbourhoods. Demographic changes in East Central Europe are discussed with respect to their convergence towards, or divergence from, general European trends. A central aspect of these debates is the question of whether the processes and phenomena linked with the second demographic transition concept are also applicable to East Central European conditions. One of the major arguments within this approach is that converging demographic patterns will, in the long term, be observable all over Europe. However, it is still true that while the interconnections between socio-demographic change and urban transformation have become increasingly important in Polish and Czech cities, this is still not reflected in research. While social scientists deal with the transformations of society mainly in a broader context (Kojder 2007, Tuček et al. 2003), urban scholars rarely refer to demographic change and its societal premises when analyzing (area-based) residential changes.

After years of decline and population losses, inner cities are regaining their residential attractiveness. Throughout Europe, the academic literature reports on 'resurgent' inner cities, gentrification and a 'return' or 'renaissance' of city-mindedness and urban living (for example Cheshire 2006, Colomb 2007, Lees and Ley 2008). These trends are not restricted to the largest urban centres, but have also been evolving in medium-sized cities and are referred to as reurbanization by Buzar et al. (2007). Sometimes, the influx of new residents and the diversification of residential population within inner-city districts are accompanied or preceded by the refurbishment of the building stock as well as by an improvement of the quality of the residential environment. Whereas, in some places, inner-city resurgence (generally understood as the stabilization of its residential function against the background of former decline) is about to replace suburbanization as the main trend of local urban development, in others it has evolved as a process occurring simultaneously with ongoing suburbanization (Couch et al. 2005). However, inner-city resurgence in East Central European cities has rarely been analyzed to date.

For the purpose of this volume, *inner city* refers to the central parts of the urban body, including the city centre and the adjacent residential areas, which are characterized by compactness, physical and functional heterogeneity as well as by its (historical) population density. In Burgess' (1925) model of concentric zones (and based upon the case of Chicago), the inner city is specifically typified as a 'zone in transition', an area of flux, where land use and the residential population are continuously changing. In this model, the inner city was sharply distinguished from the urban centre, the central business district (CBD). The European tradition is different; many cities and towns were founded in the Middle Ages and did not grow beyond their walls prior to the era of industrialization, which thus also became the historical phase of large-scale urbanization. Accordingly, in various European languages, residents still go 'to the city' (*do miasta, do města, in die Stadt, a la ciudad, in città, en ville*, to give just some examples), meaning today's city centres. Whether or not the inner city is distinguished from the city centre depends upon the research question as well as the type of city under investigation (see, in more detail, Chapter 5).

East Central European cities are post-socialist cities. The systemic change after 1989 brought about new legal and institutional frameworks for reshaping and rearranging urban society, built environments and space. Even now, urban change in East Central Europe is often assumed to differ from general European patterns, mainly because the post-socialist transition is considered to be a specific pathway of urban development that is not the same as the urban trajectories of other European regions. Such assumptions run the risk of losing sight of the importance and impact of processes on East Central European cities that are certainly comparable to wider European processes. To explain urban change exclusively by post-socialist transition could result in a blurred picture of the respective processes, their drivers and consequences (Steinführer and Haase 2007). Therefore, in this volume, a wider explanatory approach is taken, which looks at East Central European cities more comprehensively, without undermining the explanatory usefulness of the post-socialist transition. The analysis of residential changes in inner-city areas of Polish and Czech second-order cities also allows a critical assessment of the concept of post-socialist transition as an explanatory framework for the development of East Central European inner cities. The term *East Central Europe* is by no means restricted to our empirical context of Poland and the Czech Republic. And indeed, as far as our theoretical background and the examination of demographic changes and post-socialist transition are concerned, we go beyond the Polish and Czech context and consider developments in the whole of East Central Europe. The specific research on Polish and Czech cities is therefore embedded in a broader geographical context and related to neighbouring countries, so that the use of the term East Central Europe as a 'collective name' is appropriate.

(4) While most of the research on East Central European cities has focused on the development of capital cities (see, for example, Eckardt 2006, Hamilton et al. 2005, Sýkora 2005, Stanilov 2007b), our focus is on the fortunes of those cities that rank second in the national urban hierarchies (Grimm 1994, Korcelli 1996). We look at cities that are key nodes of economy, transport and culture at the national scale as well as being major regional centres for services, administration and employment for the surrounding regions, even though they do not have the status of a capital. We call them second-order cities (see also also Hamilton 1979a: 177).[2] Many of these cities have an industrial past and possess a large, pre-Second World War built-up area. They have also faced shrinkage

---

2   Other scholars also use the terms *second-tier* or *second rank cities*.

processes in terms of population, infrastructure and economic performance for decades; sometimes these started before 1989 (Haase et al. 2008). These cities have undergone far-reaching transformations in almost all spheres of their built and social environment, such as deindustrialization and tertiarization or, as Eckhardt (2006: 28) puts it, 'changes in the matrix of urban life that are not changes in scale but in scope.' We examine the impact of European integration and political and economic globalization on these cities and argue that second-order cities are no less dynamic than capital cities and question assumptions about their 'stagnating' character (Hamilton et al. 2005: 12).

Methodological Approach and Challenges

Residential patterns are understood here as 'structures of limited range' (Kelle 2008: 76). That means they are social structures that are stable over a long period of time, but they can also change at any time, due to cumulative effects of individual reactions to macro-societal conditions (see Chapter 2). In order to analyze residential change, the interplay of individual action and macro-societal conditions needs to be considered. The individual perspective – perceptions, orientations and values, as well as residential behaviour and practices of households – demands qualitative research. By contrast, aggregate perspectives on patterns and their distributions in space and time are only detectable by statistical data provided by surveys, censuses and annual statistics. Also, parts of empirical reality always remain hidden from statistics due to incomplete registration, as well as terminological or content-related limits of the statistical data (Steinführer et al. 2010). Therefore, the most appropriate way to analyze residential change is a mixed-methods approach. Being aware of the strengths and limits of both methodological strands, we want to overcome the shortcomings of one approach by utilising the opportunities of the other (Kelle and Kluge 2001: 296–8).

A comparative approach is employed in the research, that is data of the same type are compared directly. Comparisons are made at both the temporal and spatial levels. Since the data used (from statistics, documents, interviews etc.) are already constructions of a first order, our findings represent 'double constructions' or second-order constructions (Schütz 1962). In other words, we make sense of data that are already organized in a specific way, whether as statistical categories and concepts (for example, the 'family household') or the way interviewees report on the world as they perceive it. The interpretations of these data and any comparative conclusions need to be made very carefully.

Additionally, we introduce different national and disciplinary perspectives. The consortium of authors includes scientists from Poland, the Czech Republic, England and eastern Germany, who represent urban sociologists and urban geographers, as well as anthropologists.

Given this background, our methodological approach was operationalized as follows. First, we carried out small-scale data analyses, mainly drawing on census data from 1988 and 2002 in Poland and 1991 and 2001 in the Czech Republic. Second, in order both to unravel information that is 'beyond' the statistics and to allow for a more actor-based approach, qualitative, semi-structured interviews were conducted with residents, to identify types of relationships between residents and their attitudes towards inner-city living and housing. Local experts were interviewed in order to understand the context of the respective urban regime, housing market and recent socio-demographic changes in each city. Moreover, approaches and methods such as oral history, mapping and area observation techniques were applied for single thematic fields of our research, to ensure a comprehensive view of the subject of our analyses.

A mixed-methods approach means it is possible to go beyond the knowledge reached by one particular analytical method and to work at a range of spatial scales, from the individual house to the entire city. Different perspectives on change are possible, the external viewpoint of the researcher and insider perceptions of the residents or local stakeholders. It also means it is possible to use methods from different disciplines. Finally, the full range of the authors' expertise, with their different national, disciplinary and methodological backgrounds, can be brought into play.

It is important to emphasize that we, as researchers, are not just 'neutral' observers of our research subject. The authors of this volume represent a heterogeneous consortium, including not only scholars of different countries, disciplinary backgrounds and age, but also individuals with different relationships to, as well as experiences with, the socialist past and post-socialist present in East Central Europe. Therefore, our respective emotional investments or involvements concerning this past and post-socialist transition (in terms of its political evaluation) clearly differ. Our perspectives are also coloured by the traces and shape of (1) our personal national and cultural backgrounds, (2) our methodological preferences and modes of research and (3) our existing knowledge about the issues we are working on (see also Hörschelmann 2002: 53–4). We therefore have to reflect on our positionality in various ways. The international nature of the consortium inevitably raised questions in terms of the understanding of the processes under investigation. The respective national backgrounds of the researchers led to a specific view on the urban processes in East Central Europe. For example, the east German perspective was often connected to the experiences of shrinking cities, vacancies and demolition, while the English one was characterized by the familiarity with urban decline, urban regeneration, evolving gentrification and the recent 'urban renaissance' policy in the UK. The interpretation of the results combines the different assessments of inner-city change made by the various authors. Even the understanding and use of terms such as *inner city, household, migration, residential mobility* or *urban shrinkage* were intensively debated and varied according to national contexts.

Scope and Contents of this Volume

The volume is divided into three parts. *Part I* provides the conceptual background, context conditions and methodological considerations of the research (Chapters 1–6). *Part II* is concerned with the empirical investigations of and in Polish and Czech inner cities, namely Łódź, Gdańsk, Brno and Ostrava. These case studies dealt with various topics and spatial scales, ranging from the inner city as a whole to single buildings or even individual flats, as well as from the inner-city population to the individual household (Chapter 7–12). *Part III* provides a comparative look at the four case-study cities and reflects once again on the main research objectives and questions. It relates the research to other studies and contextualizes the analyses undertaken in Polish and Czech cities, in order to embed East Central Europe into a broader European framework (Chapter 13).

While the volume represents the work of a three-year project of an international consortium, the individual chapters are written by small groups or single authors. Some chapters shed light on the development of all four case studies, whereas others deal with one or two of the four cities only. Some look at the inner city as a whole, in contrast, others set the focus on particular neighbourhoods or residential locations.

Chapter 2 presents and discusses the conceptual model the volume is built on, as well as the methodology and research methods. In the first part, Katrin Grossmann and Annett Steinführer introduce the model of residential change, in the second part they summarize current research and gaps in knowledge with regard to residential change in inner cities in East Central Europe. The third part considers methodological challenges in analyzing residential change on different scales. The final part of the chapter turns to the description of the design of the research presented, the methods applied and some problems related to these methods. Finally, the implications for the main part of the volume are drawn.

After this introduction, *Chapter 3* addresses the interconnections between demographic change and housing by focusing on private households, both as major actors in the urban space and as motors bringing about residential change. Annett Steinführer and Ray Hall examine the idea of a 'housing demography', an approach that looks at the underlying dynamics and consequences of demographic and household-driven processes from an urban perspective. Core processes of contemporary demographic change – natural change and migration, ageing and household change – are then discussed and related to the concept of the second demographic transition (SDT) and the urban context. The following section narrows the focus to East Central Europe by unravelling the specifics of demographic change in Poland and the Czech Republic during post-socialism and examining the SDT debate in both countries. The chapter concludes by identifying the main implications of the issues under discussion for the volume.

Chapter 4, authored by Annegret Haase, Antonín Vaishar and Grzegorz Węcławowicz, addresses the underlying dynamics of urban change in East Central Europe during post-socialism. It starts with a conceptualization of the terms post-socialist condition and transition for the purpose of the volume. It then introduces a model of post-socialist urban change at the overall city level and identifies the impact of the post-socialist transition on residential change as the central topic of our volume. Following this, it goes beyond the post-socialist approach and includes other general processes, such as European integration or globalization, as explanatory frameworks. It concludes that post-socialist urban transition is, firstly, not restricted to change and, secondly, its very nature consists of the simultaneousness and interrelatedness of a number of processes that have their roots in pre-socialist, socialist and post-socialist contexts, thus stressing its path dependency. While the direct impact of transition-related change has decreased in recent years, the importance of other European, global and post-transition processes for urban development in East Central Europe has increased. Finally, the authors argue for a wider explanatory framework to describe contemporary urban change in East Central Europe, without undermining the importance of the post-socialist condition.

Chapter 5 deals with the inner city as the spatial focus of the volume. First, Sigrun Kabisch and Iwona Sagan provide an overview of the historical development and specifics of European inner cities and reflect on the term with respect to its character, functions and limits. The chapter then examines the debates on regeneration, renaissance, gentrification and reurbanization of European inner cities, singling out the implications for research on residential change in East Central Europe. Finally, there is a review of the Czech and Polish research on East Central European inner cities. The chapter concludes by presenting challenges for empirical research.

Chapter 6 discusses characteristics of second-order cities in Poland and the Czech Republic, together with the major changes that have taken place there during the post-socialist period relevant to this volume. The authors, Adam Bierzyński, Maja Grabkowska, Annegret Haase, Petr Klusáček, Andreas Maas, Jana Mair, Stanislav Martinát, Iwona Sagan, Annett Steinführer, Antonín Vaishar, Grzegorz Węcławowicz and Jana Zapletalová, provide an overview of Polish and Czech second-order cities within the respective national settlement systems and introduce the four case-study cities, Łódź, Gdańsk, Brno and Ostrava, as well as their inner-city areas in more detail. In addition, general demographic and housing market developments in the four cities are discussed, which provides the context for the empirical findings of later chapters. The chapter states that, in all cities, a number of different and sometimes even contradictory processes of demographic and urban change occurred simultaneously. Yet, all four cities experienced at least a short period of population decline, mainly due to out-migration. Particularly the rise in overall household numbers, which contradicts long-term population trends, provides support for the hypothesis that second-order cities in East Central Europe also face the challenges of the SDT. These demographic trends, however, collide with particular housing markets, which are characterized by a long-term imbalance between supply and demand, as well as by a number of persistent features of state socialism, despite all the transformations which occurred during the 1990s.

Chapter 7 marks the beginning of the empirical part of the volume. Annegret Haase, Adam Bierzyński, Maja Grabkowska, Petr Klusáček, Stanislav Martinát, Zdeněk Uherek and Andreas Maas describe small-scale socio-demographic processes and their underlying dynamics in the four case-study cities. The chapter challenges the prevailing story of ageing and decline of East Central European inner cities, demonstrating that, as a result of post-socialist transition, greater diversity has developed over the last two decades. It distinguishes between processes of socio-demographic transformation and their results or spatial imprints, that is transformed, newly emerged or rearranged patterns of socio-spatial segregation and fragmentation, and also identifies and explains the characteristics and underlying dynamics of this change for the four case-study cities. In doing so, the chapter reports on population decline and repopulation, ageing and rejuvenation, household change, gentrification and displacement, studentification and ethnic diversification. Finally, the results are cross-referenced in a comparative manner; the claim of an old-new diversity of inner-city socio-demographic structures is discussed. It is argued that the 'old story' of decline and ageing is no longer totally true, nor is there a new 'grand' story to tell, but rather multiple, new 'smaller' stories that partly overlap and whose long-term significance is still unknown.

Chapters 8 and *9* change from the spatial to the actors' perspective. *Chapter 8*, authored by Annett Steinführer, Annegret Haase and Maja Grabkowska, analyzes which types of households are to be found in the inner cities of our case studies (in qualitative terms), which circumstances led them to their current place of residence and which residential patterns have evolved. In so doing, the chapter sheds light on the local housing markets through the eyes of our interviewees and in terms of the tension between new market opportunities and restrictions, on the one hand, and state-socialist legacies, on the other. Evolving housing and living arrangements in general are discussed before examining students as specific new actors in East Central European housing markets. Afterwards, selected housing careers are presented in more detail. The chapter argues, amongst other things, that there is a close interplay between households as actors and the urban environment, and that with their housing choice, households react not only to given settings, such as the housing market, but also to the conditions of their own professional and private lives and to their alterations. It concludes, furthermore, that it would be short-sighted just to claim that the evolving post-socialist residential patterns simply coexist with persisting quasi-socialist ones. Such an interpretation loses sight of the far-reaching transformations, particularly of property structures, during the transition.

Chapter 9 presents a typology of attitudes of inner-city residents towards housing and living in the inner cities of Łódź, Brno and Gdańsk. The objective is to assess the attitudes that inner-city dwellers have towards their residential location, how they came here and to what extent they feel attached to the place. The authors, Katrin Grossmann, Annegret Haase, Annett Steinführer, Maja Grabkowska and Adam Bierzyński, argue that there is a variety of attitudes, ranging from enthusiasm through pragmatism and ambivalence to frustration, and that there is no clear connection between particular household types and residential attitudes. Furthermore, the chapter reflects on a phenomenon that appeared in a number of interviews: the transitory character of the housing arrangements, that is the wide-spread taken-for-granted restriction of this type of residence to a certain phase in the lives of our interviewees. Transitory urbanites, as we will call them, are identified as important actors of current inner-city change in Polish and Czech second-order cities, and their impact on the residential function of the inner city today and implications for the future are discussed. It is argued, amongst other things, that transitory urban living may be bound to particular circumstances, such as a housing market that does not suffer from the very high demand constantly found in the capital cities of East Central Europe. Thus, transitory urbanites in the inner city arise as a new type of housing market actor in second-order cities lacking strong excess demand.

Chapter 10, by Maja Grabkowska, presents the findings of a case study of the residential flexibility of households, dwellings and inner-city neighbourhoods, focusing on the role of non-traditional households. First, the concept of residential flexibility is introduced, which is understood as the degree to which a household is able to change its use of, or movement through, the built environment, in response to altered social, economic or political circumstances. The concept is then applied to inner-city neighbourhoods of Gdańsk by discussing the question of residential flexibility or flexible residents. It concludes by confirming the assumption that the new inner-city residents value first and foremost accessibility, versatility and functionalism, but also the unique ambiance and appeal of pre-Second World War architecture.

Chapter 11, authored by Katrin Grossmann and Annegret Haase, sets out the relationship between the built environment and housing preferences. Residents in the case-study cities are asked what 'brick' (as a symbol of the mainly pre-Second World War housing stock) and 'block' (referring to wide-spread building technologies from around the 1960s onwards) mean to them. The prevailing assumption that large housing estates built in the socialist era will see degradation similar to that experienced in Western European housing estates, while inner cities will gain attractiveness and become the preferred residential location, are explored and existing attitudes towards brick or block among urban dwellers in Brno and Łódź are examined. In this way, the importance of the type of building stock and the attitudes towards it with respect to housing preferences and residential location decisions can be evaluated.

The application of an anthropological approach is presented by Jana Mair and Jana Pospíšilová in *Chapter 12*. The authors discuss the consequences of tenure change for social networks and neighbourly contacts. On the basis of survey- and interview-based fieldwork, applying the approach of oral history, the authors show what the neighbourly relations of the inhabitants of the inner cities of Brno and Ostrava are like and indicate whether, and to what degree, these relationships have been influenced by the post-socialist transformation. The chapter considers the attitudes of residents towards neighbourly relations, their interactions with each other and the degree to which these interactions coincide with their expectations. Moreover, the attitudes of inner-city residents towards their neighbourhood and the city as a whole are analyzed from an anthropological perspective. The focus is not only on the present state and recent history but also on residues and relicts from the period of state socialism, or even from earlier times.

The concluding *Chapter 13*, authored by the editors of the volume, reflects more generally upon the findings of the research presented in the volume: firstly, by referring back to the conceptual model the book is built on; secondly, by considering the contribution made by the research to a better understanding of the current and likely future development of East Central European cities and, finally, by suggesting lessons learnt for future research on residential and broader urban changes in East Central Europe.

Chapter 2

Residential Change: Conceptualization, Methodological Challenges and Research Design

Katrin Grossmann, Annett Steinführer

The city changes without cease. Homes give way to factories, stores, and highways. New neighborhoods arise out of farms and wasteland. Old residential districts change their character as their residents give way to different classes and cultures. Business districts slowly migrate uptown. Mansions of yesterday sport 'rooms to let' signs today. Tenements once rented to immigrant European peasants, now house migrants from our own rural areas.

Rossi 1980: 53; originally 1955: 1

Introduction

Change is at the heart of every city – whether obvious and spectacular or hidden under the surface of apparent stagnation. Consequently, urban scholars are predominantly interested in issues of constantly transforming urban structures, be they social, economic, demographic, physical or spatial. This volume is no exception. With the focus on residential change and its interrelations with demographic processes, our approach considers the city from a specific viewpoint: as a place of residence, a place always in flux due to changes within neighbourhoods, residential buildings and dwellings. These changes are not restricted to mobility and physical movements of people but are much more diverse. How residential change comes about, at which scales and with what outcomes is the basic question of the research.

This chapter presents and discusses the volume's basic conceptual model as well as the methodology and research design. After introducing our model of residential change, current research and gaps in knowledge with regard to residential change in inner cities in East Central Europe are summarized, before the methodological challenges in analyzing residential change on different scales are considered. We then turn to our specific research design, the methods applied and some problems related to these methods. Finally, the implications for the main part of the volume are drawn.

A Conceptualization of Residential Change

Residential change is covered by a very large body of scholarly literature, even though the concept itself is only rarely used explicitly. Residential change has been long studied by urban sociologists and geographers and refers to a key process of intra-urban change. Stories of residential change, first recounted by the Chicago School with its heuristic concepts of 'natural areas' or of invasion and succession (Park et al. 1984), continue to be told in much of the neighbourhood life-cycle literature and in residential segregation and tipping-point approaches. They also appear in the revitalization, gentrification and, more recently, reurbanization debates, as well as in post-socialist discourses about small-scale socio-spatial differentiation (for a review: Schwirian 1983, Schnur 2008).

Though definitions or explicit conceptualizations of residential change are rare, most of the literature in this field identifies one key process: residential mobility. In his groundbreaking 1955 study, 'Why families move', Peter H. Rossi showed that residential mobility is a major driver of both residential and urban change. The quotation at the very beginning of this chapter continues as follows:

> How do these dramatic changes in residential areas come about? In part, industry and commerce in their expansion encroach upon land used for residences. But, in larger part, the changes are mass movements of families – the end results of countless thousands of residence shifts made by urban Americans every year. Compounded in the mass, the residence shifts of urban households produce most of the change and flux of urban population structures. (Rossi 1980: 53)

Rossi's study has given rise to numerous empirical studies on intra-urban residential mobility worldwide, to conceptual refinements and theoretical reformulations (recent overviews are given by: Mulder 1996, 2007, Dieleman 2001, Strassmann 2001, Steinführer 2004b). However, while acknowledging the major role of residential mobility and migration, we hold that there is a lot more to residential change than just mobility, as we will discuss in more detail below.

Setting our focus on *residential change*, we are most interested in those aspects of neighbourhood transformation that are related to housing issues rather than, for example, governance or economic performance. Within the broad spectrum of users and producers of neighbourhoods, residents – particularly households – are the main subjects of the research discussed in this volume. In our understanding of neighbourhood, we follow George Galster, who defines it as 'the bundle of spatially based attributes associated with clusters of residences, sometimes in conjunction with other land uses' (Galster 2001: 2112).

Residential change is the process that occurs in and around this bundle through the interplay of people, places, local contexts and society. Via this, residential patterns form, alter and are rearranged (Figure 2.1). By *residential patterns* we mean small-scale clusters of interrelated social and physical characteristics. The physical environment and characteristics of residents are interrelated, meaning that there is no direction or determination in their relationship, but rather just an interdependence. The given – but not necessarily homogeneous – social composition of a place can impact on the physical conditions of a neighbourhood, for instance, during processes of 'incumbent upgrading', that is renovation and reconstruction activities by sitting tenants or owner occupiers (Clay 1979). Physical features can attract certain groups and represent barriers for others.

This conceptualization of residential patterns as characteristics of both *residents* and *residence* resembles other concepts discussed in urban research. For example, the idea of 'natural areas', as developed by the Chicago School, connected specific groups of people with certain places: 'Natural areas are the habitats of natural groups. Every typical urban area is likely to contain a characteristic selection of the population as a whole' (Park 1967: 62). Natural areas were regarded as spatial units with a high degree of internal social homogeneity, solidarity and mutual norms (see also Hatt 1946). Likewise, the concept of residential or socio-spatial milieus refers to small-scale interdependencies of social and spatial characteristics (Keim 1979, Matthiesen 1998).

With explicit reference to recent developments in post-socialist cities, Luděk Sýkora recently introduced the concept of socio-spatial formations (Sýkora 2009a, 2009b, building on Marcuse and van Kempen 2000, Wacquant 2008). He starts from the observation of a paradox: despite empirical evidence of far-reaching socio-spatial transformations during the 1990s, segregation indices in post-socialist cities (here: Prague) have not increased in comparison to the period of state socialism and, indeed, remain relatively low. The concept of socio-spatial formations therefore describes small-scale patterns of residential segregation (as an involuntary process) and separation (as a voluntary one), respectively. Due to the proximity and simultaneousness of contrasting processes, segregation indices might not change, even though residential differentiation processes do indeed occur. Examples of such places are gentrified neighbourhoods, socially excluded areas or ethnic enclaves (Sýkora 2009b). Even though we use other methods, we share the wish to grasp small-scale changes hidden to most statistics and indices. The major difference lies in our concept of residential patterns: we start with the neighbourhood and are not only interested in socio-spatial extremes of rather high or rather low social strata, respectively, but also in areas with mixed social and demographic structures (Figure 2.1).

The term *residents* encompasses *individuals and households* (that is the resident population) of an urban area. They exhibit certain demographic characteristics and features of social status and lifestyle that are a major object of research in urban studies. These characteristics are usually examined on the aggregate level of socio-spatial patterns that, together, form the residential segregation of a given city (Park 1915, Burgers and Musterd 2002, Dangschat 2002).

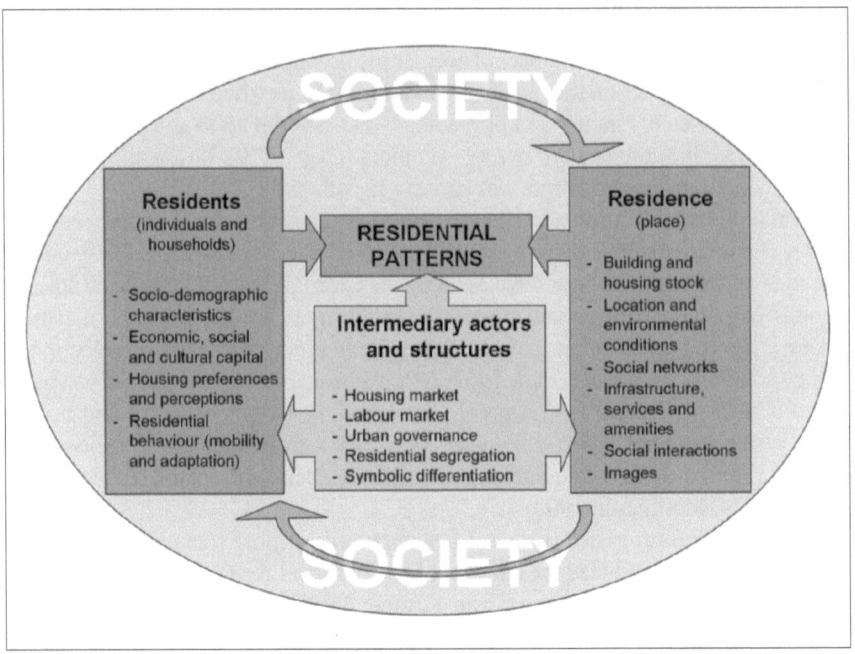

Figure 2.1 The conceptualization of residential change
Source: Authors' work (Grossmann, Haase, Kabisch, Steinführer).

In this volume, we will pay particular attention to households as actors because it is they who decide for or against a residential location, to move or stay (Chapter 3). Households have a set of *resources* at their disposal. The main focus of research has always been on economic resources, to address questions of affordability of housing and thereby the possibilities and limitations of accessing certain housing segments and urban neighbourhoods (Massey and Denton 1988). Yet, as we elaborate later in this chapter, to capture and understand residential change, one needs to look at both aggregate patterns as well as individual agency. But agency depends on more than economic resources. Therefore, we prefer a more comprehensive view of the residents' resources and include *economic, social and cultural capital* in the Bourdieuian definition (Bourdieu 1986). Economic capital relates to both income and property of individuals and households. Social capital comprises the informal ties to neighbours, friends and kin, as well as the more formal connections to schools, teachers, neighbourhood association officials, local politicians, etc. Social capital itself is highly differentiated by age and social status. Children, as well as old and socially deprived persons, are more dependent on such local social resources than people active in the labour force, who often have broad extra-local networks (Forrest and Kearns 2001: 2130–33; referring to Guest and Wierzbicki 1999). As for residential change, social capital in terms of the residents' networks might be an attractor to a certain neighbourhood, or function as a push factor to leave it. Cultural (or human) capital forms an important aspect of social inequality and is strongly related to social status and, thus, socio-spatial inequality. Cultural capital includes knowledge, skills and (formal) qualifications. A decreasing or increasing demand for certain skills and fields of expertise in the urban labour market might seriously affect residential areas, for example, when a workers' neighbourhood faces the closure of a nearby factory and the entry of IT businesses. Micro-studies conducted at small scales point to the importance of these non-economic resources (see the review by Lupton and Power 2004).

Economic, social and cultural capital, residential preferences and perceptions, along with household changes, are major determinants of *residential behaviour*. The two major modes of behaviour are moving, that is residential *mobility*, and staying, that is *adaptation* of the residential environment or adjustment of preferences (Steinführer 2004b; see also Brown and Moore 1970, Orbell and Uno 1972, Franz 1989, Littlewood and Munro 1997; more generally: Hirschman 1970). As stated above, most of the literature focuses on mobility, investigating the push and pull factors influencing individual and household decisions concerning residential location. This is also true for the research devoted to neighbourhood pathways, be it gentrification (see for recent reviews: Atkinson and Bridge 2005, Lees and Ley 2008), reurbanization (Buzar et al. 2007) or neighbourhood decline (Metzger 2000). In all of these discourses, a strong focus is on the influx of particular social groups and the displacement of others. Yet, the debate on 'incumbent upgrading' as well as the more recent one on 'residential flexibility' point to opportunities of *in-situ* change arising out of adaptations of a certain place by the residents themselves (Clay 1979, Deane 1990, Littlewood and Munro 1997, Mandič 2001, Bouzarovski 2009). Moreover, people can be socially mobile without physically moving and thus can collectively change the social fabric of a place – for example, when the inhabitants of a student neighbourhood enter the labour market after their studies, move up the social ladder, found a family and stay in the neighbourhood. The residents and their changing preferences and needs impact simultaneously on their surroundings, with their amenities and equipment. They thereby contribute to a transformation of the physical fabric and the social infrastructure of the neighbourhood. Chapters 10 and 12, in particular, focus on modes of such *in-situ* change and thereby highlight the importance of this aspect of residential change.

The latter issue leads to the right-hand part of Figure 2.1, the *residence*. While the literature on residential change pays substantial attention to shifts in the social composition of a neighbourhood (for example, Kirschenbaum 1983), the role of the place is rarely considered. The residence is made up, first of all, by the physical fabric; the *building and housing stock* of a given neighbourhood. A neighbourhood comes into being when new homes are constructed and occupied. And, starting from the very first house warming, the residence is subject to change (Galster 2001: 2116). The building stock can deteriorate, be rebuilt, upgraded or demolished. *Environmental conditions* such as exposure to noise and pollutants, the air quality or vicinity of and access to recreational areas might also change. A developing branch of research is concerned with environmental justice, the question of whether or not the access to environmental goods and the burden of environmental problems are equally distributed (see the review by Elvers et al. 2008).

Infrastructure like public transport, schools or other services is often crucial for residential location decisions. *Services* and infrastructure development depend on administrative decisions and private investment, but also on demand. The recent debate on declining numbers of residents in certain urban neighbourhoods in post-socialist eastern Germany clearly shows the problems caused by declining demand for the technical infrastructure (Hummel and Lux 2007, Moss 2008). School closures, closed post offices or shops and empty kindergartens convey an air of decay and decline to a place (see also Franz 1989: 192–204).

But residence is more than just the buildings, shops and roads. It is also a socially networked place where people interact, both intentionally and by accident. The presence of others provides opportunities to build up, reactivate and ignore *local social interactions*. Finally, it should not be forgotten that there are some 'soft' factors which contribute to the characteristics of a place. Among the most important are the *images* of a place as seen from the inside by the residents themselves or from the outside by non-residents. Images impact on the fortunes of a neighbourhood and its residents, as the debate on neighbourhood effects and stigmatization shows. Negative images especially stick to a neighbourhood, even after extensive physical or social change has occurred (Hastings 2004, Grossmann 2010).

While all of the aforementioned features contribute to the fortunes and long-term trajectories of urban neighbourhoods, residential places need to be embedded in broader contexts, both spatially and socially. Residents and residence are brought together – or kept apart – by *intermediary structures and their actors* (see Figure 2.1 above). These include the logic and structures of local *housing markets* and their submarkets, as well as the existing, transforming and newly evolving patterns of residential segregation, which are formed, in their turn, by processes of socio-spatial differentiation, including their *symbolic representations*. Significantly shaped by national housing traditions and legal frameworks, local housing markets might be more or less fragmented and thus open or closed to the housing demands of specific social groups and household types. Existing patterns of socio-spatial and symbolic representation provide opportunities and restrictions to households as actors in housing markets and shape perceptions of 'good' and 'bad' places. But also broader context factors such as *urban governance* and *local labour markets* play a part. Salient intermediary actors are, for example, investors, real estate agents, planners and municipal decision makers, but also civic actors like neighbourhood initiatives which influence pathways of neighbourhood change in general. Research on urban governance, for example, looks at the interplay of a variety of municipal, private and civic actors in revitalization, gentrification and regeneration processes, as well as their significance for neighbourhood development.

Finally, change is also induced by more general shifts in the *societal context*, that is the macro-scale. For this volume, the post-socialist transition and demographic change are the societal changes we have chosen to investigate. We aim to establish the links between these macro-processes and changes in residential patterns (see Chapters 3 and 4).

In order to make our argument clearer, we will briefly discuss an example. The concentration of childless dual-earner couples in their early or mid-professional careers, colloquially known as 'dinkies' (double income no kids, sometimes also understood as no children yet), in prestigious inner-city areas would be a residential pattern by our definition. These couples set their priorities on a professional career; they are thus able to afford a more expensive consumer lifestyle than couples with children but also than singles and other one-person households. Dinkies often have substantial economic resources and a high degree of (formal) qualifications that are part of their cultural capital. They are highly mobile, moving to where the demand for their professional skills takes them. The development of this household type is directly linked to changes in society, such as the transformation of the labour market in favour of services and the IT sector, the altered position of women in many western societies and changing attitudes towards children and family foundation (van Gils and Kraaykamp 2008). With regard to residential change, dinkies, together with yuppies and other lifestyle groups, as well as other 'new' household types (see Chapter 3) have been decisive actors in gentrification processes in many inner cities worldwide. And besides transforming the socio-economic make-up of cities, they also change the character of the place, the residence. Certainly, the residential location decisions of these groups are mediated by local housing markets, by developers and urban governance, but such restrictions are relatively less important for these households. Their choice is strongly influenced by pro-urban (or even 'anti-suburban'; Ley 1996) attitudes that are conditioned by both pragmatic reasons (such as the desire for a central location well connected to transport and infrastructure) and aesthetic preferences for individual housing and non-standardized dwellings. Chosen housing areas often represent the lifestyle that such households can afford and pursue. These areas exhibit particular physical and cultural features, for example, certain types of housing, shops, pubs and places for recreation, as well as an ascribed symbolic meaning of being stylish or 'in fashion'. These features, together with the generally strong financial position of dinkie households, are strong pull factors to move to a particular neighbourhood. On the meso-scale of the respective area, the individual location decisions of many dinkies influence and contribute as much to the residential change of these areas as to residential segregation of the entire city – regardless of whether the actors personally desired such a development.

Taken together, residential change occurs by the interplay of the various described components. It can be induced by general societal shifts, by the cumulative consequences of individual behaviour as well as by changes in the local contexts. Processes such as gentrification, reurbanization, studentification or suburbanization as outlined here, are specific forms of residential change (see Chapter 5 and 7).

Residential Change in East Central European Inner Cities: What is Known?

Urban transition research on East Central and Eastern Europe has devoted some of its efforts to the issue of residential change – although the term itself is only rarely applied (as one of the few exceptions: Ruoppila 2006). Nevertheless, the overwhelming bulk of work on the processes of socio-spatial differentiation (and, out of this, on the evolving patterns of residential segregation) in East Central European cities has also considered the influence of the post-socialist transition on residential patterns: that is, on the causes, courses and consequences of post-socialist residential change. A majority of contributions have addressed the shifts in *society* and *intermediary structures* that need to be taken into account to understand and explain residential change (Figure 2.1). In the following, we will review and summarize some of the findings of this earlier research.

From the very onset of the transition, numerous scenarios have been formulated concerning the impacts of the societal shifts on intra-urban socio-spatial structures (Friedrichs and Kahl 1991, Marcuse 1991, Musil 1993, Szelenyi 1996). Most prominent among these were predictions concerning the functional specialization of the city centres (commercialization), a new phase of urbanization (by intensified suburbanization), inner-city change (particularly small-scale gentrification) and transformations of the post-Second World War housing estates (generally expected to be substantial social degradation). About two decades later, it is clear that most of these scenarios were correct, with the notable exception of the last one: large housing estates today are, for the most part, still stable social areas that have, however, undergone processes of residential change and socio-spatial differentiation (Węcławowicz et al. 2003, Bernt and Kabisch 2006, Barvíková 2009, Szafrańska 2009a). This is also relevant for inner cities, since the expected massive outmigration of middle-class households from declining large housing estates has not taken place in most East Central European cities.

In comparison to the slow pace of transformation in the large housing estates, residential changes in other parts of the city and at the urban fringe appear to have been much more pronounced in the course of post-socialist transition. The most important socio-spatial processes in East Central Europe at the urban scale have been the commercialization of the city centres in connection with the displacement, or at least a steep reduction, of their formerly strong residential function (thus a 'deresidentialization'), the suburbanization of housing (and retail activity), selective building and social up-grading (gentrification) of some inner-city areas, as well as the decline of others (for example, Sýkora 2000a, Sýkora et al. 2000, Węcławowicz 2001, 2004, Kovács and Wießner 2004, Kährik 2006, Ouředníček 2006, Hirt and Stanilov 2007).

In retrospect, it is also necessary to stress that both the legacies of the old system (or its persistent patterns; Steinführer 2006, Sýkora 2008) as well as a number of unintended consequences of certain transition processes were underestimated or not perceived at all in terms of their effects.

While there is indeed a large body of studies on residential change in East Central and Eastern Europe, three shortcomings of the previous research were apparent when we started our work. These were:

1. a predominant focus on capital (and other large) cities,
2. a relatively greater focus on other functional and residential zones than on inner-city areas, particularly on the city centres, suburban areas and, to some extent, large housing estates, and
3. a neglect of the role of demographic processes or, to put it more precisely, of changing living and housing arrangements (which themselves are affected by broader societal changes).

In the following discussion, these points will be elaborated and linked to other chapters in this volume.

1. In general, the extremes of *socio-spatial differentiation* – social exclusion and upgrading – can most easily be found in the metropolises. In cities like Prague, Warsaw or Budapest, economic pressures are high. In addition to the post-socialist transition, capital cities face pronounced effects of globalization and 'Europeanization' (see Chapter 4). Hence, different transition processes overlap here in a much stronger way than for cities in the lower ranks of the national settlement structure. Correspondingly, there is a large literature on urban change in East Central European capitals (for Prague: Sýkora 1994, 1999, Ouředníček 2006; for Warsaw: Węcławowicz 1993, 2002b; for Budapest: Kovács 1994, Douglas 1997, Kovács and Wießner 2004),[1] but this does not apply for the cities of the next level like Gdańsk, Ostrava or Debrecen. Some studies are available for Poland (Smith 1995, Riley et al. 1999, Węcławowicz 2001, 2002b, Kotus 2006, Parysek and Mierzejewska 2006) and Hungary (Sailer 2001, Kiss 2004). The centralist structure of the Czech Republic is also mirrored in the state of research: apart from Prague, Brno alone has a relatively large knowledge base (Mikulík and Vaishar 1996, Sýkora et al. 2000, Steinführer 2003, 2004b).

We hold that the capitals, as the most important political, economic and cultural centres of the particular countries, have their own paths of development. By their very nature, they fulfil specific functions and are attractive for regional migration as well as for immigrants from abroad. The internationalization of the property and capital markets is most distinctive in the capital cities and

---

1 It is also worth noting that most contributions and anthologies on 'the' post-socialist city focus almost exclusively on the capital cities (for example, Kostinskiy 2001, Eckardt 2006, Stanilov 2007a).

influences their spatial development. The polarization in income, the degree of retail and residential suburbanization as well as the excess of demand in the housing markets of Warsaw, Prague or Budapest have no counterpart in the other cities of their respective countries (Kostelecký et al. 1997, Gawryszewski et al. 1998, Kovács and Wießner 1999). Thus, we expect that in second-order cities the characteristics and dynamics of people and places and, resulting from this, the *residential patterns* are less pronounced or less polarized than in the capital cities. But we also argue that these different characteristics and dynamics bring about distinct residential patterns – a hypothesis which, however, cannot be tested systematically in this volume.

2. Residential areas in inner cities are vitally important for the urban viability and local identity of East Central European cities because of their location, historical importance, functional mix and specific urban fabric. Depending upon the specific urban structure of a given city, inner cities might include the functional city centre or not. However, decisive for our understanding of the 'inner city' is its residential function (see in more detail Chapter 5). When we started our research in 2006, we found the inner city to be a place of residential change that had been largely neglected by urban transition research (see the argument in Steinführer and Haase 2007). This situation has changed since then (for example, Temelová 2007, Beim and Tölle 2008, Kovács 2009).

As places of *residence*, inner cities in East Central Europe had faced decades of disinvestment following the Second World War, with resulting physical decay, demographic ageing, population decline and partial concentrations of socially deprived groups. This situation did not change overnight after 1989, although inner cities in Poland and the Czech Republic attracted some selective investment from the 1990s onwards. Less attractive areas especially are still in need of repair and refurbishment (to a greater degree in Poland than in the Czech Republic). But the central location of these districts is likely to attract further capital spending in the future. Some locations are hotspots for speculative investments by developers or for state-led upgrading of the housing stock. The environmental conditions differ according to the character, history and location of specific neighbourhoods (Vaishar and Zapletalová 2003). Historically, more prestigious addresses have a higher share of green spaces in their gardens and nearby parks, whereas less prestigious areas are often located close to industrial sites with less access to green spaces within and around the blocks. Traffic- or industry-related air pollution and noise are concentrated in the inner cities. Yet, they are also places with good infrastructure, well connected to public transport facilities, well equipped with facilities for daily shopping needs, schools and kindergartens, public institutions, as well as cultural and leisure facilities.

Inner-city *residential change* in the post-socialist period is interrelated with various other societal trends and processes at different scales. We consider demographic change, particularly its related transformations of housing and living arrangements (Chapter 3), and the post-socialist transition, to be the most important, both of which impact decisively on the urban scale (Chapter 4). How they affected and still affect both *residents and residence* of the inner city will be one key question throughout this volume. Existing research points to a number of parallel and even contradictory processes: there is social mixing versus the existence or development of ethnic clusters, ageing of the population versus selective gentrification, privatization in favour of long-term tenants versus partial vacancy, physical decay versus luxury renovation and so forth. Research on second-order cities is also comparatively sparse with respect to these topics. Urban transition research has not only focused on the capital cities, as discussed earlier, but also on processes in the capital cities which can be called 'spectacular', that is processes of dramatic changes at both extremes of the socio-spatial continuum. For example, processes and outcomes of gentrification, flagship or 'historicized' revitalization (such as that of former Jewish neighbourhoods) are relevant here (Sýkora 1996a, Więcław 1997, Murzyn 2006, Temelová 2007). Moreover, social exclusion plays a substantial role in the description of inner cities. In East Central Europe, socio-spatial marginalization is often connected with high proportions of Roma population in certain neighbourhoods, particularly in the Czech Republic and in Hungary (Ladányi 1993, Burjanek 1996, Růžička 2006, Kovács 2009). However, while appreciating the necessity of such studies, we also hold that these processes describe only some development options of inner-city neighbourhoods.

3. In this volume we pay particular attention to overarching demographic processes, especially to changes in household structures, but even more to the consequences they have with respect to various socio-demographic and socio-economic changes at the meso-scale of the inner city. We assume that inner cities are both home to, and attractive for, a variety of household types, living arrangements, social classes, age and ethnic groups, because of their mixed constructional and property structures as well as their location. Research questions in this context are, for example: can we observe processes of revaluation and displacement similar to those in Western Europe? What variations in households, social structure and generations are to be found in those areas? How did residents come there and how do they change residence?

In order to answer these questions, we have chosen a micro- and meso-scale approach. The focus is deliberately on the inner city (middle level in Figure 2.2). We also pay attention to even smaller scales of analysis – such as buildings and flats – and ongoing residential processes there (see particularly Chapters 10 and 12). Tenure change (for example, via the privatization of the municipal housing stock) or adaptation (of a dwelling to the changing needs of the sitting tenants or to various needs of newcomers) are two examples which impact on residential change at both the micro- and meso-scale. On the other hand, processes in the entire city and the region must also be included because the inner city is not an independent unit. Rather it is influenced by residential processes in other areas (for example, by suburbanization or commercialization) and is itself impacting on such tendencies. Also, overall demographic processes at the urban scale (population growth or decline) play a part in inner-city residential change (Figure 2.2).

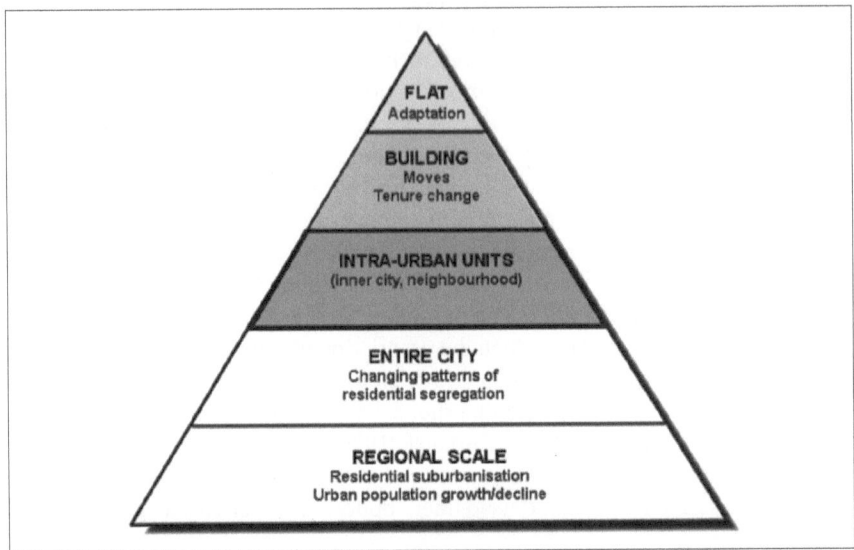

Figure 2.2 Socio-spatial embedding of inner-city residential change
Source: Authors' work.

Bridging these scales in a conceptually *and* methodologically convincing way was a major challenge. How we solved the problems of actors and structures as well as of different spatial levels will be explained in the following section.

Methodological Challenges: Integrating Methods and Perspectives

In our research, we wanted to understand how the changes in the residential patterns of Polish and Czech inner cities come about. This research provokes a number of methodological and technical questions. What kind of research strategies and what data are needed? How can different data sources and types be combined and compared? And how can one tackle residential change as a process? How can research in two countries and four cities be carried out by researchers from four different national and cultural backgrounds and from different disciplines?

The main methodological challenges were:

1. to describe and analyse aggregate socio-demographic patterns over time and to understand the behaviour and perceptions of households and individuals,
2. to transfer demographic knowledge based on the national scale to processes found at the urban scale,
3. to conceptually define and empirically delimit the inner city in specific case-study cities,
4. to relate the results to different spatial scales – from the flat and house via the inner city to the entire city level – and to interpret them within the specific local context,
5. to tackle change while investigating a specific point in time, using quantitative and qualitative data related to different time slots,
6. to compare – or at least relate and discuss – the findings from different cities in different countries and
7. to draw conclusions extending beyond the specific local case.

Mixed Methods as a Key to Analysing Residential Change

Urban research in and on Poland and the Czech Republic is largely based on census data, which is the only regular data base with small-scale information on housing and the population. In Poland, censuses date from 1921 (that is the Second Republic which is not identical with the Poland of today)[2] and, in the territory of today's Czech Republic, from 1869. They allow for a description of aggregate patterns at selected moments in time. This quantitative information is provided at different spatial scales, with the smallest units in Poland being *rejony statystyczne* or *rejony spisowe* and in the Czech Republic *základní sídelní jednotky* (in 2001) and *urbanistické obvody* (in 1991), respectively. These small-scale data also cover information on the socio-demographic and socio-economic characteristics of the residents as well as their housing situation. With the help of these data, we can consider the years 1988 in comparison with 2002 for Poland and 1991 in comparison with 2001 for the Czech Republic, describing the composition of residents of a place in a cross-sectional perspective and in comparison to an earlier census. Some census information can be backed up with registry data, available annually for the period after 2001/2002 but giving less detailed and almost no small-scale information, with respect both to social structures and to housing.

---

2   Even before 1921 there were censuses in Poland which, however, had not existed as a state between 1795 and 1918. In its Austrian, Prussian and Russian territories, from the second half of the 19th century onwards, censuses were carried out more or less regularly, but with different methods and at different moments in time. Thanks to Adam Radzimski (Leipzig/Poznań) for his support in clarifying this issue.

Yet, to understand perceptions and behaviour of residents, statistical data – be they from censuses or registers – are not appropriate. Instead, surveys or qualitative approaches are able to shed light on these issues. Surveys, unlike censuses, can look at collective specifications of preferences, choices and actual behaviour as well as their interpretation by the actors themselves. But to understand the motivations for individual action, to uncover the underlying, general structures of individual preferences, of orientations and decisions, qualitative research is needed. For example, in order to detect *in-situ* change of the physical environment, qualitative data might be most helpful. Moreover, such data help to cover census-free periods which, for our research, was a challenge for the years after 2001/2002.

In addition to the research challenges, or perhaps fortunately, this volume has been written by sociologists, geographers and cultural anthropologists who all use different methodological approaches. Thus, the investigations presented here combine a number of views on urban development, housing and demographic processes – with the major line of demarcation being that a quantitative, statistically-based approach was combined with a qualitative, phenomena and actor-oriented perspective. The data base is therefore quite rich, as the subsequent section outlines in more detail, but it creates another challenge: how does this fit together, how can one make sense of diverse data, collected in different cities and how can one really integrate the distinct approaches?

The resulting paradigm of *integrative or mixed methods* claims that, in order to capture both macro-societal patterns and individual agency, a methodological approach that combines qualitative and quantitative data and procedures is most appropriate (for a review see Morgan 2007, Greene 2008, Kelle 2008). That is, one can look simultaneously at aggregate patterns and see how social actors utilize their action spaces to form these aggregate patterns. In this way we can understand how actors 'may change culturally defined patterns of behaviour not only for themselves, but new patterns may also emerge if other members of their social group take over these patterns' (Kelle 2001: paragraph 39). This idea 'to link social behaviour to wider social systems' in order to 'meaningfully make interpretive sense of localized phenomena' (Greene 2008: 7) qualifies as a core argument of mixed methods as a new research paradigm. Instead of just aiming at a more detailed, enriched and colourful picture of the phenomenon under investigation, the understanding of the changes in social structures at the macro-level, rooted in individual action patterns and the agency of actors, is a theoretically-informed claim for the importance of mixed-methods approaches in social research.

Even though mixed methods are not new to the field of urban studies – we only have to think of the investigations of the Chicago School, or the community studies that always combined aggregate with qualitative data – the traces of the dualism of quantitative and qualitative methods are also present in the urban research community. The subject of this book – residential change – however, demands an understanding of the changes in both aggregate structures and individual action patterns, both in general and in particular. This need arises not just from the complexity of the subject, but also from the dynamics it possesses. From the very beginning of our research, we aimed at mixing methods to gain a fuller picture and a deeper understanding of the processes in the field. We were aware of the limitations of the census in providing a description of residential patterns. For instance, an empty-nester household consisting of two married adults whose children have moved out is listed in both the Polish and the Czech censuses as a family household. Young families with dependent children also appear in the same category of 'family household'. For the issue of residential patterns, these two household types make a critical difference. Moreover, the variety of housing and living arrangements is much larger than any census could ever capture (see also Chapter 3). Therefore, a self-contained qualitative study was carried out. The qualitative work became even more important when we realized that, after the last census, dynamic changes had occurred, as the empirical analyses in Chapters 7–12 show.

The basic problem that needs to be solved, however, is how the conclusions from both types of data can be brought together or, in other words, how social scientists can make sense of the findings from mixed-methods research. Udo Kelle, one of the voices calling for an integration of qualitative and quantitative data, claims that the basis on which qualitative and quantitative research can be integrated is not a pragmatic but a theoretical one (Kelle 2001, 2008). They can be integrated in describing and analysing what he calls "structures of limited range". These are social structures that are 'stable over a long period and, in still unpredictable ways, versatile' (Kelle 2008: 76). Household structures, for instance, are social structures that are both stable and at the same time flexible. To grasp change, Kelle employs the concept of agency, the individual competence of action that an actor develops over time. He considers individuals as actors who reproduce social structures through habits and routines. But when they are confronted with problems induced by macro-societal conditions, they use their action space and alter their routines. Together, the cumulative, collective effects of these individual actions change social structures (ibid.: 68–76).

This idea of structures of limited range provides a good template for the methodological concept of the work presented in this publication. Residential patterns are certainly social structures of limited range. They might be stable for a certain period, but they are always potentially in flux. Looking again at Figure 2.1, we see that both routines and change are in the arrows between the elements. The socio-demographic structures of a neighbourhood, for instance, change with individual mobility: people leave the neighbourhood, others enter it.

Existing residential patterns hold restrictions as well as opportunities for residential behaviour. As explained in Chapter 3, the use of flats is already limited by their floor plan. The number, size and structure of available flats influence household formation and housing arrangements. But the physical structures might be changed through the adaptation of flats and houses to meet the shifting needs of residents. It is the residents who for a long time just reproduce residential patterns through their habits, their everyday life and routines. Then, stimulated by shifts in societal or local context, individual behaviour changes. The collective result is a change of the aggregate social and physical characteristics of a place. By utilizing their action space, be it small or large, people gradually establish new residential patterns that might again be stable for some time.

Of course, residents act within a local and societal context that shapes their action space, for instance through the given housing markets, urban planning decisions, local labour markets, neighbourhood initiatives or greater shifts in society. Restrictions might be severe or people might be totally satisfied – if so, there will be little change.

The Challenge of the Cross-cultural Investigation and Positionality

Cross-national or, to be more precise, *cross-cultural research* has a long tradition in both housing and urban research. Since the 1990s, and very much influenced by international funding schemes (for example, the Framework Programs of the European Commission), the internationalization of social science research has intensified and has led to countless studies and papers striving towards insightful cross-cultural findings and comparisons. Yet, the methodological and conceptual challenges of such undertakings need to be solved from scratch for each research project. The path towards meaningful results is difficult but it can indeed lead to a process of learning about the two (or more) subjects of research, and not just the 'other', non-familiar one (Steinführer 2005). In a paper on *Meaning, science, context and confusion in comparative housing research* from 2001, Michael Oxley attaches particular significance to small-scale, in-depth approaches:

> Rather than study whole countries and whole housing systems, there is much that might be gained from micro-scale housing studies that, for example, focus on towns, sets of households or individual housebuilders or landlords, comparing these between countries. [...] It would in most cases involve primary data collection. We might learn much about the whole by studying a part and putting this in context. (Oxley 2001: 103)

One of the starting points of the research presented in this volume was the specific transitional path of eastern Germany, with its experience of deindustrialization, high unemployment, strong decline in birth rates, out-migration and subsequent 'urban shrinkage'. Initial research questions and the methodological design were thus created with very specific perspectives, which also implied certain expectations concerning the possible results of the field research. However, among the researchers involved, this view did not remain unquestioned and led to a continuous process of reflection on the appropriateness of specific research questions, concepts and methods. In the literature on cross-cultural research, these problems are discussed under the keyword of *equivalence*. It is considered as a methodological criterion which, in cross-cultural investigations, is as important as the classical criteria of validity and reliability. Two central notions of equivalence were elaborated by Timothy P. Johnson, following his analysis of a total of 52 different equivalence labels (Johnson 1998: 3–10):

- the equivalence of the meaning of certain concepts in different national and cultural contexts (interpretative equivalence) and
- the equivalence of the measures and procedures used in empirical research (procedural equivalence).

Interpretative equivalence relates to questions of the meaning of research concepts in different cultural settings and national research traditions – and, very often, the use of the same wording actually conceals differences in meanings. Procedural equivalence is a major criterion with regard to methodology and methods. It also implies adequacy and appropriateness of certain instruments in different contexts – it is, for example, not necessary to put the same question in different languages (and, indeed, it might even mean something different, depending on the language). Instead, it is important to ask context-adequate questions (Harkness and Schoua-Glusberg 1998). These two types of equivalence are to be understood as ideal types which can never be reached completely in practice, but which are a standard to be reflected upon.

This leads to the problem of *positionality*, that is the connection between researchers' backgrounds and the questions they ask, as well as the way they conduct research (Hörschelmann 2002). Even though positionality cannot be eliminated, its effects can be limited. This is all the more important in cross-cultural research settings, since there is usually some kind of hegemony in the origin of concepts and research questions. For instance, the male bias is one of the most discussed issues in social science in general. In urban research, many concepts and questions derive from hegemonic western perspectives. Their appropriateness for post-socialist contexts is questioned, just as the label 'post-socialist' itself has its explanatory limits (see Chapter 4).

The experience of the transition in East Central European countries, the researchers' personal familiarity with rundown inner-city residential areas in post-socialist cities as well as the experience of population decline and housing vacancies in eastern Germany, shaped awareness and influenced research questions and conduct. We discussed terms, compared concepts and thought carefully about uniform ways to delimit the inner cities. We even agreed on the use of terms and concepts in an (internal) project dictionary. The aim of such undertakings is, again, equivalence. Thus, equivalence is not only a measure of methodological quality but also a way to limit positionality.

Research Design, Applied Methods and Problems Faced

Research Design and Applied Methods

The research was organized around a core study on inner cities of Polish and Czech second-order cities (Chapters 7–9) together with some additional smaller, 'satellite', studies on specific topics such as residential flexibility, perceptions of specific types of building stock and symbolic differentiation, as well as sociability and the transformation of single residential buildings (Chapters 10–12). The core study investigates residential change in the inner cities of Łódź, Gdańsk, Brno and Ostrava, using the research methods explained below.[3]

The first step was to conduct statistical analyses, based mainly on spatially-resolved data from the last two censuses in Poland (1988, 2002) and the Czech Republic (1991, 2001). The data were analyzed on three spatial levels: the city as a whole, the inner city (as defined in Chapters 5 and 6) and selected neighbourhoods within the inner cities. The inner-city neighbourhoods were chosen to capture varieties of socio-spatial or demographic processes within a given local context and are thus not directly comparable across contexts. In addition, we also included local and regional register data (mainly demographic indicators), as well as national population data, to contextualize the local situation.

Secondly, interviews were carried out with residents of the inner cities as well as with local experts. The resident sample was gathered as a selective sample that covered certain types of households, reflecting our specific interest in demographic issues. About half of the interviews were carried out with representatives of non-traditional households, that is living and housing arrangements that depart from the core family (see Chapter 3 for more details). Interview partners were found using a snowball system.

The semi-structured interviews covered the housing biography of interviewees, their perception of the current place of residence, housing preferences and future plans. The recorded interviews were transcribed and then coded inductively. On this basis, we developed types of attitudes towards the inner cities in relation to residential preferences, perceptions and decisions (see especially Chapters 9 and 10). In Brno, further interviews were conducted outside the inner city in order to obtain more insights into other residential areas, particularly the large housing estates. Some of these interviewees were former residents of the inner city.

The perspective of the residents was rounded out by information obtained in expert interviews. These aimed at gaining a better understanding of historical and current developments in the respective city, major challenges faced by local governance and town planning, the housing market situation and patterns and processes of residential segregation. These experts were representatives of the local or district governments and neighbourhood organizations, as well as developers, real estate agents, urban planners and the like (Table 2.1).

---

3 The methods used in the 'satellite' studies are explained in the respective chapters.

Table 2.1 Overview of the semi-structured interviews conducted in the case-study cities

	Łódź (PL)	Gdańsk (PL)	Brno (CZ)	Ostrava (CZ)	TOTAL
Interviews with inner-city residents	16	20	17	16	69
Interviews with residents from outside the inner city	–	–	13	–	13
Interviews with local experts	9	9	12	9	39

Source: Authors' work.

The statistical data gained from the censuses and the population registers, on the one hand, and the information gathered via the interviews, on the other, are completely independent. They differ in their methodological background, were collected for different reasons and by different means. They also represent different time slots. The last censuses were conducted in 2001 (Czech Republic) and in 2002 (Poland), respectively, while the qualitative data were collected in 2007 and 2008. Additionally, many of the cases within the qualitative dataset do not appear in the official statistics, be they census or register data. This point will be elaborated in more detail below.

Problems Faced in the Empirical Phase

When we began our research in Poland and the Czech Republic in 2006, we were half-way between two censuses. This caused a certain dilemma: we knew that the available census data would already be outdated, however – as already stated above – in both countries it is the only data set that provides spatially-resolved data for intra-urban units. So we had to look for additional and alternative data. Among them are, most prominently, register data collected by the regional statistical offices (GUS in Poland and ČSÚ in the Czech Republic). These allow for long-term analyses of both natural population change and migration. Yet, in both countries, the register practices are subject to strong criticism. In Poland, the population registers are regarded as significantly incomplete. This is especially true for students who – according to some studies in Polish university towns – register fairly irregularly, except if they live in halls of residence. A study conducted in Poznań in 2005 revealed that about 39 per cent of all students studying in that city were not registered (Marcinowicz 2006: 49). A further major problem relates to the inadequate coverage of migration: most international immigrants are excluded from the official statistics (Kupiszewski and Bijak 2006, Bijak et al. 2007). Moreover, it is assumed they also cover just 20–30 per cent of the actual international emigration (Śleszyński 2004b, Mykhnenko and Turok 2008: 324, 333).

In the Czech Republic, there are two main data sources available, which differ greatly in numbers. The population records compiled by the Czech Statistical Office (ČSÚ) are based on census data and are regularly updated with data on deaths, live births and migration. Yet, and just as in Poland (and elsewhere), these data become increasingly erroneous over the course of time, in particular due to the inadequate coverage of international and domestic migration. While the ČSÚ data are based on population balances, another data source (ISEO registry, kept by the Czech Ministry of the Interior), takes the individual as a starting point. It contains every single citizen with Czech nationality registered in the place of his/her permanent residence – no matter whether this person is also *de facto* living there or, in some cases, is still alive. Quite naturally, the different methodological approaches, a slight under-enumeration of the population in the last census, gaps in data from some municipal offices, an uncertain number of Czech citizens living abroad but still registered at home and poor quality statistics recording foreigners, lead to significant differences between the two data sources. By 1 January 2006, for example, the difference between the ČSÚ database and the ISEO registry on the national scale was almost 227,000 persons (see also Steinführer et al. 2010). In subsequent years it decreased due to register corrections. Neither the Czech Statistical Office nor the Ministry of the Interior can judge which data base is more appropriate to calculate the actual population total. There are indications of both over- and underestimations. Not least for this reason, some municipalities have established their own registers (among them Brno), which again differ from the previously mentioned data bases.

Implications for the Volume

Residential change in East Central European second-order cities is the core process this volume is concerned with. A few points need to be taken into account:

- Residential change always involves both people (residents) and place (residence), as well as their interactions, mediated by the local and societal contexts.
- Up to now, both urban and housing studies are, to a large degree, restricted to questions of residential mobility. Non-moving behaviour is termed *immobility* and taken as evidence for lack of change. Based on our model of residential change, we also pay attention to *in-situ* and 'non-spectacular' changes, which are brought about by remaining at a certain location rather than by mobility.
- Demographic changes are regarded as major drivers of residential change. But it remains a challenge to determine the dependent and independent variables. Is demographic change one cause for urban dynamics, among others, or are demographic features always the outcome of broader societal processes (as is suggested by Bürkner 2008)?

- We pay deliberate attention to the context, that is, the framework conditions of residential change at different scales – be it the inner city itself, local housing markets or existing patterns of residential segregation. The perspective of the residents needs to be completed by information from intermediary actors.
- A mixed-methods approach is indispensable in order to understand the opportunities and limits of particular data sources.
- Several methodological problems arise from the research setting (cross-cultural investigation) and the data base. Solving them means paying particular attention to the criteria of interpretative and procedural equivalence, as well as of adequacy and positionality.

Chapter 3

Housing, Households and Demographic Challenge in Urban Space: Conceptual Considerations and Context Conditions in East Central Europe

Annett Steinführer, Ray Hall

Introduction

That demographic processes always have a spatial dimension has long been known to urban scholars. Nineteenth and 20th century urbanization was the story of rapid population growth as a result of migration, a dynamic process which had profound consequences for the urban fabric, social heterogeneity, patterns of socio-spatial differentiation and local governance. Both increase and decline in the size of the urban population impact in the short and long term on the demand for housing in terms of numbers, type of amenities and other characteristics. But it is not only the numbers of people that matter; qualitative changes in the population are also highly significant for urban space. As discussed in Chapter 2, one facet of demography and demographic change is of particular interest in this volume: private households, their housing practices, needs and decisions, together with the ways they bring about residential change, which also always implies urban change.

With the increasing interest in demographic issues in recent years, the interactions between urban populations and household change as well as their impact on socio-spatial patterns and residential change need to be rethought. This chapter aims to contribute to such a rethinking, both in general terms and with a specific focus on East Central Europe. There are five sections, the first of which examines the conceptual links between housing and demography. The second introduces the idea of the second demographic transition (SDT), to provide a broader conceptual background. The third part examines the urban dimensions of demographic change. The fourth contextualizes demographic change in East Central Europe while the final section discusses the implications of the issues raised in the chapter for the volume as a whole.

Housing and Demography: Conceptual Considerations

The Household: A Key Concept to Link Housing and Demography

Key to the relationship between demographic and residential changes is the *household*, more particularly the 'number, size and composition of households' (Gober 1992: 172; see also Rossi 1980: 20, Myers 1990a: 5). From a geographical perspective, the notion of the household carries with it the compelling idea of the interrelatedness of people and place. In this volume, we conceptualize households as *groups of people with a relatively binding internal structure, mutual economic responsibilities, a dense network of interrelations, a common (permanent) residence and the self-perception of being a household. A person living alone always forms a household and is thus a specific case.* This definition stresses the subjective meaning of the household for its members and its housing, rather than, for example, its economic function. It is therefore appropriate for the objectives of the research presented in this volume.[1] However, such a subjective conceptualization is in contrast with typical statistical (objective) definitions as, for example, by national censuses or the United Nations (UN; for an overview see Keilman 1995). In the UK, for example, a household is defined as a person living alone or a group of people who either share at least one meal a day or share living accommodation. The determining factors are thus meals and sharing accommodation, whereas the relationship between the people in the household is irrelevant. In France, all persons living in one dwelling form a household, while a 'statistical' household in Germany consists of all persons who live together and finance their living costs jointly (Housing Statistics in the European Union 2004: 98, 100). The UN recommendation is – in the exact wording – adopted by the Polish statistical definition of the household (Figure 3.1). Poland is, thus, applying the so-called 'house-keeping concept'. This approach is distinguished from the 'household-dwelling concept' which equates household and dwelling (UN ECE 2006: 106, similarly Gober 1992: 172). The Czech census, in contrast, has for some decades applied both the house-keeping and the house-dwelling concept (Figure 3.1). Irrespective of the precise wording, it is important to note that research on households in either its subjective or its objective conceptualization will bring about different, possibly even distinct and contradictory results.

---

1   Depending on the discipline, other definitions stress instead its economic, biological and sheltering functions, such as production, consumption, inheritance or reproduction (for example, Haviland 2003: 544).

Poland and the Czech Republic: Distinct Household Definitions

The Polish statistics differentiate two types of households: 1) one-person households, that is a person living alone in a separate housing unit or as a lodger together with one or several other household(s), and 2) multi-person households, that is 'a group of two or more persons who combine to occupy the whole or part of a housing unit and to provide themselves with food and possibly other essentials for living. Members of the group may pool their incomes to a greater or lesser extent.' This definition is identical with the one adopted by the United Nations.

In the Czech census statistics, up to 2001 there were three household types: 1) dwelling households *(bytové domácnosti)* consisting of all people living in one dwelling; their number was therefore always identical with the number of dwellings, 2) economic households *(hospodařící domácnosti)* consisting of the persons living together in one dwelling who share basic food expenditure, housing costs and so on; subtenants always formed separate households and 3) census households *(cenzové domácnosti)* created by people living together in one dwelling who are either blood relatives or have other relationships. In the 2011 census, a new approach will be adopted restricted to dwelling and economic households. Thus, the specific Czech tradition of applying both the house-keeping and the house-dwelling concept will continue.

Figure 3.1 Households as defined by the Polish and Czech statistics
Sources: Bartoňová 2003, Housing Statistics in the European Union 2004: 98, 100, UN ECE 2006: 107.

In cross-cultural research, the linguistic dimension of the issues under investigation must also be considered. In some languages, household carries with it the connotation of home (as, for example, in the English term as well as in the German *Haushalt* or the Czech *domácnost*). Sometimes it expresses the economic function (as, for example, in the Polish *gospodarstwo* or, again, the German *Haushalt*).[2] Finally, there are languages without a linguistic equivalent. For example, in Italy, *famiglie* refers to households as well as families but there is no proper substitute for the exact term *household* known in many other languages.

For a long time, the main reference household used to be the family, a term no less unclear and ethnocentric. Anthony Giddens defines family as 'a group of persons directly linked by kin connections, the adult members of whom assume responsibility for caring for children' (Giddens 2009: 331). For much of European history, the terms *household* and *family* were almost synonymous. For example, in English, in the past, the word 'house' could mean a kin group – people related to

---

2   In German, this economic dimension is also expressed by the obsolescent verb *haushalten* (to budget) which is directly derived from the noun. In Polish, a corresponding verb also exists: *gospodarować*.

each other – as well as a physical structure (Laslett 1974: 28). Today it is widely accepted that the family is something distinct from the household and *vice versa*:

> Household refers to the group of co-residents, people who live under the same roof and typically share in common consumption. Family is a much more ambiguous term; it refers to close kin, but the exact reference of the term tends to vary contextually. From an anthropological viewpoint, the key distinction is between a domestic group, on the one hand and a kinship system – with its attendant categories of kin – on the other. (Kertzer 1991: 156, emphasis omitted)

This is not to say that a household might not be equivalent to a family or the other way around, but we want to stress that such overlapping is just one expression of an increasing diversity of both households and families.

The existence of at least a two-generation context (that is parent/s and child/ren) is central to many definitions of the family. Yet, this notion was and still is strongly challenged by the rise of alternative forms of living together, particularly by cohabitation, same-sex partnerships without or with children and unrelated people living together (see also Waite 2005: 93). Therefore, demographers are now more cautious in defining the family and are also aware of the threat of ethnocentrism, as is obvious from Linda Waite's definition: 'In most times and places, families were responsible for the production, distribution, and consumption of commodities, for reproduction and socialization of the next generation, and for coresidence and transmission of property' (ibid.: 88). However, the so-called core (or nuclear) family, usually understood as a unit of two heterosexual adults rearing at least one common child, remains a major point of reference in household demography. Beside the two-generation structure, traditional western notions of the core family strongly relate it to kinship, marriage, coresidence and the presence of two (rather than just one) adults. In the 19th and 20th centuries, moreover, a gender-specific way of sharing the work (male breadwinner-housewife model) became central to the understanding of the core family. Today it is widely accepted that such an understanding refers to a normative conceptualization of 'how families should be' (Silva and Smart 1999: 1) and is only one of a wide variety of actual family realities and practices. However, in the debate about the 'new', 'alternative', 'non-traditional' or 'non-conventional' household types, this normative reference point of the western core family is still prevalent.

Based upon these considerations, we distinguish five basic household types (see also Table 3.1 below):

1. one-person household,
2. cohabiting couple,
3. family household with dependent child/ren,
4. other types of family household,
5. unrelated persons living together.

   With respect to residential change, it is important to note that these household types might exhibit a number of socio-demographic compositions (living arrangements) and forms of housing (housing arrangements). While the concept of *living arrangements* pays particular attention to the legal dimension of marital status, as well as the presence or absence of children, the idea of *housing arrangements* mainly relates to physically living together or apart, in one or several residences (similarly Mulder and Dieleman 2002). From Table 3.1 it becomes obvious, for example, that the wide-spread phenomenon of younger single people living alone is just one manifestation of the one-person household. Equally important are widowed and divorced persons, as well as people with a permanent partner but still living on their own (LAT – living apart together households). Not only the living arrangements, but also the housing arrangements of one-person households differ: they might live on their own, but also share a flat or a house or live as subtenants. Yet, these arrangements also depend on national housing cultures and local housing markets (as discussed in Chapters 8 and 9). Other household types include cohabiting couples (heterosexual or same-sex), as well as one- or two-parent families with dependent or adult children. The phenomenon of adult children living with their parents has become more significant recently: for example, between 1995 and 2005 in most European countries the average age at which both young men and young women leave the parental home has increased (Eurostat 2008: 26–8). This development has different reasons, including unemployment and the affordability of housing particularly in larger urban areas. Unrelated people might form one or several households, for example, as a flat or house share or in a landlord/main tenant-subtenant relationship. Flat or house sharing represents a specific housing arrangement but, under certain conditions, it can also be a household type or even a living arrangement. This depends upon how it is perceived by the actors themselves. Sharing a flat together with the housing costs, is the basic reason for such an arrangement, however, it is evident from the wide-spread *Wohngemeinschaften* in German or Austrian university towns, that living with others for companionship is another important motive (Steinführer and Haase 2009b). Thus, a single person might live in a flat share but regard him- or herself as a one-person household or, on the contrary, stress the communal living. Flat or house sharing may be unavoidable where there is high demand for housing and housing market pressure in economically prosperous agglomerations, so it is an involuntary housing option rather than a desired living arrangement. To decide upon its meaning for the individual is, therefore, not a conceptual question but an empirical one.
   Finally, Table 3.1 also points to one more problem faced by researchers when investigating households and residential change. Even though we emphasized common residence as one defining element of multi-person households, this does not mean that households always have only one place to live. As a result of commuting to places of work or education, a second residence might be needed without necessarily dissolving the original (primary) household (see the growing literature on 'multi-local housing', for example, Weichhart 2009).

Table 3.1 Basic household types and their possible living and housing arrangements

Household type	Possible living arrangements of the household members (examples)	Corresponding housing arrangements (examples)*
1. One-person household	Single Widow/er Divorcee	Living alone Flat/house share Subtenancy
	Living apart together (two persons of opposite sex with intimate relationship), married/not married Living apart together (two persons of the same sex with intimate relationship)	Two persons, each living separately (alone, in flat/house share or subtenancy)
2. Cohabiting couple	Married/non-married couple of opposite sex, without children Couple of the same sex without children Married/non-married 'empty-nester'-couple	One common residence Flat/house share
3. Family household with dependent child/ren	Married/non-married couple of opposite sex with at least one child Couple of the same sex with at least one child 'Patchwork'/step family with at least one child	One common residence One main and one secondary residence (due to commuting) Flat/house share
	One parent with at least one child	One common residence Subtenancy Flat/house share
4. Other types of family household	Married/non-married couple with at least one adult child One parent with at least one adult child Three-generation household	One common residence

5. Unrelated persons living together	Single
	Widow/er
	Divorcee
	Living apart together
	Married/non-married couple
	Married/non-married couple with at least one dependent child
		Living together with unrelated others in:
		   flat/house share
		   subtenancy

Note: * All types of collective housing (such as students' dormitories or nursing homes) are not included.
Source: Authors' work.

Table 3.1 is far from being exhaustive. A number of household dimensions are not considered, for example, the economic relationships between the members of the households, their age structure or internal division of labour, as well as their participation in the labour market. Yet, linking household structure, living and housing arrangements together mirrors one of the fundamental research interests of this volume: the interrelationships between households and the residential change they bring about.

The apparently simple, but long-neglected idea that it is households and not individuals that change residence – be it via intra-urban mobility, the change from an inner-city to a suburban location or migration elsewhere – dates back to Peter H. Rossi's *Why families move* (see also Chapter 2). Another key legacy of Rossi's work was the conceptual consideration 'that household housing needs are strongly conditioned by stages of family life cycles' (Rossi 1980: 25). Although this idea was evident in Burgess' model of concentric zones as early as the 1920s (Burgess 1984), a longitudinal perspective on the household and related residential change was still missing. After Rossi's work, this notion has spread to residential mobility research (for example, Gober 1990, Herlyn 1990, Kendig 1990, Mulder and Hooimeijer 1999, Mulder 2007). The German sociologist Ulfert Herlyn speaks of the 'spatial mediation of the life course' (Herlyn 1990: 7), meaning that each life phase is connected with certain (typical) housing arrangements and, ideally, also with a certain type of neighbourhood. Although there is no 'one size fits all' dwelling or neighbourhood, households need to be considered in time, since they change during the life courses of their members: they might expand, shrink, split up, or dissolve completely. As such, they require varying forms and types of housing. Such longitudinal perspectives are particularly needed in light of the second demographic transition as introduced below which brings about a higher degree of household instability and a decreased importance of 'one' standard life course (previously described as the 'family life cycle'; Loomis and Hamilton 1936, Glick 1947, Gober 1992: 174). Yet, the notion of passing through a series of life stages (childhood, post-adolescence, founding a family and rearing children, empty nest, widowhood) – always more an ideal than reflecting a real life course – lost its explanatory power in the second half of the 20th century. Table 3.2 provides an illustrative example of both a longitudinal and cross-sectional perspective on the household.

Table 3.2 Households, living and housing arrangements in time and space – one example

A household trajectory in a longitudinal perspective	The same household(s) in a cross-sectional perspective
two single partners establishing an intimate relationship (and describing themselves as a couple) without permanently sharing one and the same dwelling – after a certain time deciding to merge the two separate households into one – founding a family with one child – getting married – having a second child – one partner finding a job elsewhere and running a second household close to his/her workplace – getting divorced after their children have left home and dissolving their common household – ...	• two singles living on their own • a LAT (living apart together) arrangement: two partners occupying separate dwellings but regarding themselves as a couple • a cohabiting, non-married couple • a core family with non-married parents and one child at one common residence • a core family with two children and married parents at one common residence • a core family with two residences • two divorced persons living on their own, children founding their own households ...

Source: Authors' work.

Changing households need changing or at least flexible dwellings. This very basic statement implies that people might adjust their housing situation to a changing household structure by either moving or adapting their accommodation. Therefore, our conceptual approach (Chapter 2) and the empirical analyses (Chapters 7–12) consider both the options which can be taken by households – moving or staying – and the residential and, therefore, also urban change they cause.

A New Attempt at 'Housing Demography'

The interconnections between housing and demography are explored by different research fields. With the volume *Housing Demography,* published in 1990, the US-American geographer Dowell Myers explicitly tried to bridge what he perceived as the considerable gap between (demographic) population and (geographical and sociological) housing research (Myers 1990b):

> The close connections between housing and demography have been long neglected. Until recently, the two topics were pursued by different groups of researchers who did not communicate. [...] In the past, demographers have emphasized fertility and the structure of household groups in which the vast majority of the population lives, but they have traditionally stopped short of investigating how the search for accommodation in housing units affects those households. On the other hand,

economists, geographers, and urban planners have explored housing as a key component of urban structure, but they have rarely examined the demographic details of who lives in those housing units. (Myers 1990a: 3)

Myers (1990a: 11–12) conceptualized the new field of research as the interface between population and housing that might be approached from both the demand side (population-based inquiry) *and* the supply side (housing-based inquiry). The main dimensions of housing demography are household formation and composition, housing choices, housing construction and inventory change, as well as the resulting spatial patterns (ibid.: 13–17).

While Myers was optimistic about overcoming the gap between the two strands of research, there has been little subsequent work labelled as 'housing demography' (Gober 1992, Reed 2001, Mulder 2006) and what there is has mostly remained at the level of conceptual considerations, rather than empirical research.[3] Recently, the German geographer Olaf Schnur maintained that the 'ambitious concept of "housing demography" to date has resulted in only a few practical works and therefore remains as no more than an approach without a completed base' (Schnur 2008: 33). Sixteen years after Myers's volume, the Dutch geographer Clara Mulder again addresses the same problem when complaining about 'little research [...] devoted to the relationship between population and housing' (Mulder 2006: 402).[4] Mulder pays great attention to the two-sidedness of the relationship between housing and population. She states:

> Population influences housing via housing demand. But also, housing influences the number of people and households via the attraction or deterrence of migrants, keeping in place or pushing away the resident population, and intricate links with leaving the parental home, separation, and having children. These connections between housing and population vary over time and between places. (ibid.: 409)

Since this volume wants to explicitly link people and place, residents and residence, households and inner cities, we see it as part of the effort to further establish the field of housing demography. Our primary interest is inner-city residential change in East Central Europe – covering both socio-demographic

---

3 See also the related debate from an anthropological perspective, once called 'demographic anthropology' (Howell 1986), more recently 'anthropological demography' (Bernardi 2007).

4 This is not to say that there is no research at all on this issue. In particular, population geographers have published a number of papers on this relationship (for example, Murphy and Sullivan 1985, van Engelsdorp Gastelaars and Vijgen 1990, Clark and Dieleman 1996, Ogden and Hall 2000, Lee et al. 2001, Hall and Ogden 2003, Buzar et al. 2005, Ogden and Schnoebelen 2005, Mulder 2007, Levin et al. 2009). Also in economics there is a debate on the interrelatedness of housing markets and households (for example, Mankiw and Weil 1989 and the ongoing follow-up debate on their hypotheses, as well as van Imhoff et al. 1995).

change and changing housing realities, set within the framework of broader societal processes of demographic change and post-socialist transition, at the scale of the inner city. We conceptualize residential change in an explicit micro-meso-macro approach by analyzing the rationales, perceptions and actual behaviour of the residents and taking into account the overall local context as well as the specific housing markets.

The interrelations between demographic features (in particular household change) and housing become even more interesting in the case of rapid change, be it societal transformation (as, for example, in East Central and Eastern Europe over the past two decades, Chapter 4) or demographic change, as discussed in the next section. Changing social conditions, then, meet built structures, which are much less subject to change and thus characterized by a high degree of inertia. While inertia (or persistency) is also typical for societal institutions and social practices, it is a major characteristic of the urban fabric and the housing stock. Residential dwellings, along with technical and social infrastructure, are built at certain moments in time for specific (assumed) needs and preferences, but they usually last longer than their first-generation users. However, with changing household patterns, a de-standardization of the life course, smaller and fewer families and an increasing number of one-person households – all typical features of the second demographic transition as described below – housing demands are changing. For example, hierarchical floor plans which are so typical of 20th century architecture (see Figure 3.2) and which start from the assumption of a 'typical household' – usually a core-family household with two adults and at least one child – might be appropriate for a young family as a starter unit, but become less suitable once the child grows older or a second child is born. With greater individualism and increasing social welfare in the western world over the last few decades, the spatial demands of individuals in the private sphere have changed. Thus the dwellings shown in Figure 3.2 might today become a home to a widowed pensioner or a young person living alone rather than to a nuclear family with children.

The example provides evidence for both the inertia of built structures *and* their relative flexibility – that is the capacity of urban dwellers to change original intentions, to adapt physical structures to their needs and to keep housing stock functioning in the longer run (Chapter 10). It also highlights the need for both institutional and physical framework conditions of the housing market and the urban fabric (Chapters 6, 8 and 11) to be taken into account when analyzing the interrelationships between demographics and housing as well as their consequences for inner-city residential change.

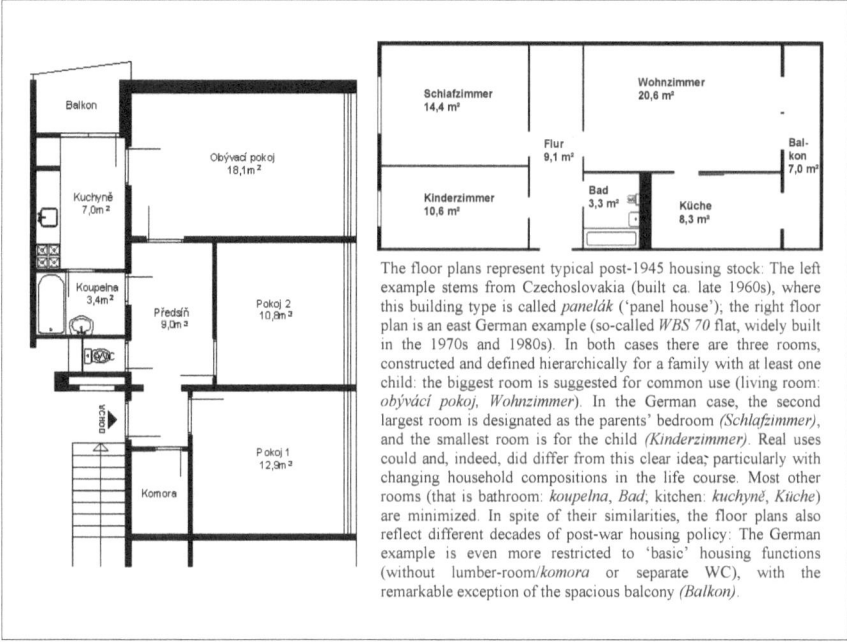

The floor plans represent typical post-1945 housing stock: The left example stems from Czechoslovakia (built ca. late 1960s), where this building type is called *panelák* ('panel house'); the right floor plan is an east German example (so-called *WBS 70* flat, widely built in the 1970s and 1980s). In both cases there are three rooms, constructed and defined hierarchically for a family with at least one child: the biggest room is suggested for common use (living room: *obývácí pokoj, Wohnzimmer*). In the German case, the second largest room is designated as the parents' bedroom *(Schlafzimmer)*, and the smallest room is for the child *(Kinderzimmer)*. Real uses could and, indeed, did differ from this clear idea; particularly with changing household compositions in the life course. Most other rooms (that is bathroom: *koupelna, Bad*; kitchen: *kuchyně, Küche*) are minimized. In spite of their similarities, the floor plans also reflect different decades of post-war housing policy: The German example is even more restricted to 'basic' housing functions (without lumber-room/*komora* or separate WC), with the remarkable exception of the spacious balcony *(Balkon)*.

Figure 3.2 Floor plans of standardized post-1945 housing stock (Czech Republic and Germany)

Source: Courtesy of Jindřich Pospíšil, Brno (left); authors' archive (right).

Core Processes of Contemporary Demographic Change

Conceptual Embedding: The Debate on the Second Demographic Transition

The term *second demographic transition* was coined in the 1980s by the demographers Ron Lesthaeghe and Dirk van de Kaa to describe the complex individual processes behind the macro-patterns of declining fertility, a sustained increase in life expectancy and a consequently ageing society, the postponement of, or complete abstention from, marriage and parenthood, as well as decisive transformations of household structures accompanying changed perceptions of family and parenthood (Lesthaeghe and van de Kaa 1986; van de Kaa 1987). In a continuous process since then, the concept has been further refined in a number of national contexts (van de Kaa 1994, 2004, Lesthaeghe 1995, Kotowska 1999b, Sobotka et al. 2003). Though not undisputed, the concept represents one of today's grand theories of demography. Some of its critics, for example, dispute both the claims of it being something decisively new and of being universal, or at least cross-European. Others see the changes as a continuation of the first demographic transition and support their claim with many contradictory findings from various country case studies. Criticisms were also formulated with respect to the assumption of converging patterns across Europe and the weak theoretical foundations of the concept (Huinink 2001, Sobotka et al. 2003: 254–5, Coleman 2004). However, even for these critical demographers, the second demographic transition is respected as 'undoubtedly the theory of the decade' (Coleman 2004: 11), whose limited coverage, with respect to time and space, needs to be taken into account.

Both the first and the second demographic transitions were characterized by fertility *and* mortality rates that were lower than in the respective previous periods. In a historical comparison with the so-called first demographic transition which occurred in Europe in the last decades of the 19th and in the early 20th century, Lesthaeghe (2007) summarizes the outstanding features of the second demographic transition in the following way:

- later marriage,
- increase in cohabitation and single living,
- increase in divorce,
- low remarriage rates,
- fertility postponement,
- efficient contraception,
- increase in the share of children born out of wedlock and
- greater childlessness.

In particular, the SDT is characterized by four factors that occur simultaneously:

- a long-term decline in fertility well below the so-called replacement level in most European countries from about the 1960s onwards,
- ageing that results from different interdependent processes: the current low fertility rates, the historical decline in fertility after the post-Second World War baby boom and the increased and ever-increasing life expectancy at birth, due to considerable improvements in welfare, hygiene and medicine,
- far-reaching changes to household structures accompanying these processes and
- international and intra-regional migration gaining importance.

At the same time, these four interrelated processes represent major aspects of demographic developments and will be discussed in a slightly different order in the following sections. After presenting the two sides of population development (natural change and migration) that are particularly important in quantitative terms, ageing and household change are discussed with a specific focus on the qualitative changes they bring about.

Population Development: Natural Change and Migration

Quantitative population development represents one of the key concerns of demographers. Population growth or decline is a result of the interplay of natural change and migration. The long-term decline in fertility below the replacement level of a total fertility rate (TFR) of 2.1 births per woman in almost all countries of the western world was defined as the 'principal demographic feature of the second transition' by van de Kaa (1987: 5). The wide range of societal causes and frameworks far exceeds the scope of this chapter. Currently, demographers distinguish between low and 'lowest-low' fertility countries where the TFR is 1.3 or lower (Kohler et al. 2002). In 1999, there were 14 European countries in the latter category. With three exceptions, all of them were in Eastern Europe. Very generally speaking, fertility decline is a result of a sustained postponement of childbearing (Billari et al. 2006), a long-term reduction in the number of children born to a woman, as well as increasing numbers of women and men remaining childless (Konietzka and Kreyenfeld 2007).

Migration, the second driver of population development, is the reason why many European countries have maintained population growth in spite of declining fertility. Today, some policy makers and demographers refer to 'replacement migration' and consider migration as a way of counteracting low fertility (UN 2001, for a critical view Keely 2001). In Western Europe, more than three quarters of the population growth between 1997 and 2006 was due to immigration, that is in terms of numbers of immigrants, but also, to a lesser extent, as a result of the higher birth rate of migrants. Moreover, immigrants contribute to a more balanced age structure of European populations (Sobotka 2008). However, migration is the aspect of population change which is the most difficult to measure and to conceptualize (Brown and Bean 2005) and, more profoundly, to make projections about. International migration is a decisive factor on the national scale, but migration also occurs within countries and regions and thus influences the population size and composition of sub-national units. Moreover, it is a multi-faceted phenomenon with a number of possible categorizations along different dimension, such as:

- legal versus illegal migration (legal dimension),
- long-term, short-term (seasonal), circular versus return-migration (temporal dimension),
- migration for economic, religious or political reasons (rationale dimension),
- migration of highly qualified versus migration of unskilled workers (socio-economic dimension),
- migration during the qualification, work or retirement phase (biographical dimension) and
- migration of individual men and women or of families with children (gender and demographic dimension).

This list is far from complete; it only highlights the broad social behaviours and motives concealed behind the term *migrant.*

Ageing

While rates of population change vary considerably throughout Europe, a consistent demographic trend across Europe is ageing. Its main determinants are low fertility rates and increased life expectancy or, to quote Uhlenberg (2005: 143): '[...] population aging is the price paid for solving the challenge of population growth'. Ageing brings about long-term consequences for population structure and infrastructure needs. It is the most outstanding feature of Europe's demography: ageing affects all regions and most of its major cities (EC 2006, Kabisch et al. 2008). The median age in Europe in 2003 was 37.7 years, compared to a global value of 26.4 years (CoE 2005: 13). However, it is important to remember that older people do not form a homogeneous group but are of different social strata, lifestyles and states of health. An important aspect of ageing is the growing number of very old people, an age group that is growing faster than all the others (Rychtaříková 2008: 101). The German gerontologist Hans-Peter Tews (1993: 17) speaks of a 'threefold ageing' comprising (a) the absolute and (b) the relative increase of older people, as well as, (c), the faster increase in the number of over-75-year-olds. The same author conceptualized what he calls 'structural transformation of age' (*Strukturwandel des Alters*; ibid.: 23–32) with the help of five characteristics:

- *Rejuvenation of older age groups*: As a result of increased life expectancy, together with an, as yet, more or less unchanged official pension age across Europe, the length of life remaining after leaving paid employment has increased significantly. Moreover, the self-perception of 'being old' has been transformed in recent decades. Today it seems more appropriate to take into account the heterogeneity of older age groups, for example, by differentiating them into several groups.
- *Living without professional duties and relationships (Entberuflichung)*: Pensioners in affluent western societies are much more disconnected from working life now than in earlier decades. Though differing from country to country – and certainly different in many countries in eastern parts of Europe – many are not integrated in the labour market at all. Tews (ibid.: 26) stresses that the very notion of the post-working phase of life is anticipated by people a long time before reaching pension age and, thus, considered normal.
- *Feminization*: Women have, in almost all societies, a longer life expectancy than men. Moreover, in many European countries, some male cohorts among the old are still significantly smaller as a result of the 20th century wars.
- *Singularization*: This relates to both voluntary and involuntary processes of

living and residing alone, be it as an ageing single or as a widow/er. Living alone at an old age is also one of the main reasons for the increase in one-person households (Grundy 1999, Hall et al. 1999).

- *Oldest-old age (Hochaltrigkeit):* Threefold ageing also leads to changing perceptions of 'the very old'. In Europe today – and significantly different from just a few decades ago – it is the 80 and over age group which is described as the oldest-old. In many cases, this part of the life cycle is linked to the negative sides of ageing, in particular to debilitating diseases and restricted ability to manage daily duties, leading to a reduced quality of life.

Ageing is therefore both an individual characteristic (that is related to individuals in their life course) as well as a characteristic of a social entity, be it a neighbourhood, a city or a society.

Currently, the negative aspects of ageing tend to be stressed both by the general public as well as by policy makers and researchers. However, it is important to remember that never in human history have so many people had the chance of living to old age while enjoying such relative prosperity.

Household Change

It was stressed above that traditionally the family, or even more specifically, the nuclear family, was seen as the standard household. However, in the second half of the 20th century, not only has the sequence and timing of family formation changed decisively, but the questions of whether to found a family at all or whether to establish alternative types of households have become more important. Van de Kaa (1987) characterized the related patterns of the second demographic transition with the following four statements:

1. Shift from the golden age of marriage to the dawn of cohabitation,
2. Shift from the era of the king-child with parents to that of the king-pair with a child,
3. Shift from preventive contraception to self-fulfilling conception,
4. Shift from uniform to pluralistic families and households. (ibid.: 11, emphasis omitted)

All of these four characteristics impact on household compositions. Most European countries have seen a diversification of household types, along with a growing importance of childless living arrangements, such as, for example, one-person households (among them: singles), childless couples (including dinkies as discussed in Chapter 2) and unrelated others sharing a flat. To describe these arrangements, different terms such as 'non-traditional', 'non-conventional' or 'new households' are used (for example, Droth and Dangschat 1985, Spiegel 1986, van Engelsdorp Gastelaars and Vijgen 1990, Schneider et al. 1998, Ogden and Hall 2004). None of them is a perfect description: 'non-traditional' and 'non-conventional' are not free from normative connotations, but neither are these household types 'new' because they are not a recent phenomenon at all. They do though differ from earlier periods with respect to their quantity, societal significance and social acceptance (Huinink 2007). These non-traditional household types – the term we generally use in this volume – are not restricted to childless arrangements. One-parent households, patchwork and step families also have to be considered (Kuijsten 1996, Silva and Smart 1999; see also Table 3.1 above). Although empirical investigations have shown that the increase in the plurality of household types is, at least for (western) Germany, overestimated (Huinink and Wagner 1998: 99–104, Wagner and Franzmann 2000), it is still true to say that the variability of household types, with and without children, has been growing since the 1970s.

Other recent household trends are, firstly, significantly smaller mean sizes and secondly, less stable arrangements, as individuals shift from one living arrangement to another several times during the course of their lives (for a review Buzar et al. 2005). These changes are particularly significant for the objectives of our volume because of their importance both for housing markets, residential and urban change, as previously discussed in the household and housing demography sections.

The Urban Dimension of Demographic Change

The processes of the second demographic transition are most visible in the urban environment. This is particularly true for migration, above all for immigration, which has always been a decisive driver of urban change. In- and out-migration are at the heart of urban change. The other demographic trends discussed above are also predominantly urban phenomena. Across Europe, childlessness and one-person households are most wide-spread in cities. Finally, in many European cities, ageing is of major concern to the urban authorities. Yet, since demographers usually work at the national scale, it is urban scholars who need to investigate the demographic dimension of local developments.

For many years *population growth* was considered to be a normal, quasi-natural precondition of urban development (Grossmann 2007). However, now that population decrease is on the agenda in many cities across Europe (Turok and Mykhnenko 2007), urban research is being challenged to consider the conditions of population development in greater depth and to investigate its local trajectories and overall trends (Steinführer et al. 2010). The related debate on *urban shrinkage* started in Germany at the end of the 1990s. Population decrease serves as its core indicator. Here, at the turn of the millennium, large numbers of housing vacancies, in particular in eastern Germany, concerned both practitioners and scholars and led to the re-emergence of an almost forgotten debate. The concept of *urban shrinkage*

(*Schrumpfung*)[5] – used for the first time in western Germany in the 1970s (Göb 1977) – relates to a significant decline in the number of urban residents over a relatively short period of time. This is caused by the interplay of out-migration (for jobs) to economically more prosperous regions and suburbanization as well as low or 'lowest-low' fertility. At the same time, the losses are not replaced either by declining mortality or in-migration. In the post-socialist eastern Germany of the 1990s, shrinkage was the main direction of urban development (Herfert 2007).

Meanwhile, the discourse on urban shrinkage spread from Germany to an international arena (Oswalt 2005, Turok and Mykhnenko 2007, Grossmann et al. 2008b). Indeed, urban shrinkage is by no means restricted to eastern Germany. Recent research shows that 40 per cent of all European cities with more than 200,000 inhabitants have been shrinking or experiencing a population decline over recent decades (Turok and Mykhnenko 2007). In particular, post-socialist Europe – including eastern Germany – forms a new 'pole of shrinkage'. Mykhnenko and Turok (2007) point out that 'over the period 1960–2005 the average East European city changed from being an engine of growth to a growth laggard, at best. Shrinkage dominates the urban landscape in the region, with three out of every four cities now shrinking, compared to two decades ago with just one in thirty' (ibid.: 46–7). They conclude that 'shrinkage rather than growth or recovery has become the dominant trajectory' (ibid.: 2). While this discourse is mostly concerned with population decline – and sometimes loses sight of its economic causes (Bürkner 2008) – the related British and American debates on urban decline or decay (Friedrichs 1993, Beauregard 2003), as well as the concept of 'weak market cities' (Brophy and Burnett 2003), are much more focused on economic pathways of deindustrialization.

Why is it necessary to reflect on urban population decline? From the example of eastern Germany in the 1990s, the following lessons can be learned:

1. Massive tendencies of both economic and population decline may pave the way to a self-perpetuating downward spiral.
2. Although decreasing population numbers lead to reduced densities of usage, services and technical infrastructures need to be maintained for the remaining population – therefore, while absolute costs might decline, they increase in relative terms.
3. Planning instruments, governance practices and allocation procedures are all related to situations of urban growth and not to shrinkage. Therefore, while urban practitioners are faced with the constant need to make decisions, there is no 'best-practice' toolbox available to them.

---

5   A major problem of the concept relates to the fact that the metaphor of shrinkage (or 'contraction', Mykhnenko and Turok 2007: 46) does not best describe the actual development since, in spatial terms, the cities are not 'shrinking' but sometimes even continue to grow.

4. Residential segregation has not been eliminated, but to a large part follows the same logic as under conditions of urban growth (Grossmann et al. 2008a).

A central question addressed in this volume is whether or not the experience of urban shrinkage and its consequences might also be transferred to East Central Europe.

As discussed earlier, *ageing,* as a major demographic process of contemporary European societies, is an overarching aspect of urban life in many cities across the continent (EUROCITIES 2007, Kabisch et al. 2008). However, there is relatively little research devoted to ageing in the urban space (Bunchandranon et al. 1997, Peter 2009). Just like other facets of demographic change, ageing often creates an imbalance between available housing supply and the actual housing and services required in residential areas. In addition, the five facets mentioned above that, according to Tews (1993), characterize the specifics of current ageing tendencies – rejuvenation of older age groups, living without professional duties and relationships, feminization, singularization and oldest-old age – need to be seen in the light of housing needs. Older people, who no longer take part in the labour market, not only spend more time at home and in their neighbourhood compared with people of working age, but also in comparison with old people of previous generations because of their increased life expectancy. With the growing social differentiation of older people, their housing and service requirements also become more differentiated and are not just restricted to residential care homes. The desire of older people to continue to live in their own home means that dwellings are needed that are in good physical condition, with the right facilities and without barriers to living at home, for example, with step-free access, non-slip treads and suitable bathroom equipment. Self-determined living may be achieved either by living alone or sharing a dwelling. Finally, one has to consider that future older inhabitants may well have residential patterns that are different from those of today's older people because of their higher mobility and longevity (Kabisch et al. 2008).

While *household transformations* take place at all spatial levels, an additional observation is that cities, and especially inner cities, lead the way in changing household compositions. Across Europe, the share of one-person households is highest in the largest cities and it is even higher in their inner cities (Urban Audit 2001). Other new types of household, such as cohabitation, same-sex unions and unrelated others living together, are also spreading faster in cities, particularly inner cities, than elsewhere. This diversification of living and housing arrangements transforms the affected neighbourhoods and carries with it a new dynamic with respect to residential change. In line with these changes, urban experts have recently started referring to reurbanization as a small-scale process and in qualitative rather than quantitative terms, when discussing the relationships between demographic, residential and urban changes, particularly in European inner cities (Buzar et al. 2007, Steinführer and Haase 2009b, Haase et al. 2010, in more detail Chapter 5).

Contextualizing Demographic Change:
The Situation in East Central Europe

Principal features of Demographic Change in Poland and the Czech Republic during Post-socialism

After 1989 and the onset of the transition period, Poland and the Czech Republic, like all other countries in post-socialist transition, experienced a dramatic decline in fertility that fell well below replacement level. In spite of many demographic changes occurring simultaneously, the fertility decline was, and still is, regarded as the major process of changing demographic behaviour across the region (for an overview see Kučera et al. 2000, Philipov and Dorbritz 2003). Although the downward trend in the number of births had started earlier (Rychtaříková 1999: 20, Okólski 2006: 106, 109), all countries in transition experienced an accelerating decline in fertility, along with increasing deferment of first marriages, greater numbers of children born out of wedlock and growing divorce rates (a trend which in Poland started much later than in other post-socialist countries). In many cases, individual decisions in favour of fertility and marriage were postponed and, thus, fertility rates were particularly low during the 1990s before starting to rise again (in Poland from 2005, in the Czech Republic from 2000 onwards). As a result of negative migration balances (Poland), or a lack of sufficient replacement migration (Czech Republic), there was a slight fall in national population numbers after 2000 and in the 1990s, respectively (CoE 2005; see also Tables 3.3 and 3.4 below). Future population projections are negative for both countries and assume that ageing will gain greater importance (Philipov and Dorbritz 2003: 184–6, GUS 2004, Kretschmerová and Šimek 2004, Matysiak and Nowok 2006).

As a result of the dramatic population changes after 1989, the demographic debate in both countries can be described as a crisis discourse. In the public, and to some extent the academic debate, the dominant judgements about the decline in birth rates and marriage rates, as well as the growing numbers of extra-marital births are pessimistic. Ageing is seen to have some positive aspects but, in its societal consequences (for example, for the pension systems), it is regarded as negative. The decline in infant mortality and the downward trend in mortality in general are seen positively. For the Czech Republic, the drop in the number of abortions is also considered to be one of the few positive features of demographic change (for the Czech debate Steinführer 2011).

In a cross-European demographic report produced by the Council of Europe, Poland appeared three times among the countries with extreme features. It was one of the ten countries with lowest dependency ratios, lowest mean age of women at first marriage and lowest divorce rates (CoE 2005). Thus, the type of demographic change characteristic of the second demographic transition is, according to these rankings, not very apparent. However, Table 3.3 provides evidence that demographic changes are also taking place in Poland – among them, most prominently, the postponement of important demographic decisions (such as getting married or having children), declining birth rates (which started to decline later compared to the other countries in transition), ageing, as well as improving life expectancies for both men and women (in more detail Kotowska 2001, Strzelecki 2003, Okólski 2006, Kotowska et al. 2008). In particular, the recent increase in divorce rates has resulted in public debates.

Table 3.3 Selected demographic indicators for Poland, 1980–2005

	1980	1990	1995	2000	2005
Population on 1 January (in 1,000)	35,734.9	38,073.0	38,284.0	38,654.0	38,157.1
Natural population balance	342,595	157,377	47,025	10,320	-3,902
Migration balance	-21,159	-15,814	-18,223	-22,200	-12,878
Rate of natural increase	0.96	0.41	0.12	0.03	-0.01
Rate of net migration	-0.06	-0.04	-0.05	-0.05	-0.03
Total fertility rate (TFR)	2.26	2.05	1.62	1.34	1.24
Infant mortality (per 1,000 live born children)	25.5	19.3	13.4	8.1	6.4
Life expectancy at birth (men)	66.0	66.5	67.6	69.7	70.8
Life expectancy at birth (women)	74.4	75.5	76.3	78.0	79.4
Crude marriage rate (per 1,000 population)	8.6	6.7	5.4	5.5	5.4
Crude divorce rate (per 1,000 population)	1.1	1.1	1.0	1.1	1.8
Mean age of women at birth of first child	23.4	23.3	23.8	24.5	25.4
Mean age of women at first marriage	22.7	22.6	23.1	23.9	25.5
Extra-marital births (in %)	4.8	6.2	9.5	12.1	18.5
Proportion of population at post-working age (men: 65+, women: 60+) to population at pre-working age (0–15)	45.9	49.4	59.4	71.1	87.4

Note: Thanks to Adam Bierzyński for his support in compiling the data. The index in the last row is widely used in Poland and differs from the one in Table 3.4.
Source: CoE 2005; GUS Regional databank; authors' work.

Post-socialist changes in demographic behaviour occurred more rapidly and profoundly in the Czech Republic, compared with Poland (Table 3.4). In the 2005 Council of Europe ranking, it was in the top ten of either the highest or lowest values five times, namely among the countries with the lowest percentages of the age group 0 to 14 years, the lowest dependency ratios, the lowest total female first marriage rates (below age 50), the lowest total fertility rates (TFR) and the highest divorce rates (CoE 2005). Table 3.4 provides the respective data for the period 1980 to 2005.

Table 3.4 Selected demographic indicators for the Czech Republic (and the Czech part of Czechoslovakia, respectively), 1980–2005

	1980	1990	1995	2000	2005
Mid-year population (in 1,000)	10,326.8	10,330.8	10,333.2	10,272.5	10,234.1
Natural population balance	18,264	1,398	-21,816	-18,091	-5,727
Migration balance	1,856	624	9,999	6,539	36,229
Rate of natural increase	0.18	0.01	-0.21	-0.18	-0.06
Rate of net migration	0.02	0.01	0.10	0.06	0.35
Total fertility rate (TFR)	2.10	1.89	1.28	1.14	1.28
Infant mortality (per 1,000 live births)	16.9	10.8	7.7	4.1	3.4
Life expectancy at birth (men)	66.8	67.6	69.7	71.7	72.9
Life expectancy at birth (women)	73.9	75.5	76.6	78.4	79.1
Crude marriage rate (per 1,000 population)	7.6	8.8	5.3	5.4	5.1
Crude divorce rate (per 1,000 population)	2.6	3.1	3.0	2.9	3.1
Mean age of women at birth of first child	22.4	22.5	23.3	24.9	26.6
Mean age of women at first marriage	21.7	21.4	24.6	26.4	28.1
Extra-marital births (in %)	5.6	8.6	15.6	21.8	31.7
Abortions (per 1,000 population)	8.4	12.2	6.0	4.6	3.9
Ageing index (65+/0–14)	56.9	58.3	72.5	85.5	97.0

Note: Where data are contradictory, CoE (2005) data are used. The ageing index (last row) is widely used in the Czech Republic and differs from the one in Table 3.3.
Source: ČSÚ internet sources (http://www.czso.cz/).

Household data are not monitored on an annual basis nationally but are collected by the census held every ten years. They are thus rather limited at the national but even more so at the local level. This has also resulted in relatively few studies of household change in the post-socialist period. Some analyses use the European Value Survey, which also provides information on households (Rabušic 2001, Lesthaeghe and Surkyn 2002, Sobotka et al. 2003). However, such data are not available for smaller geographical units such as cities or intra-urban units. These data limitations are important for our methodological design (see Chapters 2 and 7–9). Even so, there are typical features of the second demographic transition in the national data. According to the census, the number of households in the Czech Republic increased between 1991 and 2001 (depending on the definition of household) by 3 to 6 per cent with a continuing long-term trend towards smaller households (Bartoňová 2003: 269). The same is true for Poland, where household numbers increased by 11 per cent between 1988 and 2002 while the population number grew by less than 1 per cent (Kotowska 1999a: 215, GUS). Based upon the European Value Survey, Sobotka et al. (2003: 262–5) argue that behind the statistics much more fluid behaviours are taking place, which are in line with the second demographic transition. The almost universal pattern of early marriage and early family formation seen in the late 1980s has been transformed into a range of different behaviours (ibid.: 258, 262–5, Okólski 2006: 132–6). According to Sobotka et al., 'Czech society has gradually become tolerant towards certain forms of non-traditional family behaviour' (Sobotka et al. 2003: 259), hence cohabitation, postponement of marriage and childbearing outside marriage, as well as childlessness, are seen much more often than was the case during state socialism. In Poland, changing attitudes towards new household arrangements, such as cohabitation, have contributed towards actual changes in demographic behaviour. These transformations are as yet scarcely detectable statistically in Poland and fertility remains strongly linked to marriage, but demographers expect a future rise in cohabitation and childbearing outside marriage (Mynarska and Bernardi 2007, Kotowska et al. 2008: 818).

The Second Demographic Transition Debate in and about East Central Europe

A central aspect of the demographic debates about East Central Europe is the question as to whether the concept of the second demographic transition is applicable there. As mentioned above, the convergence assumption, implying that similar demographic patterns will be observable all over Europe in the long term (van de Kaa 2004: 8–9), are central to the second demographic transition debate. Therefore, unsurprisingly, in order to describe and explain population processes after 1989, the concept has been applied to East Central European countries (Kotowska 1999b, Rychtaříková 1999, 2000, Rabušic 2001, Sobotka et al. 2003, Coleman 2004, Surkyn and Lesthaeghe 2004, Kotowska et al. 2008).[6] Whether the trends described here can be interpreted as evidence that post-socialist countries are indeed experiencing the second demographic transition and therefore as a verification of the all-embracing character of the process, is part of that ongoing controversial debate. Since demographic changes occurred in East Central Europe much more rapidly and dramatically than in Western European countries, some scholars have argued that they are better described as a demographic 'shock' (Rychtaříková 2000). If this is the case, changing fertility behaviour can be seen as a way of coping with the challenges of the post-socialist crisis rather than as an adaptation to western models of rational choice (Fialová and Kučera 1997: 99–100, Rychtaříková 1999: 28, Kotowska 2001: 59–63, Dorbritz and Philipov 2002: 427–8, 454–8). By contrast, others argue that the changes are a result of changing values as part of the second demographic transition (Rabušic 2001, Sobotka et al. 2003, Surkyn and Lesthaeghe 2004). The Czech sociologist Ladislav Rabušic goes one step further and completely rejects the notion of a demographic crisis. In his opinion, the demographic changes in the last two decades can be interpreted as a profound modernization of both Czech society and family behaviour (Rabušic 2001; see also Steinführer 2011). Polish demographers point to the ambiguity

---

6   There was also an earlier (that is pre-1990) debate on the SDT concept in Eastern Europe (see, for example, vol. 30 of the Czech journal *Demografie* in 1988). A systematic treatment of the discourse and a comparison with the post-1990 debate would far exceed the scope of this chapter.

of the ongoing processes: Kotowska et al. (2008) stress the importance of the economic and social transformations after 1989 in bringing about demographic changes rather than cultural and ideational changes. In the western discourse, however, great significance is attached to the latter in order to explain the second demographic transition.

Answering the question as to whether or not East Central European countries fit into the model of the second demographic transition is not part of our research agenda. Even so, it is likely that there are many reasons for the demographic changes that have taken place during post-socialism, including shifting values, postponements of certain demographic decisions, due to the rapid transitions taking place, as well as changing patterns of behaviour caused by a new social, political and economic environment. Furthermore, some of the main demographic trends (declining birth rates, as well as the diversification of household types and smaller household sizes) started well before the post-socialist transition and accelerated after 1989. Although several population indicators suggest trends similar to those in Western and Northern Europe, it is still too soon for a final assessment (similarly Steinführer and Haase 2007). For the future, projections suggest that Eastern Europe will continue to experience population decline, low birth rates, accelerating ageing and out-migration (Philipov and Dorbritz 2003: 165–87). So, while we cannot provide a definitive answer to the question about how to interpret the demographic changes related to the second demographic transition in East Central Europe, it is clear that more than just the post-socialist explanatory background has to be taken into account (see also Chapter 4). Moreover, in this volume we are primarily concerned with some of the consequences of the demographic changes for urban development and housing rather than with demographic change on its own.

Conclusion

Demographic processes never occur in isolation but are always socially constituted and embedded at different societal levels and spatial scales. In spite of a different institutional framework compared with Western Europe, Polish and Czech societies have, in the past decades, also undergone processes of individualization, pluralization of lifestyles, changing gender roles, shifts in professional careers, the prolongation of adolescence and postponement of post-adolescence, as well as a redefinition of traditional concepts like family and marriage. This modernization process became more dynamic and, at the same time, changed decisively with the breakdown of state socialism at the end of the 1980s. The new societal institutions of the post-socialist phase again impact on demographic decisions and their timing. They have lead to a new demographic régime which is characterized particularly by a postponement of demographic decisions, such as having children or getting married, but also by a greater diversity in individual biographies and corresponding living arrangements. It is a key hypothesis of this volume that this diversity will be reflected in the urban space, on the one hand and that it contributes to residential change in inner-city neighbourhoods of Polish and Czech second-order cities, on the other. Population and household development are the major concerns of our volume, while ageing and immigration are only dealt with in passing. The actors we are mainly interested in are households which we define in a way that is not simply restricted to a statistical understanding but takes into account people's self-perceptions about their way of living and their housing. With this perspective, we want to contribute to a further refinement of the scientific approach towards housing demography, which we combine with qualitative and small-scale approaches from urban geography, urban and housing sociology as well as anthropology.

Chapter 4

The Post-socialist Condition and Beyond: Framing and Explaining Urban Change in East Central Europe

Annegret Haase, Antonín Vaishar, Grzegorz Węcławowicz

Introduction

The dismantling of state socialism and its replacement with a market-oriented system was accompanied by fundamental changes in urban space and societies in East Central European cities. The systemic change brought about true qualitative political, economic and social change. It also accelerated transformations that had already started earlier, for example, urban decline and out-migration in old-industrialized cities, the decline in birth rates and processes of population ageing (Haase et al. 2007: 156, Mykhnenko and Turok 2008: 311). The breakdown of state socialism led to the transformation of the institutional and legal regulation systems of the cities and affected nearly all spheres of urban space and society, including the housing and real estate market, urban land use, as well as the realms of actors, strategic policies and planning. The direction of these transformations was, as Harloe (1996: 11) stated in the mid-1990s, 'clearly [...] from socialist cities, but what to is much less certain'.

Over the last two decades the post-socialist transition has become one of the central topics of urban research in East Central Europe. A vast body of scientific literature has emerged and a variety of theories have helped to explain and understand the manifold expressions of post-socialist urban transition (Andrusz et al. 1996, Kovács and Wießner 1997, Enyedi 1998, Kostinskiy 2001, Altrock et al. 2005). Different strands of debate have developed covering a spectrum between modernization and catch-up approaches that saw post-socialism as a period of adaptation to western structures (Gowan 1995) at one extreme and approaches that emphasized path dependencies and the (continuing) divergence of post-socialist cities away from western patterns at the other (Musil 1993, Pickles and Smith 1998, Kawka 2007, Sýkora 2008). This chapter will not contribute to these debates. The authors argue instead that it is the simultaneousness, interdependence and overlapping of different processes which make the changes in East Central Europe and its cities unique. These cities, with their historical legacies and characterized by the dramatic speed of change, have – in contrast to their western counterparts – experienced another trajectory of development after 1989 that has led to patterns specific to them. In addition they have been affected by other macro-trends, such as European integration and globalization, which have attenuated the impact of the post-socialist condition on their development (Enyedi and Kovács 2006: 17, Schlögel 2006: 473).

Set against this context, this chapter deals with the post-socialist condition as an explanatory framework for East Central European urban development. It starts from the assumption that looking at post-socialist cities two decades after the 'fall' of state socialism is a good opportunity to re-examine some of the long-held assumptions about the pathways, characteristics and the underlying dynamics of their development. It sets the scene for the way the post-socialist transition is understood in this volume but also for a critical reflection on the concept of the post-socialist transition and its explanatory power for East Central European urban change.

The chapter is structured as follows: in the first section, the terms *post-socialism* and *transition* are introduced and conceptualized for the purpose of this volume. The second introduces a systematization of post-socialist urban change and considers the impact of the post-socialist condition on urban development in East Central Europe. The third and final section discusses the explanatory value of the post-socialist condition for East Central European urban development and reflects on the interlinkages of post-socialism and other macro-processes that impact on urban change.

Theorizing Post-socialist Transition

> We have argued that the nature and meanings of post-socialism are distinctive but that, to be fully understood, connections must be made to other parts of the world, where similar processes are being implemented and experienced in different ways.
>
> (Stenning and Bradshaw 2004: 255)

The term *post-socialist transition* is associated with two different ideas. On the one hand, it refers to the political and societal macro-process of the systemic change from state socialism to any kind of subsequent regime. On the other hand, it serves as an umbrella term for all the transformations that have been observed in post-socialist societies since 1989. Given this context, what do the authors of this volume understand by post-socialist transition? The following section deals with the conceptualization of post-socialism and transition for the purpose of this volume. We first introduce our understanding and use of these terms. We then briefly detail the debates on post-socialism and transition to embed our view. Finally, the urban context is specifically considered.

The terms *post-socialism* and *transition* have been the subjects of numerous debates and are often used synonymously. Whereas some scholars speak of post-socialist societies or – in our case – cities,[1] others use the term 'in transition' (for example, Burawoy and Verdery 1999, Hamilton 1999, Berdahl 2000) or 'after transition' (Sýkora and Bouzarovski 2011). There are, however, some scholars who argue for a deliberate distinction between the two terms. For example, Stenning (2005a: 113) marks transition as a passage or period and post-socialism as a wider concept with a longer duration than the transition itself. She argues that there should be a clear distinction between the two terms and underlying concepts because, otherwise, if they are identical, '[...] post-socialism must end with the "end of transition", and in this way it becomes a temporary, transitional category with no power beyond a limited historical moment'. In addition to *post-socialist,* the term *post-communist* is also used,[2] or other terms like 'Soviet-type system', 'centrally administered or command economy' and so on (Kornai 1992: 10). At this stage, one must not forget that transition research is always 'charged' by the emotional investment and involvement of its scholars, maybe more than other debates. The choice of different terms is, therefore, not just casual or pragmatic but relates to the researchers' personal experiences of state socialism. This chapter dissociates itself from any emotional or ideological debates about these terms. Ultimately, any choice of terms remains a matter of semantics and – as long as the meaning of a term is clearly defined – these semantics do not require further attention (see also Kornai 1992: 10). Therefore, for the purpose of this volume, we speak of *state socialism* when we refer to the time before 1989, in order to set out what the system in that period was like, and about *post-socialism* when referring to the time after 1989, to highlight both the break (there is no longer any state socialism) and, at the same time, a kind of continuity (there are many legacies that have endured).

Our Understanding of Post-socialism and Transition

The research presented in this volume builds on two robust definitions of post-socialism and transition that both highlight the close relationship between the two terms and, at the same time, their distinctiveness. We define post-socialism in a twofold manner. *First,* we understand it as the qualitative change from state socialism to a democratic, market-oriented system. It remains a relevant category that draws attention to the commonality of (socialist) experience that exists within and between certain regions and countries (see also Flynn and Oldfield 2006: 20). *Second,* we also use the term *post-socialist condition* to describe all the processes, the emerging formations and re-configurations, changing attitudes and so on, that represent changes experienced at the macro-level in daily life in, for example, urban

---

1   For example, Bridger and Pine 1998, Stark and Bruszt 1998, Kostinskiy 2001, Hann 2002, Hörschelmann 2002, Stenning 2005a, 2005b, Tsenkova and Nedović-Budić 2006, Stenning and Hörschelmann 2008.

2   For example, Musil 1993, Sakwa 1999, Sýkora 2000a, 2000b, Staniszkis 2001, Blokker 2005, Hirt 2005, Gentile and Borén 2007, Sýkora and Bouzarovski 2011.

environments. The explanatory value of post-socialism resides in the fact that it may explain things that developed after the breakdown of state socialism but are directly linked to this past and its reality. Or, as Stenning and Hörschelmann (2008: 322) state: 'The practices and policies of post-socialism are distinctively inflected by the socialist past and narratives of the past.' This is what is discussed in this volume for Polish and Czech inner cities. We have to explain how post-socialism, conceptualized in a procedural, path-dependent manner, affects residential change and to what extent the post-socialist condition (still) serves as an explanatory framework to understand this change (see also Outhwaite and Ray 2005: 19–24).

With respect to transition, we conceptualize it as a process whose direction is largely defined but whose end and characteristics are open and may vary from context to context; we do not equate transition with modernization (Blokker 2005). Transition is referred to as a 'broad, complex and lengthy process' (Sýkora 2008: 284) that describes a process from state socialism towards forms of democracy and market economy, but underlines the general openness of this process, as well as its complexity and uniqueness (Eyal et al. 1998, Stark and Bruszt 1998), a 'multi-facetted transition from a mixed form to [...] a mix of multiple formal and informal institutions which can be complementary to or conflicting with each other' (Heilmann 2002: 284, cited in Segert 2007: 3). In this understanding, transition is both evolutionary and path-dependent since it leads, at the same time, to new, emerging structures and to the intersection of old and new structures (see also Pickles and Smith 1998: 13). Transition is driven by a number of transformations, that is reconfigurations of various political, institutional, economic, social and urban structures that shape a newly-evolving logic of a 'society after transition' (see also Stark and Bruszt 1998: 7, Burawoy 1999, Stenning and Bradshaw 2004, Sýkora and Bouzarovski 2011: 3). We offer both a wider and more inclusive perspective of the post-socialist condition, with the accentuation of the term *transition* for the systemic change and refer to the various changes and rearrangements as transformations that go on when the systemic change has already come to an end. In this vein, transition marks the phase of post-socialism that describes the change from the old to the new system. While this phase, and thus also transition, represents a finite process, the transformations brought about by transition will remain a relevant category to explain the nature of the post-socialist condition beyond the transition period.

Figure 4.1 shows that transition, as it is conceptualized here, can be divided into different phases[3]: It starts with the early years which brought the more or less abrupt breakdown of state socialism. A second phase ('early' transition) was characterized by the restructuring and re-organization of the state, society and economy. Regardless of how long this phase lasted, the next phase describes a

---

3   See also Sýkora (2008), who conceptualizes post-socialist transition as a two-step process 'from socialism towards capitalism', undergoing, firstly, a phase of revolutionary change and, secondly, a phase of evolutionary adaptation before the impacts start to develop increasingly from global processes or from within the newly-established capitalist societies.

kind of consolidation and adaptation to the newly-established political, legal and institutional regime ('developed' transition). Going one step further, the model indicates that, after consolidation, the transition phase in a strict sense may have an end or, to put it differently, arrive at a stage where its explanatory value becomes more and more questionable or less and less convincing, in comparison to other explanations. This is what clearly distinguishes it from post-socialism: while transition comprises the process of change from state socialism to any kind of non-socialist system, post-socialism relates to the evolving system and patterns after the breakdown of socialism that will still endure when the transition approach, in terms of (rapid) change and its consequences, will be of limited explanatory value. Post-socialism and transition do indeed have, in this regard, different temporalities, which make it necessary and useful to treat them as two different approaches. While the scope of the research presented in this volume relates to the period from 1988–2009, on-site research was conducted only from 2006–2009.

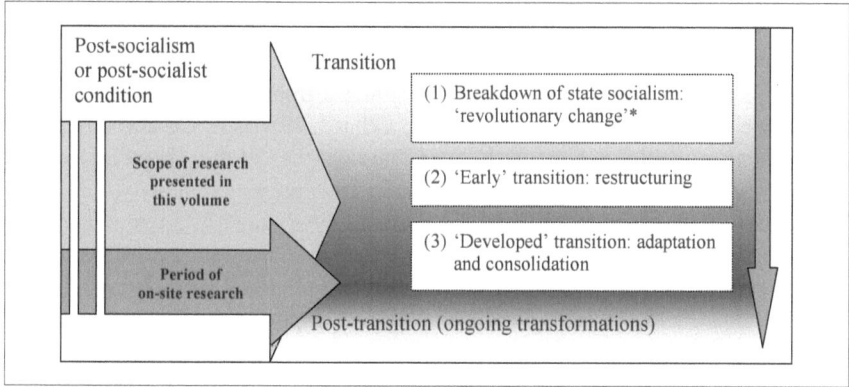

Figure 4.1 The interconnections and distinction between post-socialist condition, transition and transformation

Note: * The term 'revolutionary change' is cited after Sykora (2008).
Source: Authors' work.

Terms and Debates

Post-socialism has been analyzed as a framework or concept by a number of authors.[4] These analyses reflect on the nature of the theory and concept itself, covering a number of different disciplines (for example, economics, political

---

4   For example, Grabher and Stark 1997, Pine and Bridger 1998, Burawoy 1999, Hann 2002, Hörschelmann 2002, Humphrey 2002, Verdery 2002, Lowe 2003, Bradshaw and Stenning 2004b, 2004, Beyer 2004, Stenning 2005a, 2005b, Stenning and Hörschelmann 2008.

science, sociology, anthropology, history, geography). Post-socialism, an inherently multi-disciplinary and multi-scale issue, rests on the understanding that there is (still) enough that these countries and societies have in common that differentiates them from what we call 'capitalist' countries and societies. Post-socialism has been polarized between scholars who coined the term to represent a process of modernization or catching-up to western structures and standards and others who underlined the variety and distinctiveness of post-socialism(s) as well as its path dependency from pre-socialist and socialist patterns and their underlying dynamics (Hamilton et al. 2005: 11). Sýkora (2008: 289–90) even claims that post-socialist transition is not only path-dependent but also path-shaping, in so far as old path dependencies are overthrown and new ones, impacted by the post-socialist societal regime, are built. Old and new path dependencies can be contradictory, complementary or indifferent (ibid.: 292). Path dependencies, hence, are assigned a considerable significance among transition scholars, as Stark and Bruszt (1998: 6–7) state: 'If some hold that the postsocialist present is determined by its future (that is intelligible, only in terms of a predesignated future), others argue that its present is condemned by its past.'

A further 'bi-decennial' controversy relates to, as mentioned above, the question whether post-socialist societies are undergoing a transition or a transformation (mentioned, for example, by Pickel 2002, Eckardt 2006). To expand thoroughly on this debate would go beyond the scope and purpose of this chapter. However, and leaving aside much of the related polemics and rhetoric often involved (Sakwa 1999: 120–121, Polanska 2008), the debate can be divided into two strands: the economy-related discussion that speaks of a transition from the centrally-planned to a more market-oriented economy (see also Stenning and Hörschelmann 2008: 314) and social science discussions that criticize this approach for being teleological and 'neglectful of the 'embeddedness' of the diversified transformations taking place' (Hamilton 1999: 136).

Apart from this differentiation, transition itself has been shown to be a 'highly problematic term' (Pine and Bridger 1998: 2). Criticism has related mainly to the neoliberal, normative or evolutionist approach of the concept and to its teleological determination of the goal of change, that is market capitalism. Whilst some authors underscore the development of specific, path-dependent new structures, consisting of elements from both socialist and pre-socialist tradition as well as elements of western societies (Steinführer 2001: 218, Beyer 2004: 75), others point especially to the persistence of socialist patterns (Hann 2002: 5, 11, Verdery 2002: 21, Lowe 2003: XVI). Pickvance (2002: 195–6) emphasizes that transition was often used as a 'temporary label' to be applied until post-socialist societies had achieved democracy and established a market-economy although, among economists, it is questionable whether the majority of post-socialist states can be seen as part of a modern capitalist system (Lane 2007: 33). In this respect, the category *post-socialist* can still be seen as providing a useful conceptual framework. Hence, as Humphrey (2002: 15) suggests, the time has not yet come to lay it to rest.

The variety of approaches to define post-socialist transition in an overall societal context is reflected also in the debates led by urban scholars. In the early 1990s, most of the urban scholars dealing with the post-socialist transition in Eastern Europe – 'perhaps in both the East and the West' (Hamilton et al. 2005: 11) – assumed that the shift from state socialism and the centrally-planned state economy towards democracy and a market-oriented economy would result in a (faster or slower) convergence of Eastern European cities to western structures. Socialist cities were thought to be rapidly overtaking the neoliberal version of the western blueprints and becoming both 'Europeanized' and globalized (Sýkora 2008: 286). Such thinking proved to be not only naïve but also lacked understanding of the power of the past for post-1989 developments in the whole of Eastern Europe. In the meantime, it became obvious – not only with respect to urban environments – that the neoliberal assumptions about a pure replacement of institutions and mentalities, as well as about a simple convergence towards western-style, post-fordist modernization, were not suitable for an adequate interpretation of the transition processes (Grabher and Stark 1997, Pine and Bridger 1998: 3–7, Burawoy 1999: 303, Hann 2002, Bradshaw and Stenning 2004a: 12–14, Bradshaw and Stenning 2004: 10). In other words, transition, instead of leading in a straightforward way to a new system, now seems to have generated complex and highly differentiated systems of adjustment, both at the regional or national, as well as at the local level (Pickles and Smith 1998: 20).

As stated above, at least since the late 1990s, it has been widely accepted among urban scholars that post-socialist change is not a process leading to just one predestined target-state in the future (Harloe 1996). It rather represents a set of processes with considerable regional and local inflections (Hemmnet 2003). Transition depends on many factors, including systems of local legal, institutional and governance frameworks, policy choices and modes of privatization and distribution (Andrusz et al. 1996). The idea of transition implies a certain starting point (the socialist city) and a certain end point (the post-socialist or capitalist city) and describes the way and manner cities go from one to the other.

Among urban scholars, there are two different approaches about how to understand transition – as an evolutionary change or as a path-dependent change. While evolutionary change approaches claim that old, socialist structures have been replaced by newly established, western-style or capitalist ones, the second approach argues that the long-term imprint of past periods on present urban change in Eastern Europe must be taken into account (Sýkora 2000a, 2000b, 2008, Hamilton et al. 2005: 11). Path dependency approaches emphasize that it is first and foremost the contradictions between the new, 'capitalist' rules of transition and the existing socialist urban environment that bring about and push forward restructuring. The last two decades mapped out constantly diverging scenarios for different regions and countries in Eastern Europe (Nedović-Budić and Tsenkova with Marcuse 2006: 10). But most post-socialist cities embody a variety of socialist and even pre-socialist patterns, which brings us to the conclusion that state socialism represents a part of European modernity and that the post-socialist transition is therefore, a part of the global restructuring of post-modernity (Bodnár: 2001). While built structures, social relations and cultural practices show a high degree of continuity (path dependency), other areas, like the housing and property market or the economic sector, are characterized by fundamental, rapid change. In this way, it is legitimate to characterize post-socialist cities as continuously undergoing an urban transformation that incorporates much of the past (Sýkora and Bouzarovski 2011: 5).

Post-socialist transition consists of several dimensions of change. It comprises the change from socialism to capitalism at the national and local level and, at the same time, the integration of socialist states into the global capitalist order (Sýkora 2008: 286). Subsequently, post-socialist societies have undergone various types of transformations through the interaction of different components and processes. But, despite this, some overarching directions of change can be identified that allow for the inclusion of the post-socialist condition under the umbrella term *transition* (see also Haase and Steinführer 2009):

- the shift from state socialism towards forms of democracy and market-oriented economy but underlining the general openness of this process, as well as its complexity, while not following western textbook models or the forecasts of western scholars,
- interactions between legacies and impacts of structures, institutions, networks, behaviours and mind-sets that either stem from the pre-socialist period or are related to the socialist decades or represent newly-established, that is post-socialist, developments[5] and
- the creation of specific patterns, because path-dependent developments cannot be identified simply as convergence to, or divergence from, western patterns; in most cases, we see heterogeneous forms of rearrangements and re-organization of older patterns together with more recent ones.

The longer the debate on transition continues, the more the extent to which the post-socialist condition will determine the future of East Central European cities is discussed with respect to what comes after it. Not surprisingly, there are a range of arguments. While Sakwa (1999: 127) proposes that post-socialism will have ended when there is no longer any attempt to learn from it, others argue that post-socialism still represents a relevant category since the legacies of state socialism are not swept away over night and there is, subsequently, still a number of comparable structures in different national and local contexts within the former state-socialist world (Stenning 2005a, Flynn and Oldfield 2006, Czepczyński 2008). Sýkora described the 'post-communist city' in 2000 as a still unfinished project (Sýkora 2000b). What is more, post-socialism is even treated by some

---

5   Pickles and Smith (1998: 11–12) underscore that one has to recognize that any reconfiguration of political, economic and social systems, patterns or structures occurs in a concrete milieu that provides the context within which these structures are reworked and that impacts on the transformation that is taking place.

scholars as a complex term relating explanatory horizons of the past, the present and even the future whose explanatory value is not restricted to Eastern Europe. 'In short, post-socialism cannot be reduced to neoliberal economic restructuring, nor just to the legacies of socialism (and pre-socialism), nor indeed to the passage of "transition". It is all of these' (Stenning and Hörschelmann 2008: 325). At the same time, we have to remember that the debate on post-socialist transition has also changed and enriched western debates. In a certain sense therefore, we are all post-socialist now. The debate on the transfer of concepts 'from West to East' or path dependencies of East Central European cities shows that we have to consider more carefully the 'repercussions of transition in the West' (Dunford and Smith 2004) and how the uniqueness of post-socialist transition is developing as part of the interconnection with other macro-processes at the European and global level (Hamilton et al. 2005).

Urban Change in East Central Europe during Post-socialism

It is often at the urban scale that the large-scale discourses of post-socialism meet the everyday strategies of change, survival and success, practised by individuals and institutions, where the abstract processes of change are made real and lived.

(Stenning 2004: 104)

Post-socialist urban change in East Central Europe has to be seen as a continuum – it is not just the replacement of one urban regime by another – but a transformation of structures, actors and representations. The most radical change was during the transition but change is continuing and will impact on the future fortunes of East Central European cities. Kostinskiy (2001: 451) calls post-socialist cities 'cities in flux': He refers mainly to the fact that post-socialist cities faced very complex developments in less than two decades: a number of breakdowns ('shocks') and crises but also processes of resurgence and even unexpected successes and a regaining of strength. Some scholars even argue that 'the potential of East European cities has been boosted by a "new conventional wisdom" within the international policy community, identifying [these] cities as engines of growth and cohesion' (Mykhnenko and Turok 2008: 312).

Today, one can clearly speak of a divergence of Eastern European urban trajectories and fortunes. Apart from the post-socialist transition, other macro-processes such as European integration and globalization have increasingly been impacting on the development of Eastern European cities. Set against this background: what does the post-socialist city mean today? What are the premises and foundations of the related debate that have to be taken into account in order to explain current changes in East Central European inner cities? This section examines the debate on the post-socialist city. It introduces a systematization of post-socialist urban change, its transitions, urban responses and consequences.

The importance of post-socialist change for residential change, as one component of urban development in particular, will then be examined in more detail.

Systematizing Post-socialist Urban Change

Many scholars have proposed various models of, and definitions for, the post-socialist city. Sagan (2000a: 70–71, 2000b) points to the marketization of socialist urban space and society as the fundamentals of the post-socialist transition. As a consequence, the labour and housing markets changed and a 'new urban order' emerged, characterized by commercialization, segregation, suburbanization and polarization, as well as by the effects of external (European, global) processes (Węcławowicz 1993, 1996, 2002b, Kaczmarek 2000). Enyedi (1998: 30–33) sees the main characteristics of post-socialist transformation at the (local) urban scale in the reshuffling of the urban hierarchy, the transformation of urban society, a changing set of factors influencing urban development (actors, coordination, government or governance and so on) and, finally, the transformation of the city's built environment. Sýkora (2000b) underlines the specifics of the post-socialist city in terms of the difference in the time span of two transformations: firstly, the transformation of the societal foundations that produce new spatial patterns and, secondly, the changes of the urban spatial structure itself. Kaczmarek (2000: 55) underscores the interconnection of the production of new patterns of spatial and symbolic differentiation as well as the changing behaviour of urban residents within the urban space, a view that comes close to that of the research presented in this volume. Finally, Stanilov (2007c: 8) points to the juxtaposition of complexity and chaos:

> On the urban scene, this post-socialist/post-modern condition has been reflected in a chaotic pattern of development, generated by the retreat of central authorities, the appearance of a multitude of new players and the frivolous application of patterns of development "borrowed" from the West. The once monolithic structure of the socialist city has been shattered in multiple fragments, pulled in different directions by various economic, social, and political interests, yet somehow it is holding together, brimming with energy suddenly released after half a century of comatose existence.

When looking at urban change during post-socialism, it is important to consider the institutional and legal framework at different spatial levels that imprint on the local sphere, the international, national, inter-urban and the local, intra-urban scale (Musil 1993: 900, Tsenkova 2006: 24). The urban system (at different levels) serves as the primary channel linking the spheres of economic, institutional and legal activities with the cities (Beaverstock et al. 2000, Bourne and Simmons 2003, Tsenkova 2006: 23). Cities form a component of the urban system, linking local realities to overarching trends and processes of change, both in terms of social constructs and products or as the results of market and consumption processes (Tsenkova 2006: 23).

The systematization of post-socialist urban change introduced in this section and Figure 4.2, focuses on East Central Europe and, more specifically, on Poland and the Czech Republic. In this framework, the Polish and Czech cases represent societies that underwent a 'relatively quick transition from the socialist to a "mixed" model with some remnants of state control', which means that we find structures here that are no longer socialist but are nonetheless different from western structures (Tosics 2005: 72). The housing sector can serve as an example at this stage. While privatization reshaped parts of the local markets in Polish and Czech cities, a municipal sector with fixed rents and social housing still exists (see also Chapter 6).

The systematization of post-socialist urban change shown in Figure 4.2 considers the different spatial levels mentioned above and the overall institutional change as the context in which post-socialist urban transition and its transformations take place. The model conceptualizes a 'triple transition': (1) the transition towards democracy (political systemic change), (2) the transition towards market orientation (economic systemic change) and (3) the transition towards decentralization which Tsenkova (2006: 23) describes as 'the quiet revolution [...] of the devolution of power and responsibility to local governments'. Each of the transitions was 'echoed' at the urban scale by a number of processes or changes:

- *transition to democracy*: change from government to governance, development of a civic society, change in urban society,
- *transition to market orientation*: social change, housing market change, economic change, spatial/settlement system change (all undergo processes of differentiation),
- *transition to decentralization*: empowerment at the local level, participation, bottom-up procedures.

These processes led to a number of consequences for the individual domains of urban space and life that shape the general 'trajectory of urban change in postsocialist cities' (Table 4.1, Tsenkova 2006: 47). Across the entire post-socialist urban world, political and institutional decentralization have taken place which have strengthened local autonomy and legitimized the new democratic rule, these have also occurred at the urban and even intra-urban (that is district) scale. Cities have been transformed from being 'state events' towards cities 'as autonomous entities' (Schögel 2006: 474). As a consequence, many cities are now faced with a large number of public, semi-public and private bodies, a lack of necessary coordination mechanisms and confusing or contested areas of responsibility (Tsenkova 2006: 39).

As for the settlement and ownership structures, deregulation has led to polarized urban areas and housing markets as a consequence of restitution, privatization and re-organization of housing ownership. In this vein, Tsenkova (2003: 193) speaks of the creation of 'nations of home owners' with levels of home ownership over 80 per cent. Both renewal and decline can be found close to each other. While the share of brownfield and derelict land increased within the boundaries of the city, new land was used for retailing and for new investments within the suburban zone.

With respect to their economic structures, post-socialist cities have undergone profound changes and deindustrialization processes (decline in secondary and rise in tertiary sector) resulting in restructuring, decline of traditional businesses and a differentiation in the economic fortunes of the cities which has impacted on urban consumption and production patterns (Tsenkova 2006: 31). The reduction of the workforce, rising unemployment and a sharp decrease in wealth and purchasing power were new phenomena of economic change that harmed an urban society that had already undergone a fundamental re-organization of social classes, groups and hierarchies with respect to political and social reputation, income and wealth.

Demographic and residential change juxtaposed social differentiation, new forms of exclusion, fear and crime at many places in the cities. New actors like private property owners, developers and the '*nouveaux riches*' brought about new decision-making rules and hierarchies. Fragmentation also became more obvious within the urban space. Marketization led to an increasing land price gradient from the city centre towards the periphery as well as to an increasing differentiation of residential areas and housing markets.

Finally, the deregulation, privatization and marketization of the infrastructure and of urban services like retailing, water, electricity, sewage and public transport, brought about growing inequalities in access to and provision of these fundamental goods to the urban population. Rising diversity, inequalities and imbalances, hence, may serve as a general description of the urban post-socialist reality.

It has to be said that many processes such as the decline in fertility were already underway before 1989, although they did not receive much attention, and that the development of suburbanization was hindered by the centrally-planned economy. In this vein, Table 4.1 represents a 'simplification' but serves to make clear the general direction of transitional shifts that led – in addition to new change – to an acceleration or strengthening of all processes or phenomena that were already in place before 1989.

As a result, the dimensions of change described in Table 4.1 led to a number of re-configurations of urban structures that are listed in the lowest part of Figure 4.2: the re-organization and rearrangements as a result of post-socialist transition and restructuring, the juxtaposition of changes and persistencies and the emergence of convergent and divergent trends towards western or capitalist patterns. This coexistence is reflected by a juxtaposition of patterns within the urban space. These include a consolidation of new, transition-related patterns that, because of their characteristics, are affected by both national and local specifics; a continuity or persistency of socialist modes of distribution, action and decision-making; and a continuity or 'regaining strength'[6] of pre-socialist patterns under changed conditions that are closely related to the historic heritage and development of the

---

6   'Regaining strength' refers to the fact that 'return' may in fact represent continuity: As early as 1979, Hamilton (1979b: 221) described the inherited and unequal pre-socialist patterns in socialist cities. The continuity of patterns mentioned for the case of poor housing areas in inner-city Łódź – despite the Holocaust and the change of the ethnic composition of the population – was already noted by Majer (1988).

particular city and its built and cultural heritage (Sýkora 2008: 287–8, Steinführer and Haase 2009a: 403).

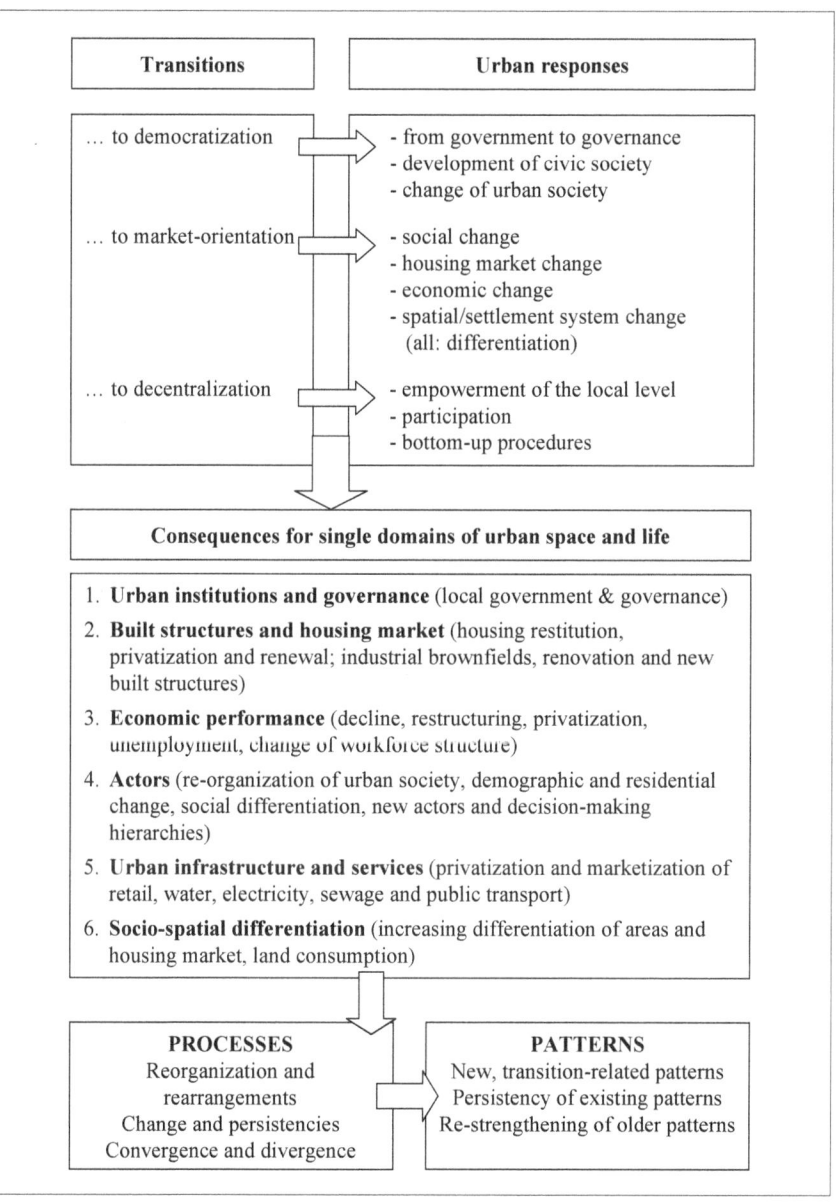

Figure 4.2 Systematization of post-socialist urban change
Source: Authors' work (based on our own research and Tsenkova 2006: 47).

Table 4.1 Major consequences of urban change during post-socialism

Domain	'from' (socialist)	'to' (post-socialist)
1. Urban functions, institutions and governance	central government decision-making, little local autonomy, top-down procedures	local government and governance, locally-based decision-making systems, bottom-up procedures
2. Built structures and housing market	neglect of the old built-up stock, large-scale housing estates built according to central norms, housing provision by the state, mix of tenure types, equalized low rents	deregulation, restitution, privatization, commercialization of housing stock, renewal and decline, industrial brownfields and newly-built structures, suburban housing market
3. Economic performance	centrally planned economy, regulation at the state level, control of markets through income and price policies, planned mono-structures	deregulation of markets, growing competition, decline of manufacturing, restructuring, unemployment, change of workforce structure, increase of services/tertiary sector
4. Actors	comparatively low social differentiation of urban society, specific (socialism-related) hierarchies, relatively egalitarian distribution of wealth and income, 'hidden' differences, relatively high safety standards, top-down organization of social decision-making	re-organization of urban society in terms of social classes, groups and hierarchies with respect to political/social reputation, income and wealth, demographic and residential change, social differentiation, forms of exclusion, new actors and decision-making hierarchies
5. Urban infrastructure and services	state-provided infrastructure and services, universal access, strong support of public transport, investments in water and sewage systems	deregulation, privatization and marketization of infrastructure and services: retailing, water, electricity, sewage and public transport, growing inequalities in access and provision
6. Socio-spatial differentiation	small land price gradient, little differentiation of housing areas, 'hidden' differences, predominance of mixed-use areas, no CBD, almost no suburbanization and lower land consumption	increasing land price gradient and differentiation of areas and housing markets, segregation and fragmentation, reuse of the city centre/CBD, brownfields and derelict lands, suburbanization and new land consumption

Source: Authors' work (based on our own research and Tsenkova 2006: 47).

Beyond Post-socialism: Other Macro-trends and their Explanatory Power for East Central European Urban Change

> I see no obstacles to the integration of Central Europe into the European urban system. Central Europe has never been completely isolated from European urbanisation processes during the period of state socialism. In addition, the state socialist 'detour' lasted only a little longer than forty years. [...] Integration will expose Central European cities to strong external competition. This competition may overwhelm them; on the other hand, it may accelerate a long delayed process of modernization.
>
> (Enyedi 1998: 33)

To what extent do post-socialism and the post-socialist transition explain inner-city residential change today? Which other processes are important and how do they interact with the post-socialist condition?

Cross-referencing recent research concerning the diverging trajectories and fortunes of post-socialist cities reveals that there is 'more than a gentle hesitation' (Eckardt 2006: 17) as to whether East Central European cities are not simply transforming from one system to another or whether they are also impacted by other processes in a European or even global framework. As one of an emerging community of scholars, Węcławowicz (2000: 28) argued for a wide embedding of post-socialist urban change by identifying it, together with European integration and globalization, as one of the 'fundamental transformations of the late 20th century'. In other words, although the debate on the post-socialist city continues and will probably do so for many years, there are other contexts that play an important role with respect to East Central European urban development, too. In the following, we argue for an integrative view. We briefly list some of the most influential processes or contexts and show how they are impacting on East Central European inner-city residential development.

Globalization

Although we are dealing with local urban contexts here, their fortunes depend on and are embedded within global developments or macro-trends. According to Sýkora (2008: 290), global processes are reflected by local specifics. During the last two decades, post-socialist cities have been increasingly influenced by internationalization and globalization, while, simultaneously, the transition itself led to a fundamental change of the political, economic and societal structures within the cities and their national contexts (Sýkora and Bouzarovski 2011: 7). But – and this might, according to Sýkora et al., be a 'key feature of post-socialist change' – while the transition is fundamental for becoming internationalized, many decisions about the fortunes and misfortunes of the cities are made by actors that form part of the global game (ibid.: 10).

Globalization is, apart from the post-socialist context, today perceived as the most powerful factor that re-configures economic and cultural relationships between nation states, regions and even continents (van Kempen et al. 2005b: 3–5). Moreover, globalization has underscored that international competitiveness has become more deeply entwined with territorial development (Atkinson and Rossignolo 2008b: 7). Globalization makes us think that the world is becoming smaller or, perhaps more realistically, that most spheres of political, economic and further activities are becoming more interconnected and interspersed. For the urban realm, according to Hamilton et al. (2005: 3), globalization means 'a process which is diffusing, deepening, and accelerating the functional integration, competition, cooperation, dependency, and interdependency of cities and their regions, across international borders, continents, and oceans'.

As a result of these processes, the role of cities – both at the global but also at the regional level – has become increasingly important. They are viewed as forerunners and hot spots of global change in terms of economy, commerce, communication and knowledge and also of societal and demographic change. At this stage, both convergence and divergence between (urban) regions worldwide have increased. The impact of globalization on local processes and decision-making in cities has to be recognized. According to Atkinson and Rossignolo (2008b), both socio-demographic change and post-socialist transition are some of the challenges for the present and future of urban Europe.

Although the influence of globalization on the urban world is not the same everywhere, since 1989 it has assumed rising importance in the former socialist cities. The most recent literature demonstrates that East Central European urban development is being discussed in a global context (Węcławowicz 2003: 147, Hamilton et al. 2005, Pichler-Milanovic and Dimitrovska Andrews 2005, Sýkora 2005, Stanilov 2007c). This is primarily true for the capital cities, particularly those offering good locations and favourable legal and institutional conditions for the globalizing economy. In the globalized world, large cities are representing centres of national as well as international power and entrances through which investments, information and innovations flow onto the territories of the corresponding state and its neighbours, as centres of progressive services, centres of concentration of information and its dissemination by media and also as cultural centres (Taylor and Walker 2001).

It is, however, mainly capital cities where this process of globalization really takes place in this way.

> Globalization and leadership in restructuring national economies is usually creating significant divergence between i) capital cities and their capital city regions on the one hand, where the effects of reforms and restructuring are most marked and ii) second- or third-order cities where change is or may be less marked and more narrowly confined. (Hamilton et al. 2005: 12; see also Beaverstock et al. 1999, Stanilov 2007c: 12)

For second-order cities, that is those which are the focus of this book, above all European, as well as national, regional and local policies, strategies and funding schemes are decisive for the re-organization of their economic, social and infrastructural performance – maybe even more than globalization. Other instruments, for example, those related to government decisions, subsidies for the housing market or urban infrastructures and amenities or housing allocation regulations, are those that determine the development trajectories of cities. They can make a difference to the success or misfortune of a city both on the national and the international level (van Kempen et al. 2005b: 4).

European Integration

There is an ongoing debate about whether East Central European cities are characterized more by global effects than by regional-international forces, such as the process of European integration. After years of pre-accession activities and the entry of Poland and the Czech Republic into the European Union (EU) in 2004, the weight of 'Europeanization' plays an ever increasing role in the fortunes of the post-socialist cities and their regions. This process includes both the premises and objectives of urban and regional policy of the EU and also the related instruments and funding schemes. In comparison to globalization processes, Europeanization is, of course, less characterized by market forces than by the strategic objectives of the EU.

In 2004, Poland and the Czech Republic – together with other countries in East Central Europe – joined the EU. Although Stenning (2005a: 114) argues that post-socialism or the post-socialist attributes of the respective cities will 'not disappear after EU accession or the achievement of EBRD 4+ ratings[7] across the board', the last few years have clearly led to a shift of focus when speaking of the political, economic and social challenges in East Central Europe. On the one hand, the EU context has become the most widely used and accepted context of comparison for those former socialist countries, societies, cities and regions that are included in the European integration process (Altrock et al. 2005: 9, van Kempen et al. 2005b). On the other, the new member states have been integrated into European programmes and funding schemes which have impacted on their strategic politics and planning priorities and have led towards a new understanding of how to tackle and coordinate urban challenges (Stanilov 2007b, Reiter 2008). The objective of these initiatives, strategies and instruments was for them to become the common

---

7   Founded in 1991, the European Bank for Reconstruction and Development (EBRD) uses the tools of investment to support market economies in 30 countries from central Europe to central Asia. Its mission is to support the former state socialist countries in the process of establishing their private sectors. EBRD provides project financing for banks, industries and businesses, both new ventures and investments in existing companies. It also works with publicly owned companies to support privatization, the restructuring of state-owned firms and the improvement of municipal services.

foundations or premises of reference and cooperation in the political discourse and practical policy making related to the future of the 'European city'.

The EU has seen a fight for an explicitly urban dimension to its regional policy during recent decades. Given that cities are seen by EU officials and institutions as key drivers of European and national development and competitiveness, a corresponding policy framework was shaped comparatively early (Reiter 2008). The EU's urban policy has changed its focus several times during this period: whereas for the 1980s and 1990s, the 'socially sustainable' city and its 'European character' was the first priority, the Lisbon Agenda (2005) set the focus more on the competitiveness and economic performance of cities and urban regions. Recent documents suggest an attempt to meet both objectives – cohesion and competitiveness. Since officially urban policy does not belong to the areas of responsibility of the supranational bodies, the EU has less influence on its shape and priorities (Frank 2005: 307), a situation that has remained nearly unchanged up to the present, although initiatives like the Eurocities network try to foster Europe-wide cooperation and dialogue between cities and their regions. More and more, the agenda of policies derives from the original idea of the maintenance of the 'European character' of the European city and aims to strengthen their position in the globalizing international arena (from the 'welfare' towards the 'workfare' city).

If such approaches are followed, there is the danger that all other aspects, with the exception of economic competitiveness (for example, demographic, social, moral and cultural issues), could be of interest only insofar as they are 'capitalizable'. Moreover, there is a concern that it will be the 'big competitive cities', that is, selected capitals, that benefit most from an economy-driven funding, while smaller or economically weaker cities will fall behind, which will, in turn, further widen the gap between richer and poorer urban regions.

East Central Europe as an Integral Part of Europe ...

Apart from the process of European integration, East Central European urban development is generally ignored by the academic debates on the 'European city' both by western and also by regional urban scholars because of its status of 'being in transition'. As part of the EU it is obvious that urban development in the post-socialist countries must be integrated into a broader context which is, first and foremost, European, and that treating them as exceptional is no longer justified. It is striking that the recently intensified and partly controversial debate about the European city (Hassenpflug 2002, Le Galès 2002, Kazepov 2004, Siebel 2004, Frank 2005, Bernt 2008) generally still leaves out the former socialist part of Europe. Yet such exclusions are short-sighted because East Central European cities are already facing or will soon face, structural challenges similar to their western counterparts, such as increased land consumption, the proliferation of inner-city brownfield sites, social and regional polarization and demographic and household changes (Steinführer and Haase 2007). East Central European cities are influenced by both general Europe-wide developmental trends and, in its current form, an ongoing post-socialist transition (Węcławowicz 2003: 147–50). Recent studies on population trajectories of European cities under 200,000 inhabitants (Mykhnenko and Turok 2008, Turok and Mykhnenko 2007) clearly demonstrate that East Central European cities show both comparable and distinct patterns with respect to other European cities.

... and the European Debates on Urban Development

The recognition of the 'Europeanist character' of East Central European cities and their fates leads to the following 'dilemma': if one perceives the region as an integral part of Europe, one has also to integrate it into all the theoretical debates on urban development, such as those on urban decline, regeneration, revitalization, gentrification, urban renaissance or resurgence (see Chapter 5). A closer look at the processes that occur in East Central European cities, however, shows that a simple application of these 'textbook models' has its limits or, to put it more critically, helps little (in some cases more, in others less) in explaining urban change and its appearance within the urban space in East Central Europe. This is, more or less, the conclusion of research carried out by both western and urban scholars from the region (Standl and Krupickaite 2004, Bernt and Holm 2005, Sýkora 2005, 2009a).

Given this context, one might argue that looking at western concepts helps to identify commonalities and differences in East Central European developments. As a result, the emergence of hybrid forms seems to be the most realistic and – at the moment – clearest assumption that can be made for the cities. Hybrid structures might sound disappointing for many who argue that, 20 years after the transition, one should at least have arrived at some clear statements about what is happening in the post-socialist world. However, such an assumption underlines that there is a lack of conceptual clarity and a further need for discussion among transition scholars. It is to be hoped that this 'fuzzy' picture will challenge urban research in the future to look for appropriate terms and approaches to explain urban change in the later stages of post-socialism. This is not about distinguishing East Central European urban development 'by force' from western trends simply because it is post-socialist and, therefore, must always be different. It is, rather, about focusing attention on the specifics that exists for the post-socialist context and that have been described in this chapter. We cannot give an ultimate answer within this volume but we hope we can close more chapters in the 'catalogue of today's and future challenges' than we open.

Conclusion

Post-socialist transition has been discussed for two decades. It is now at a stage where it seems legitimate to scrutinize its explanatory power as well as generally to challenge its use as a helpful term for current developments both generally and with respect to urban contexts. Research has not resulted in a unanimous understanding and use of terms describing the systemic change from state socialism to a successor regime. For the purpose of this volume, we have deliberately decided to use the term *transition,* while being aware of its theoretical pitfalls and 'temporal scope'.

In our conceptualization, post-socialist transition is, first, not restricted to change but also refers to persistencies. Second, its very nature consists of the simultaneity and interrelatedness of a number of processes that have their roots in pre-socialist, socialist and post-socialist contexts which refer to its path dependency. While third, the transition itself, as a temporally restricted period, might have come to the end of its explanatory power, the post-socialist condition and its related manifold transformations – from which residential change represents just one – will continue to be a relevant reference context for current and future change in East Central European cities. In this vein, we argue for a wider explanatory approach that looks at these cities from a more comprehensive point of view. The explanatory power of the post-socialist transition is no longer sufficient to explain East Central European urban change completely. Approaches developed in the west and being applied and tested for their scope elsewhere, such as catch-up development, divergence or convergence are still less appropriate to fulfil this function. The recognition of the complexity of the current development and the fact that it brought about other, hybrid forms of formations, arrangements and relationship mechanisms, leads to the conclusion that, to date, the spectrum of terms and explanatory approaches has to be challenged again, not in order to overload the specifics of East Central European or the post-socialist condition but to come closer to what is really happening in the respective societies. One of the main challenges for future research will be, therefore, to analyze more closely the interplay of the post-socialist condition and other macro-trends that shape the fortunes of East Central European cities. In this volume, an attempt is made to apply such an integrated view to inner-city residential change.

For the analysis of inner-city residential change in Polish and Czech cities the post-socialist condition has to be considered as an explanatory background, although it is impossible to properly distinguish which impacts relate to it and not, for example, to others, such as demographic change or globalizing housing or property market structures. Often, it is impossible to say what has influenced a particular process of change. Our point here is not to opt for a new ranking of explanatory frameworks but to state clearly that post-socialist urban development needs to be looked at in an embedded manner. Since the post-socialist transformations have become interspersed with European and global developments, it is more than ever necessary to understand their interplay, rather than to analyze which impact can be ascribed to a particular context or macro-process.

Residential change, as it is analyzed in this volume, represents a good example of how local transformations and overarching change create a reshaped, rearranged pattern of socio-spatial structures within Polish and Czech inner cities. Our analysis helps not only to understand what impact the post-socialist condition might have for the current development of these cities but it also provides a broader framework in which to discuss urban systemic (political and economic) change *per se* and how such change can be interpreted with particular reference to local settings and pre-conditions. It relates today's development to historic legacies as well as to future options, based on current constellations shaped by the post-socialist condition. 'Complete' answers to the many questions raised cannot be provided; this would require an in-depth analysis which is far beyond the intention and scope of this chapter.

Chapter 5

The Inner City in Focus

Sigrun Kabisch, Iwona Sagan

Introduction

After years of decline, European inner cities are increasingly regaining their multi-functionalities including residential use. They have become more and more attractive for residence for different groups of people with specific socio-demographic characteristics, housing preferences and residential behaviour. With regard to these processes the economic, social and built characteristics of inner cities have changed over time which in turn has been reflected in the corresponding scientific debate.

The following chapter will give some insights into the debate about, and the understanding of, the term *inner city*. We begin by taking a historical perspective on both the development of the inner city and how it has been seen. Two approaches can be taken when examining the inner city: a quantitative, data-driven approach and a more qualitative approach, concerned with identity, tradition, place attachment and urbanity. Western debates on urban regeneration, resurgence and other concepts are referred to in order to embed our approach into the overall debate on different urban development paths concerning residential patterns in the inner city. We then examine the development of the inner city in East Central European cities, their pre-conditions rooted in the period of state socialism and the opportunities and obstacles to their development during the period of post-socialist transition. In conclusion, we emphasize that the inner city is key to understanding the city as whole.

The Inner City in its Historical Perspective

The term *inner city* refers to the centrally located parts of a major city or metropolitan area. However, the meaning and understanding of this term has been evolving over time. Today, it is difficult to use the term *inner city* without defining how it is understood and described for the purpose of a particular study. The variety of use and understanding of the term stems from its role in the historical development of a city. As a core part of the city, the inner city has always been shaped and reshaped in a sequence of development stages. Therefore, in both the morphological and the socio-demographic structures found in the inner city one can trace the social, economic and political changes experienced by the city.

To explain the socio-spatial structures of a particular inner city, the historical perspective is indispensable. One of the first scientific approaches to the inner city was developed by the Chicago School in the early part of the 20th century. Although the models and theories were developed almost a century ago, and referred to industrialized and growing American cities, they provide a sufficiently broad scope of interpretation and generalization which is still relevant to contemporary urban issues. In the concentric zonal model of a city, developed by Burgess (1927), the notion of inner city is used and defined in a precise way. It refers to two out of five identified zones. A 'zone in transition' surrounding the central business district (CBD) contains old factory complexes, interspersed with areas of deteriorating tenement neighbourhoods. This zone is the main area of the city which absorbs those new migrants to the city who are not able to compete economically for the better residential locations. It is an area with a permanently high degree of mobility and change. In this zone, ethnic minorities are often housed in poor conditions. As a result, the zone in transition can become an area of high crime rates and social disorganization. Beyond the 'zone in transition' lies a belt of more settled working-class housing which was also identified as part of the inner-city area.

The concept of invasion-succession theorized the process of inner-city changes. As new commercial enterprises are established and new immigrants arrive to residential areas around the CBD, the 'zones in transition' expand into the zone of workers' homes and impel longer established groups to migrate outwards towards the suburbs. As the immigrants move up in socio-economic status, they in turn move outwards and are replaced by newer immigrants. Thus, a non-random spatial pattern emerges, with lower socio-economic status groups located centrally. The Chicago School's perspective highlights the continuing state of change and transition characterizing the inner city. Selecting the inner city as a focus of residential change and demographic challenge of contemporary East Central European cities firmly locates this study in more general traditions of urban research.

Gradually, and particularly from the 1960s onwards, the term *inner city* began to be associated with urban areas suffering substantial economic and social difficulties and became a synonym for an area of poverty, a ghetto or a slum, with less educated and more impoverished residents and an area of crime concentration. In the British context, during the Thatcher era, MacGregor (1990) described the inner city as a 'public issue which represents a constellation of social worries to do with urban poverty, squalor, ill-health, deprivation, decay, crime, social disintegration and social polarization. The core issue is that of urban poverty' (ibid.: 64–5). To date associating the inner city as an area of decay and deprivation is common in North American and British publications. In other countries, inner cities have developed in a much more diversified way.

Many Western European industrial cities and especially their core areas have experienced changes after the Second World War similar to those which occurred in rapidly industrializing North American cities (Petsimeris 1998). The economic prosperity and its accompanying increase in employment were caused primarily by an expansion of manufacturing in the period 1951–1971. The growing economies of Western European countries stimulated the inflow of immigrants to the cities from other European and non-European countries looking for work and better living conditions. The growing number of incomers started to change the socio-spatial structures of the cities. As in North American cities, the immigrants' first destination was usually the inner-city area. Consequently, the social heterogeneity of these areas increased. Analogous to tendencies described by the Chicago School, filtering-down processes, which deepened social segregation, caused the downgrading of large parts of the inner city. However, such negative social processes have not affected continental Western European inner cities to the same extent as North American or many British cities. Numerous long-settled inner-city districts in cities such as Paris, Rome or Vienna often remained the residential areas with the most expensive rents, inhabited by the wealthiest and most influential citizens. In the tradition of continental Western Europe, possessing a well maintained apartment, conveniently located near the city centre, was as valuable as possessing a family home on the outskirts. This attitude has helped to keep the middle class in the inner parts of the city and, by the same token, to provide its members with good living conditions in densely populated central districts.

After the Second World War, East Central European cities developed differently from their western counterparts. The totalitarian system which was imposed on these countries changed the social and economic conditions of development, strongly influencing the growth trajectories of cities. The politically and ideologically based economy separated the industrialization process from that of urbanization (Sagan 1995). It slowed and deeply affected the development of cities which did not experience a period of intensive growth and social diversification as a result of foreign migration characteristic of Western European cities. The critical issues for reshaping socialist cities were the lack of market rent of land and the lack of economic competition for attractive locations. These systemic constraints resulted in the creation of areas specific to socialist cities (Hamilton 1979b, Szelenyi 1983, Smith 1996). In the most general scheme, it is possible to identify:

- a central area, usually consisting of a historic medieval or renaissance city core, surrounded by an area with a mixture of services, industrial and housing land use,
- a transition area with a domination of pre-Second World War built-up housing stock,
- new large housing estates,
- a peripheral suburban area, which included open land, still in agricultural use and low density, one-family private housing.

It is important to note that the areas did not follow a concentric zonal model, rather they created a type of mosaic pattern (Węcławowicz 1988, 2003). The lack of market rent of land and the domination of state-subsidized public transport altered the traditional socio-spatial structures which were typical for cities in market economies (Sagan 2000a). Despite the heavy losses during the Second World War, the historic urban fabric survived in the central and transition areas and physically delimited the area of the inner city. Since they were a symbol of the pre-socialist era the old workers' districts, as well as the former middle-class housing stock, received little attention and inadequate financial support (Friedrichs 1978). This resulted in a pervasive physical deterioration of the inner cities. Residents who could move away did so in order to live in a better urban environment. As a consequence, the processes of social segregation were induced not by social factors, which occurred in the non-socialist world, but by the sheer physical dilapidation of the housing stock. Studies undertaken by East Central European and western scholars, confirmed residential differentiation and segregation during the state-socialist period (Pickvance 2002). The inner city which Pickvance termed the inner commercial, housing and industrial districts from the capitalist period, has always played a distinctive role in urban life and transformation. It represents the centre of power in political, administrative, economic and cultural respects and hosts the institutions whose influence reach far beyond the city's boundaries (Enyedi 1998). The residential areas adjacent to the city centre traditionally provide residence for the political and economic élites of the city or for working-class populations, in the form of a mixed industrial-residential type. Their physical structure is diverse, ranging from very simple to valuable housing stock. Nevertheless, the final result of the process remained the same: inner-city districts of post-socialist cities, just like those in the west, urgently needed regeneration and revitalization.

From this review of the differences and similarities in the development of inner cities, it is evident that there is no one definition of the inner city precise enough to cover all cities. Depending on the tradition and regional perspective, some include the central district and others do not. In the European research tradition, there is a tendency to consider the central district which quite often not only includes the CBD but also the historic core of the city, including residential areas, as a central part of the inner city. North American studies, based on a different urban development history, refer to the inner city as the area developed around the CBD. To identify the inner-city areas in East Central European cities for the purpose of this study, a morphological criterion, based on the density of pre-Second World War housing, has been used. As discussed above, the pre-socialist urban structure is the critical element for describing this part of the city. Since central districts fulfil the chosen criteria, they have not been excluded from the research. Our flexible approach, based on identifying the most typical inner parts of the cities under examination, helps in the investigation of ongoing residential change (see Chapter 6).

The Characteristics of the Inner City

To provide an overview of the main characteristics of the inner city, we use quantitative as well as qualitative features. This allows us to produce a comprehensive description, encompassing the heterogeneity and complexity as well as the density and compactness of the inner city. The *quantitative* approach starts from administrative borders and statistics. It contains the geographical area, the physical building structure and demographic as well as social statistics.

Looking at several studies dealing with the inner city in Europe shows that there are no clear criteria for defining the inner city. In his study, Mangen (2004) compares inner-city development in five Western European cities, located in different countries. For him, a decisive factor is the availability of data provided by administrative statistical offices. For instance, in Hamburg, one of his five cities, he used a total of 12 neighbourhoods in the city core *(Kerngebiet)* and adjacent locations to the west. For Marseilles, the three central arrondissements were involved (ibid.: 52–3, Appendix 208–215). In his initial words on mapping the inner city, he emphasizes that the conceptual and functional equivalence is problematic. Beyond the crude population statistics, there are considerable differences in the quality and the range of the material generally available.

The *geographical location* is also a quantitative delimitation. It is important to stress that the inner city is not just the city centre. Rather, interlinkages between the centre with its residential offers and its adjacent residential districts are a decisive feature of the inner city in terms of jobs, leisure time opportunities and cultural and educational facilities, all of which are essential preconditions for compact and lively inner-city areas. A further quantitative characteristic is the *structure of buildings* in terms of architecture and diversity of use. The latter includes public, administrative, commercial and cultural buildings which are typical of a city centre, as well as tenements and a housing stock built in different historical periods. In East Central European cities at least in those not destroyed during the Second World War, the housing stock is dominated by pre-war buildings. The reason for the existence of many unique and architecturally valuable buildings in these cities is the completely different urban development over decades. Whereas Western European cities often pursued a policy of mass demolition and construction of completely new buildings in keeping with the architectural style of the time, in East Central European cities the inner-city building stock attracted little interest. Thus the traditional housing stock which survived the Second World War, although often in a bad state of repair, has undergone physical rehabilitation since the 1990s. The renovation and rehabilitation processes have helped to preserve the old housing stock which offers now a unique place of residence and contributes to the remarkable cityscape of many East Central European cities.

In contrast to the quantitative features, the *qualitative* approach includes the *local identity* of the residents, *local history* and traditions as well as the *flair and the uniqueness of the place* (Murzyn 2004: 258–9). To choose inner cities for residence reflects the residents' awareness of specific functions, facilities and opportunities as well as their interplay. These locations are formative parts of the whole urban fabric in which *place attachment* is a distinctive characteristic. The special features of place attachment, including residential satisfaction can be determined by using interview results on, for example, the perception and evaluation of residential environments, housing conditions and relocation decisions. Here, it is not only the household type which is the key to understanding perceptions and dynamics, but also housing and living arrangements – that means the way people are living together – socially and flat-sharing.

The inner city constitutes a distinctive part of the built urban heritage involving public as well as residential buildings. With their functional mix of housing, commerce, work, education, leisure time facilities and further urban amenities, inner cities are a symbol of *urbanity*. This includes complexity, density, change and expansion during the period of industrialization from the early 19th century onwards. And today the inner city represents the main characteristics of urban life in its concentrated forms and diversities.

The end of the Fordist model of economy and the end of industrial city prosperity resulted in the shift towards a service-based economy and brought a new value to inner cities. Inner cities are the scene of some of the most spectacular revitalization investments, which often make the particular city known worldwide as an example of successful urban policy. The urban renaissance, broadly discussed in the literature (Brühl et al. 2005, Porter and Shaw 2008), symbolized the rebirth of the whole city by bringing back the meaning of its inner city.

Finally, the *appropriate linguistic expression* for the inner city as our specific research area has to be analyzed in the East Central European context. An analysis of the scientific literature and official documents shows that there is no single, defined term for the inner city. In English, the term *inner city* is neither unambiguous nor neutral (Burjanek 2009) and the same holds true for the different translations. In Polish, the terms *śródmieści* (inner city), *obszary śródmiejskie/wewnątrzmiejski* (inner-city areas) and *obszary starej zabudowy* (old built-up areas) are common. In Czech, the area might be called *vnitřní město* (inner city), *střed města* (city centre), *střední čtvrti* (inner-city quarters) or *čtvrti staré zástavby* (old built-up quarters). In the Czech research literature, one can also find *jádrové části velkých měst* and *centrálně položené městské části* (core parts of large cities and centrally located urban areas, respectively), *subcentrální části velkoměsta* (sub-central areas of the city) or even *centrum města* (city centre; examples taken from Vaishar and Zapletalová 2003: 24, Mulíček and Seidenglanz 2004: 37). Nevertheless, these terms include the geographical perspective in terms of location, which is the decisive feature for the naming of an area. In some expressions, the old or traditional building stock is also included.

Western European Discourses on Inner-city Development

There are several debates concerned with changing social, demographic as well as built structures and their relationships within the inner city. The following concepts will be included here: urban renewal, revitalization, regeneration, resurgence, renaissance, gentrification and reurbanization. Table 5.1 provides an overview.

**Table 5.1 Important urban development concepts: Content and original
context**

Concept	Content	Original context
Urban renewal	Process of essentially *physical change* on a local level, related to neighbourhoods and housing estates, action against urban decay and social deterioration of the inner cities, based on large projects of private investors (for example, waterfront development)	US
Urban revitalization	Focus on reconstruction of the physical environment in the historical city core, *new uses for old buildings*	US
Urban regeneration	*Comprehensive and integrated political perspective* on problems, potentials, strategies and projects within the social, environmental, cultural and economic sphere; a governmental activity with strong spatial focus; longer-term and strategic purpose; implementing policies in existing urban areas rather than developing new urbanization; addressing the problems of inner-city areas facing problems of imbalance and decline; a normative concept	UK
Urban resurgence	*Broad process of revitalization and repopulation* of inner cities, relates to a previous context of long-term decline	UK
Urban renaissance	Describes the *physical and social 'refurbishment', 'repopulation' and 'revitalization'* of the inner city; relates to city-mindedness as a housing preference and lifestyle but also to all processes of upgrading (physical, symbolic, price-related) taking place in inner cities	UK and US
Gentrification	*Upgrading* of residential areas by in-migration of Young Urban Professionals and affluent people in connection with urban renewal and revitalization, causing *displacement* of the traditional local population	UK
Reurbanization	Trends of *back-migration* into cities after a period of suburbanization, urban liveability, mixed-use and sustainable community development in the inner city, demographic rejuvenation in quantitative terms; relative growth and diversification of population and household numbers; increased residential attractiveness for socially mixed population stimulating *households' decisions to stay in the city*	UK and DE

Source: Authors' work.

All of these concepts imply processes of urban development resulting in
improvement and stabilization of living conditions in the inner city. Whereas

urban renewal and urban revitalization (Couch 1990) are usually directed towards a range of measures concerning the built environment, concepts such as regeneration, resurgence, renaissance, gentrification and reurbanization can imply a more integrated development. These include built and social aims dedicated to the improvement of the physical housing conditions as well as of the liveability for different groups of residents. Furthermore, they suggest cooperative approaches, such as public-private partnerships. Nevertheless, there is no strict separation between the concepts describing inner-city change but rather they overlap with each other, albeit with different focal points.

The starting point of all the concepts was the dramatic change of social structure, which in turn was closely related to the built structure in the inner-city areas during the last few decades. These locations have become the main target of policies of intensive urban renewal and revitalization.

From the late 1980s onwards, signs of a return to urban living have been detected in a number of countries (Kujath 1988, Cheshire 1995). In the framework of urban regeneration, a more comprehensive and integrated political perspective on problems, potentials, strategies and projects developed (Roberts and Sykes 2000, Couch et al. 2003, Kühn and Liebmann 2009). It included the social, environmental, cultural and economic spheres, reflecting the complex structure of inner-urban viability. Processes of urban regeneration and resurgence seem to have affected a large number of European cities, as well as cities elsewhere in the world (Cheshire 2006, Cheshire and Gordon 2006, Turok and Mykhnenko 2006, Buzar et al. 2007, Colomb 2007, Buzar et al. 2010). In addition to resurgence, the 'urban renaissance' is associated with a more general 'revival' of cities (Storper and Manville 2006: 1247). In broader terms, these concepts have been defined 'against a context of previous decline', although they have been distinguished 'from simple growth as such' (Cheshire 2006: 1232). Urban resurgence today forms a part of complex changes taking place within most European cities. It describes a number of patterns, drivers and related processes and pathways for upgrading affected inner cities or inner-city neighbourhoods. Several dimensions of inner-city resurgence have been identified: first, the demographic rejuvenation of inner-city areas in quantitative terms, including the relative growth of population and household numbers, as well as the diversification of household structures (Ogden and Hall 2000, Buzar et al. 2007); second, the qualitative aspects of inner-city resurgence, which involve increased residential attractiveness and socio-cultural revitalization (Seo 2002) and third, a long-term stabilization of the residential population, social mix and housing functions of the inner city, including the emergence of 'balanced neighbourhoods' (Cheshire 2006).

Additionally, broader societal trends, such as the professionalization and feminization of the workforce and the development of work-centred household types, a large proportion of single households and the emergence of new household compositions, all with distinct urban preferences, have changed both people's perception and the social fabric of the inner city. This has resulted in gentrification, which forced displacement of the traditional local population (Slater et al. 2004,

Atkinson and Bridge 2005, Lees and Ley 2008). Inner-city resurgence may, for example, have fostered the displacement of lower social classes from the inner city, while new segregation patterns have arisen on the urban scale, transferring social problems to other places (Lees and Ley 2008). Gentrification is one of the most widely discussed issues in the context of inner-city changes. The causes of the process have mainly been investigated from production and consumption points of view. From a production perspective, gentrification is driven by the logic of capital inflow to the inner parts of the city. The consumption perspective stresses the social exchange issue. Gentrifiers are bringing new styles of life and new residential requirements which are changing the inner-city habitat. The western invasion-succession model of gentrification assumes that artists are usually the first wave of gentrifiers in a run-down inner-city area. They are followed by middle-class groups such as liberal public sector workers (Ley 1996, 2003) who introduce specific lifestyles to inner-city neighbourhoods on a broader scale. While many gentrification analyses focus on new social and symbolic structures (Lees 2008), the demographic dimension of these processes has often been ignored. At that point, reurbanization comes into play. The fact that reurbanization's vision of 'urban liveability, mixed-use and sustainable community development' uses a 'language [...] familiar to many gentrification scholars' (Colomb 2007: 13) has resulted in the concept being subjected to a series of theoretical critiques, whose proponents fear that the term has been co-opted by urban developers as a discursive method of camouflaging the adverse social impacts of gentrification. Some of the key contributions within this vein of work argue that the demographic aspects of reurbanization are being quietly brought into the urban revitalization debate in order to remove *class* from mainstream conceptualizations of gentrification (van Criekingen 2010, Slater 2008) or from urban studies in general (Bürkner 2008). They suggest that gentrification can be 'stretched out' to comprise many of the constituent dynamics of reurbanization, without compromising the original, class-based meaning of the term.

However, opposite arguments suggest (Buzar et al. 2007, Smith and Butler 2007) that the concepts of reurbanization and gentrification set out to describe different sets of processes. Reurbanization, as we understand it, refers to a much wider range of dynamics, compared to gentrification, since its main purpose is to capture the rising overall 'liveability' and 'survivability' of compact inner-city areas for a variety of social classes and demographic groups (van den Berg et al. 1982, Jenks et al. 1996, Williams et al. 2000, Haase et al. 2010). In conceptual terms, gentrification, with its focus on the politics of displacement and selective urban upgrading, cannot be equated with the broader, multi-directional process of reurbanization, which takes place over the entire inner city and is principally characterized by increasing in-migration as well as decreasing out-migration rates after a long phase of decline (Lambert and Boddy 2002, Buzar et al. 2007). Moreover, reurbanization entails a wide variety of migration flows into the entire inner city, often involving marginalized social groups, such as international migrants and working-class families with children, that cannot be considered as 'gentrifiers', according to the conventional understanding of the term. In this context, one of the decisive driving forces of reurbanization is the often neglected interplay of housing markets, urban development and population change (Haase et al. 2010).

East Central European Inner Cities During and After the State-socialist Era

In contrast to inner cities in Western Europe that have undergone processes of urban regeneration and resurgence over the last 20 to 30 years, cities in East Central Europe are associated with built-up areas in public ownership and poor physical condition which are in general decline and suffering from both physical and demographic ageing (Szelenyi 1996: 306–7, Billert 2004). While in both the gentrification and the reurbanization literature in the western debates, the influx of new residents and distinct household types as well as the diversification and rejuvenation of the inner-city population are highlighted, it is widely agreed that East Central European inner cities are inhabited by predominantly an older and less affluent population who either cannot afford better housing conditions or who were assigned the flats during the state-socialist period. Up to now therefore, it is the legacies of the previous system that have been stressed together with their persistence under new societal conditions (Haase and Steinführer 2009).

To obtain a clearer picture, it is necessary to look back to the period of state-socialist city planning when inner cities became hybrid areas. They consisted of pre-socialist built structures, as well as socialist intrusions. The latter included socialist architecture for public and residential buildings together with poor maintenance of the old stock which was regarded as bourgeois housing stock and a replacement of the old residents with new ones.

The huge housing crisis in the immediate post-Second World War period meant socialist governments were looking for different, quite often provisional, solutions. One was to subdivide large flats in nationalized tenement houses or one-family private properties into several units for more than one household. As a result, inner-city areas were repopulated and redensified. The inner-city living conditions significantly worsened in that period. The solution to the housing shortage reached by the socialist governments was a decision to develop industrial housing construction. The prefabrication of flats and the unit construction system (*blokowisko* in Poland, *sídliště* in Czechoslovakia) led to an enormous increase in the number of newly built apartments, so achieving the political aim of 'solving the housing question'. Since these locations provided the open space necessary for the technical equipment such as cranes as well as space for large scale building, the newly constructed prefabricated housing estates were concentrated on the outskirts of the cities. The large housing estates inevitably sucked out the younger and more affluent residents from the old inner-city housing stock since they provided new and better living conditions.

The following decades were characterized by an ongoing housing shortage, as household numbers rose, a shortage which could only in part be met by the provision of flats in growing new housing estates at the urban fringes. The inner cities underwent step-by-step downgrading processes (Wießner 1997), reflected by a poor state of repair and an increasing number of uninhabitable flats; so neglect of maintenance led to underutilization (Sýkora 1993: 284). Ageing in place contributed to ageing neighbourhoods as a result of the generally low residential mobility during state socialism and the non-existent influx of younger urban dwellers who preferred the newly built housing estates on the outskirts of the cities (Matějů et al. 1979, Cieśla 1984).

It is well-known from many studies that East Central European inner cities after 1945 were characterized by a systematic neglect by the authorities and a steady decline of the building stock as well as the social status of the inhabitants. Scholarly works from both Western and East Central Europe from the 1970s and 1980s report on the deterioration of the housing situation and a concentration of poor, old and socially weak people within the city centre and the adjacent old built-up areas.[1] As a consequence, inner-city reality was strongly associated with population decline. Sowa refers to the 'depopulation of the inner city' of Kraków during the late 1980s, a characterization that applied also to inner-city Łódź, most of the Silesian industrial cities but also to old-industrialized cities such as Brno and Ostrava where old built-up workers' districts had already deteriorated in to 'ghettos of poverty' before the beginning of the post-socialist transition (Szczepański and Ślęzak-Tazbir 2008).

During the 1980s, a new awareness of the architectural value of the old housing stock and its importance for place attachment and urban identity developed. Some central places and shopping malls underwent renovation. But there was no change in the overall political decisions of the time; the inner cities, and especially the adjacent housing areas, remained undervalued and suffered from continuing decay.

Following the political and socio-economic changes in the European state-socialist countries during the 1990s, the inner cities changed their appearance in terms of their built and residential characteristics. Accompanied by an extensive change of ownership, gradually renovation took place and the local economy became more service sector based. Thus there were new opportunities in terms of economic function, urban amenities and living and housing conditions resulting in housing mobility for a variety of residents and household types. The prevailing picture has become more differentiated. In particular, accelerating suburbanization from the late 1990s onwards again reinforced the processes of inner-city areas' decline and ageing. Privatization and general access to cheap credit, coupled with a conspicuous lack of planning controls on land conversion, have fuelled a

---

1   For Łódź: Hamilton and Burnett 1979: 283–90, Majer 1988: 85–6; for Kraków: Sowa 1988: 71–4; for Warsaw: Dangschat and Blasius 1987; for Prague: Matějů et al. 1979.

boom in suburban residential and commercial development. Blocked for decades, people's desire for better living conditions, with the one-family house in the countryside idyll at the very top of their list of wants, has been the main force shaping intra-urban migration movements. Wide-scale out-migration from inner-city areas, encouraged by suburban housing opportunities, has overlapped with more general demographic trends in terms of a fall in birth rates and an ageing population (see Chapter 3). As a result, the process of urban shrinkage observed in some western cities for decades has now begun to affect severely post-socialist cities released from central command and control systems (Zborowski 2009). East Central European inner cities seemed therefore to continue to follow their long-term trend accompanied, at least in Poland, by an ongoing dilapidation of the inner-city building stock. There are estimations that of the pre-Second World War housing stock in Poland 20 per cent (or 340,000 dwellings) need to be torn down, due to their bad state of repair (Billert 2004: 47).

Set against this background, several signs of resurgence within East Central European inner cities represent an opposite trend to this story of decline and ageing. According to some scholars, there are, seemingly in accordance with western patterns, more young and, to some extent, more affluent people who are moving to the inner city. In the case of Prague, Burcin and Kučera (2006: 181) noticed a deceleration of population decline in the inner city for the period 2001–2004. Sýkora (2005) found evidence of neighbourhood change by gentrification (and gentrification-like) processes in some major urban centres of East Central Europe. However, he argued that this development is not primarily linked to positive sentiments towards these neighbourhoods, but rather to a 'utilitarian demand for housing in convenient and pleasant locations close to places of work for professional élites' (ibid.: 105). For Polish inner cities, Parysek (2005: 109) claims that processes of reurbanization, that is a 'redevelopment of the central parts' of larger cities are reflected in increased housing densities and a number of renewal projects. Buzar and Grabkowska (2006) describe a number of social practices of residential mobility and spatial flexibility in the inner city in Gdańsk, which have led to a socio-spatial transformation of this urban space (see Chapter 10). Housing biographies might reveal that dilapidated, but nevertheless still architecturally impressive pre-Second World War tenement houses have become attractive for the children of families who, decades ago, escaped to 'modern' large housing estates. Especially, some selected areas have experienced an upgrading in built and symbolic terms, for example, former Jewish quarters. Such areas are at the focus of urban development strategies and public attention.

Regeneration urban policy, which has become one of the main issues in the post-socialist urban development debate, is mainly concentrated on large projects targeted at particular, limited areas within inner-city districts. Looking for financial support, urban authorities compete for European Union funds to secure their plans. Thus, the proposed projects have to meet the aims and expectations of the funder. EU policy is aimed at individual projects for improving both economic and social conditions in terms of revitalization as well as the historic and cultural protection of unique places.

One of the most important targets which seem to absorb the attention of local authorities in their regeneration plans is brownfield regeneration. By contrast, the continuing depopulation of inner-city districts suggests that the improvement of housing conditions is seen as less important. Another specific attribute of the regeneration strategies in East Central European inner cities is that they are strongly linked to the trend of commercialization of public space. Local authorities, struggling for funds and investment capital, are much more exposed to investors' activity. The weak position of planning bodies and the lack of an effective system of public control and participation in the preparation of spatial development plans often results in the transformation of public space into commercial areas. Commercialization has been especially driven by the development of offices and multi-purpose commercial centres (Sýkora 1999, Kotus 2006).

In the post-socialist urban context, observed processes of gentrification take the form of 'oases of wealth' in the midst of spaces of physical and social decay. They are limited to selected plots. These processes lead to the deeper fragmentation of the inner-city urban fabric (Sýkora 2005). Selective gentrification, at least in part, results from the lack of coherent laws enabling planned and controlled revitalization. In particular, current legislation means that crucial regulations are atomized to a substantial number of different, and in reality, incompatible laws, acts and statutes. In addition, the unknown legal status of many of the inner-city plots significantly complicates regeneration processes. This is especially true for the Polish situation where, unlike the other East Central European states (Blacksell and Born 2002), the law of property restitution to former owners is still absent. The processes described above result in the polarization of the inhabitants of the inner-city areas. As already mentioned, selected parts of the inner city have been subjected to the initial stages of gentrification, leading to a growing isolation of the wealthy social strata. Parallel to selective regeneration, the steady deterioration of vast amounts of the pre-Second World War housing stock as a result of the lack of local government investment has not yet been stopped. These areas, often owned by local authorities, run the risk of becoming spaces of physical and social blight. Strategies of regeneration hardly ever include specific plans for adaptation and incorporation of indigenous inhabitants into newly regenerated places. Quite often, the revitalization projects of inner cities have a tendency to limit the regeneration policy to the typical successful model approach. Concentrated as they are on brownfield regeneration, they usually propose turning post-industrial buildings into cultural or commercial centres (for example, the shopping and entertainment centres Manufaktura in Łódź, Vaňkovka in Brno). Thus, they are much more place- than people-oriented initiatives. Although activities aimed at social environment improvement are present in the revitalization programmes, they tend to be rather more rhetorical than practical in nature.

The problems of regeneration and revitalization of inner cities require a flexible and comprehensive policy, oriented towards both top-down and bottom-up processes. As current studies demonstrate (Grabkowska 2007), while top-down projects may be highly effective for the physical regeneration of the areas, bottom-up processes are of critical significance for social and demographic regeneration. Thus, to achieve a desired resurgence of a place, mixed strategies should be applied. Limiting the activities to one-size-fits-all projects reduces the effectiveness of urban development policy.

Conclusion

Summing up, we can state that East Central European inner-city development reflects the new trends of, and responses to, current urban development. We argue that the inner city encloses the city core and its adjacent residential areas. Consequently, the inner city is the central part of the city, with public and commercial functions and which also has a residential function. It is the central district, without administrative borderlines, in which traditions, identification and place attachment of the local people have been defined. Nevertheless, the inner city is part of the city and needs to be embedded into the whole urban fabric. It should not be separated from the other parts. Rather, the closer the links between the inner city and the other urban areas, the stronger the cohesion of the whole city.

There are very different preconditions for development in Western European countries and East Central European states, but similarities in urban resurgence can increasingly be observed. Despite these similarities, the local contexts remain quite different, which leads to different quantitative and qualitative results. Thus we should not neglect the specifics of the post-socialist character of East Central European cities. Only on the basis of detailed analysis is it possible to draw appropriate conclusions about the prospects for future development and to make appropriate practice-oriented recommendations. In that sense, the main challenges stemming from the debates to date are the maintenance of the unique character of the inner city in physical terms, as well as its viability, based on a socially and demographically mixed residential population. This requires considerable efforts in the reconstruction of buildings and infrastructure to ensure they meet present requirements, but also maintain affordable rents and fees. It provides the basis for residential change which attracts different household types in demographic and socio-economic terms.

After decades of domination by standardized prefabricated large housing estates, the pre-Second World War tenement houses in the inner city have become an increasingly attractive housing environment. Together with a variety of urban infrastructure and services, the inner city possesses appropriate conditions for new life in old buildings. The observed first trends could be effectively supported and strengthened by a regeneration policy which allocates some development funds to such areas. The ongoing processes are not free from the negative effects of gentrification and displacement. But, due to the limited strength of the gentrifying social class in post-socialist cities, the penetration of middle- class households into lower class inner-city neighbourhoods results in a continued fragmentation of the inner-city urban fabric.

In a debate about changes in the post-socialist inner cities, many fundamental questions are still open:

- What are the main motivations for living in the inner city? Is inner-city living a pragmatic choice or rather a lifestyle choice?
- Who lives in the inner city? To what extent does the persistent socio-spatial model of the post-socialist city with a declining and ageing core surrounded by better-off suburbs still represent reality?
- What opportunities does the inner-city housing market provide in response to the changing preferences and needs on the demand side?
- Is there a new pattern of social networks and neighbourly relationships in inner-city areas?

The following chapters provide answers to some of these questions based on empirical research results derived from statistics, surveys and in-depth interviews.

Chapter 6

Łódź, Gdańsk, Brno and Ostrava and their Inner Cities: Urban and Demographic Development during Post-socialism

Adam Bierzyński, Maja Grabkowska, Annegret Haase,
Petr Klusáček, Andreas Maas, Jana Mair, Stanislav Martinát,
Iwona Sagan, Annett Steinführer, Antonín Vaishar,
Grzegorz Węcławowicz, Jana Zapletalová

In this chapter, second-order cities in Poland and the Czech Republic are introduced. Their characteristics are discussed, together with the major changes that have taken place there during the post-socialist period relevant to this volume. It therefore integrates the theoretical and methodological sections introduced in Chapters 2–5 with the empirical, case-study based research presented in Chapters 7–12.

The chapter is structured as follows: it starts with an overview of Polish and Czech second-order cities within the respective national settlement systems. The second part introduces the four case studies, Łódź, Gdańsk, Brno and Ostrava, together with their inner-city areas in more detail. Finally, we take a city-wide overview of the two dimensions of residential change identified in Chapter 2: residents and residence. This includes a discussion of the demographic and housing market developments in the four cities, which provides the context for the empirical findings of later chapters.

Polish and Czech Second-order Cities in the Context of Post-socialist Change

The relative paucity of research on the impacts of transition on non-capital cities of East Central Europe led us to select four such cities for our investigations of the significance of ongoing urban population decline and household change for inner-city residential change. This volume focuses on selected second-order cities which fulfil a significant role in the urban hierarchies of Poland and the Czech Republic. Two are old industrial cities in the narrow sense (Łódź and Ostrava), and two are multi-functional cities with a long history as centres of trade (Gdańsk and Brno). Their position in the national settlement systems as well as the changes that have been taking place in these types of cities in Poland and the Czech Republic are discussed in the following sections.

Specifics of Second-order Cities

We understand second-order cities (which are sometimes called second-tier or second-rank cities) to be those urban centres in a national settlement system that rank second after the capitals with respect to size, function and economic importance (Grimm 1994, Korcelli 1996). Although population size is a major criterion, it is not the only one. Second-order cities are nodes of the economy, cultural centres and national transport hubs as well as major regional centres for employment, trade, services, administration and education. Many of these cities have a long industrial tradition and have, since the onset of the post-socialist transition – and in many cases long before – been experiencing deindustrialization, accompanied by high unemployment and out-migration as well as a restructuring of the economy towards the tertiary sector. This includes an increasing significance of the service sector but also a growing importance of administrative and educational functions.

The polycentric and relatively equal distribution of urban places in *Poland* is considered to be one of the assets of the spatial organization of the country (Figure 6.1). Throughout the post-Second World War period, the number of cities slowly increased as a result of administrative changes or population growth and, in 2007, this number had risen to 891 cities. During the forced industrialization period between the 1950s and the 1970s, some cities grew more quickly due to extra investment in housing in selected agglomerations by the central government. The significance of urban places as a concentration of the working class and of industrial production had the greatest priority at this time. Yet, policies were also implemented which attempted to limit the growth of the largest cities, for example, by applying restrictions with regard to migration for permanent residence to Warsaw and other large cities.

In the post-socialist period, the economic recession, together with unemployment, became the main differentiating force of spatial change. Depending on their economic performance, Polish cities can be classified as winners or losers in the socio-economic transformation. Moreover, cities face the challenges of competitiveness on both global and European scales. Currently, their economic and political position is relatively weak. European integration has brought and intensified the competition for investment, jobs and locational advantages in the regional and urban hierarchy on the continental scale. On the list of the 74 Metropolitan European Growth Areas (MEGA; DG Regio 2004), there are eight Polish metropolitan centres: Warsaw as a first-rank metropolis and Gdańsk, Katowice, Kraków, Łódź, Poznań, Szczecin, Wrocław as lower-rank metropolises. The Union of Polish Metropolises, however, promotes a larger network of twelve urban centres and includes Bydgoszcz, Lublin, Białystok and Rzeszów with those mentioned above. It is assumed that these twelve Polish cities together can contribute substantially to the European urban network as more important players.

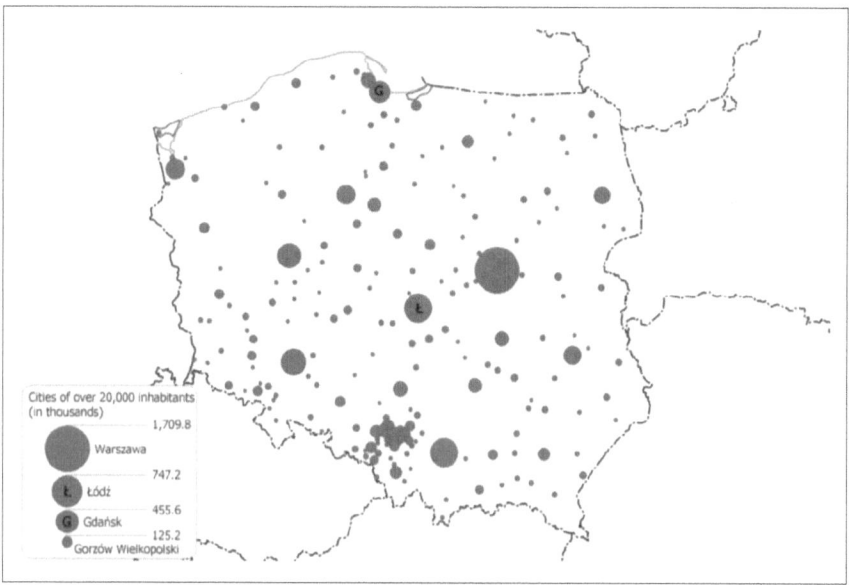

Figure 6.1 Poland's polycentric urban system
Source: Authors' work.

Not only the relatively equal distribution, but also the population size of the Polish national urban system and its geopolitical location create important assets. Nevertheless, the transport and communication links between the major urban centres within the country, as well as to the European core area, are under-developed. Improvements of the transport system and the increase of spatial accessibility will significantly shape the development of the major Polish cities and their place in the European urban hierarchy (Korcelli 1997, Węcławowicz 2002b).

In contrast to Poland, the *Czech Republic* is dominated by its capital city: Prague is both the largest and the most important Czech city (Figure 6.2). It is also the only Metropolitan European Growth Area in the country. Prague is both functionally, and with respect to its size, large in terms of the national settlement system. This imbalance has deep historical roots. Prague was the capital of a larger country, Czechoslovakia with Slovakia and Transcarpathian Ukraine between 1918 and 1938, and with Slovakia between 1945 and 1992. Before the existence of an independent state, Bohemia and Moravia, which form the two major historical territories of today's Czech Republic, were integrated into the settlement systems of the large Austro-Hungarian Empire.

When taking population size alone into account, Brno, Ostrava and, with some restrictions, Plzeň, can be considered as the Czech second-order cities. Olomouc, České Budějovice, Liberec, Hradec Králové, Ústí nad Labem and Pardubice would then be third-order cities. However, when other indicators are also included, such as administrative and juridical functions as well as education, research, transport, commerce, tourism, culture and sport, prices of land and quality of life, as well as functions within the European settlement system, the description of the urban hierarchy is somewhat different: Prague, unambiguously, remains the most important centre of the Czech urban hierarchy. Brno clearly ranks as number two. Ostrava, although the size of its population is similar to Brno's, ranks together with smaller regional centres such as České Budějovice or Liberec in various functions. Ostrava's problems result from its remote geographical position, structural difficulties associated with a specific population composition (see below) as well as its heritage of heavy industry and environmental pollution.

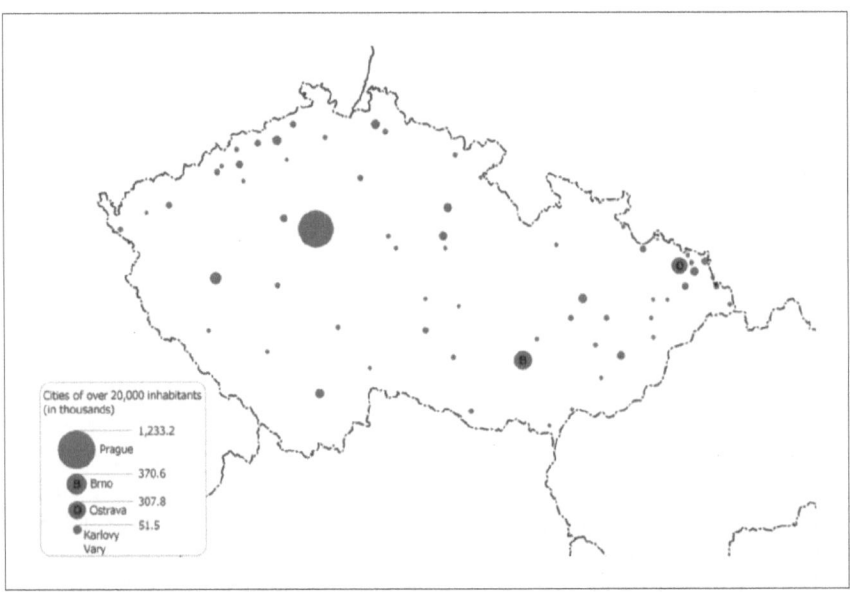

Figure 6.2 Urban centres in the Czech Republic
Source: Authors' work based on work of Śleszyński (2010).

Table 6.1 Population development in the largest Polish cities 1988–2008

	Warsaw	Łódź	Kraków	Wrocław	Poznań	Gdańsk	Szczecin	Bydgoszcz	Lublin
— Population according to the population register (as of 31 December; in 1,000)									
1988	1,655.0	854.0	746.4	639.1	588.3	464.1	410.3	377.9	340.3
2001	1,609.8	786.5	740.7	634.0	572.0	455.5	415.6	383.2	354.0
1988=100	97.3	92.1	99.2	99.2	97.2	98.1	101.3	101.4	104.0
2002*	1,688.2	785.1	757.5	639.2	577.1	461.7	415.1	372.1	358.4
2003	1,689.6	779.1	757.7	637.5	574.1	461.0	414.0	370.2	356.6
2004	1,692.9	774.0	757.4	636.3	570.8	459.1	411.9	368.2	356.0
2005	1,697.6	767.6	756.6	635.9	567.9	458.1	411.1	366.1	355.0
2006	1,702.1	760.3	756.3	634.6	565.0	456.7	409.1	363.5	353.5
2007	1,706.6	753.2	756.6	632.9	560.9	455.7	407.8	361.2	352.8
2008	1,709.8	747.2	754.6	632.2	557.3	455.6	406.9	358.9	350.5
1988=100	103.3	87.5	101.1	98.9	94.8	98.1	99.2	95.0	103.1

Source: GUS (several years), authors' work.

Note: * Revised, based on census data from 20 May 2002 (not directly comparable with 2001 data).

*Residents and Residence: Challenges for Polish and Czech Second-order Cities
Arising from Demographic and Housing Market Changes on the National Scale*

Polish and Czech second-order cities have faced a number of challenges during post-socialism, the major period of reference of this volume. From the perspective of residential change, the focus in this section is on the following issues and processes: population development shifting between decline and growth, accelerated demographic ageing, changes in households, as well as far-reaching housing market transformations.

Demographic Change: Population Development, Ageing, Household Change

Population development   In Poland, cities with more than 100,000 inhabitants became the major population gainers as a result of the forced industrialization in the post-Second World War period. Urban growth slowed down significantly in the 1980s as a result of economic recession, which reduced rural-to-urban migration. This was also true for second-order cities. During post-socialism, the majority of the largest Polish cities have either stagnated or experienced a population decline (Table 6.1). Beside losses caused by out-migration, they have shown the greatest decline in natural population growth of all urban settlements (Kupiszewski et al. 1998: 274, 288–9). In spite of the general economic stagnation during post-socialism, other large cities that were not dominated by declining industrial activity emerged as winners in the socio-economic transformation. Nevertheless, these cities also experienced population decline, particularly as a result of suburbanization (Węcławowicz 2005).

However, the official population statistics remain ambiguous. The last census in 2002 provided evidence that some of the largest cities, characterized by booming labour markets (for example, Warsaw, Kraków and Poznań), experienced population growth in the inter-census period, contrary to the calculations based on registrations, which showed that the population was in decline over the same period (Śleszyński 2004a, Steinführer et al. 2010 and Table 6.1). For example, in Warsaw, around 65,000 more inhabitants were counted in the census, compared to those that were actually registered. In other large cities, the numbers were lower, but still significant (for example, 18,000 in Kraków and 7,000 in Poznań). It is to be expected that, for the period after the 2002 census, the underestimation of the urban population will gradually increase (see also Bijak et al. 2007 for the case of Warsaw). For the long term, however, population projections are negative and suggest a decline of more than 20 per cent for many large cities (GUS 2004b).

In the Czech Republic, the major urban centres registered population increases between 1970 and 1991. In contrast, in the 1990s, all the larger cities, with the exception of Olomouc and České Budějovice, experienced population decline (Table 6.2). Besides negative natural change, negative net migration at the regional scale, as a result of suburbanization, increasingly played a key role in population decline. Thus, from the mid-1990s onwards, out-migration exceeded in-migration.

Prague and Ostrava were major exceptions: Prague because this was only the case between 1998 and 2001, and Ostrava because it is the only large city in the Czech Republic whose migration balance has, so far, been negative throughout the entire transition period. While fertility rates started to increase in most cities from the late 1990s onwards, out-migration (particularly through suburbanization) was greater and kept population development negative (Čermák 2005). Yet, as a general tendency, the register data of the ČSÚ point to a slowing down of the population decline in recent years and, in 2007 (still with the exception of Ostrava), even suggest a slight growth. On the national scale, this is due to positive net migration, that is, a growth in foreign immigration, after the accession to the European Union in 2004, as well as to a positive natural balance (in most cities from the mid-2000s). The latter, however, is expected to be only a short-term phenomenon.

A competing population record, the so-called ISEO registry, is run by the Czech Ministry of the Interior. Its numbers differ considerably from those of the Czech Statistical Office, as discussed in Chapter 2. Although they are based on distinctly different methodologies, both claim to represent the population of a certain territory. At the beginning of 2007, the difference between them was almost 61,000 inhabitants in the case of Prague (and 31,000 for Brno). Since 2007, the differences have been diminishing in the course of data cleansing, but they still remain considerable for some cities (Steinführer et al. 2010).

There is a basic difference between Poland and the Czech Republic in terms of job-related migration in general and, more specifically, temporary out-migration. It is an important cause of population decline in Polish cities,[1] whereas job-related migration remains at very low levels in the Czech Republic (Aleš 2001, Slany 2005, Nowak-Lewandowska 2006, Mykhnenko and Turok 2007: 33–4). It is, however, nearly impossible to estimate its scope and duration, all the more since most of the out-migrants do not register as out-going or in-coming. In Poland, incomplete registration is particularly significant in the case of international emigration, which increased markedly after the accession of the country to the European Union in 2004. The official statistics are assumed to include less than 20 per cent of the actual emigration (Śleszyński 2004b, Bijak and Koryś 2006: 28). Due to the definition of the Polish Central Statistical Office (GUS), emigrants remain in the population registers and belong to the category of 'factual residents' (*ludność faktycznie zamieszkałą*).

---

1   During recent years, temporary out-migration from Poland has outweighed permanent emigration. The majority of labour migrants go to Europe or North America. Temporary labour migrants possess a higher level of education and qualification than permanent out-migrants (Nowak-Lewandowska 2006: 171, 175).

Table 6.2 Population development in the largest Czech cities 1991–2008

	Population according to the population register (as of 31 December; in 000s)								
	Prague	Brno	Ostrava	Plzeň	Olomouc	Liberec	Hradec Králové	Ústí nad Labem	České Budějovice
1991	1,217.3	388.5	327.4	172.4	106.0	101.8	100.2	99.9	98.3
2000	1,181.1	381.9	320.0	166.8	102.7	99.2	98.1	95.5	98.2
1991=100	97.0	98.3	97.7	96.8	96.9	97.4	97.9	95.6	99.9
2001*	1,160.1	373.3	315.4	164.3	102.2	98.4	96.4	94.9	96.7
2002	1,161.9	370.5	314.1	163.8	101.6	97.7	95.8	94.5	96.0
2003	1,165.6	369.6	313.1	164.2	101.3	97.8	95.2	94.1	95.2
2004	1,170.6	367.7	311.4	162.6	100.8	97.4	94.7	93.9	94.6
2005	1,181.6	366.8	310.1	162.8	100.4	98.0	94.4	94.3	94.7
2006	1,188.1	366.7	309.1	163.4	100.2	98.8	94.3	94.6	94.7
2007	1,212.1	368.5	308.4	165.2	100.4	99.7	94.3	95.0	95.1
2008	1,233.2	370.6	307.8	169.3	100.4	100.9	94.5	95.3	94.9
1991=100	101.3	95.4	94.0	98.2	94.7	99.1	94.3	95.4	96.5

Note: * Revised, based on census data from 1 March 2001 (not directly comparable with 2000 data).
Source: ČSÚ (several years); authors' work.

Immigration is also only partly taken into account by the Polish population statistics, because only those international immigrants with an official residence permit in Poland are included (Kupiszewski and Bijak 2006, Bijak et al. 2007). The recent phenomenon of re-migration of Poles from the US, the UK or Germany to Polish cities – mainly people with higher education who had emigrated in the 1980s (Kraler and Iglicka 2002) – is again only partly reflected in the statistics.

In the Czech case, foreign immigration at the local level is only in part covered by the official statistics. Čermák et al. (1995) estimated that, in the mid-1990s, between 100,000 and 150,000 foreigners were living in Prague, but only about 50,000 were actually registered.

We can conclude that, in spite of the contradictory and partly incomplete data officially available in both countries, there is broad evidence for at least temporary population decline in Polish and Czech second-order cities during post-socialism (and partly before). However, the trajectories of the single cities are very different and need more detailed investigation.

Demographic ageing As a result of declining fertility rates, increased life expectancy, selective out-migration, mainly by families in the course of suburbanization, and, in Poland, emigration, ageing is about to become a major issue for the cities in both countries (Tables 6.3 and 6.4). In Poland, the percentage of the population of post-working age increased from 19 to 24 per cent between 1990 and 2005. Even more significant, however, is the decrease in the 0–17-year-old group (from 46 to 28 per cent) in the same period. If insufficient replacement migration occurs, this will lead to accelerated ageing at the national scale, once this cohort reaches retirement age. Yet, Table 6.3 shows that ageing is by no means restricted to urban areas. In fact, the proportion of older people in rural regions in the period investigated was always higher than in the cities   but the latter have rapidly caught up in the post-socialist period.

Table 6.3 Percentage of population of pre- and post-working age in Polish urban and rural areas, 1980–2005

	Population of pre-working age (in %)		Population of post-working age (in %)	
	Urban areas	**Rural areas**	**Urban areas**	**Rural areas**
1980	43	57	17	25
1985	47	56	18	26
1990	46	56	19	27
1995	40	53	21	27
2000	35	49	23	27
2005	28	39	24	25

Note: Pre-working age: 0–17; post-working age women: 60+, post-working age men: 65+. The distinction between these two groups is widely used in Polish statistics.
Source: GUS 2006: 56–9.

In the Czech Republic, the official data use the ageing index, which shows a rapid ageing of the urban population (Table 6.4). In Prague, for example, in 1991 there were 86 people aged 65 years and over for every 100 aged under 15; in 2001 the number had increased to 122 (and to 130 in 2006). After almost two decades of transition, Ústí nad Labem in Northern Bohemia is the only major city left where there are fewer than 100 over-65s for every 100 under-15s. In four cities (Plzeň, České Budějovice, Hradec Králové and Pardubice), the relative change within less than 20 years is more than 200 per cent.

Table 6.4 Index of ageing for the ten largest Czech cities for 1991, 2001 and 2008

City	Ageing index (65+/0–14)			Relative change 1991–2008 (in %)
	1991	**2001**	**2008**	
Prague	85.8	122.3	130.0	152
Brno	73.9	111.6	134.7	182
Ostrava	55.5	79.4	106.8	192
Plzeň	65.3	111.2	134.2	206
Olomouc	62.4	95.5	121.3	194
Liberec	62.9	90.1	106.2	169
České Budějovice	51.7	87.1	120.1	232
Hradec Králové	63.6	110.3	144.9	228
Ústí nad Labem	56.2	77.0	91.0	162
Pardubice	58.1	109.2	137.8	237

Source: ČSÚ (several years); authors' work.

Household change   As elsewhere in Europe, Poland and the Czech Republic have experienced declining household size and a diversification of household types, particularly, though not exclusively, in the transition period. Between the last two censuses household numbers grew significantly in both countries (Table 6.5). Almost all the large cities experienced a greater increase in household numbers than in total population, with household numbers increasing even when the population was declining. The only exception was Katowice in Poland, where there was a small decline in household numbers together with a much larger decline of population. The increase in household numbers was mainly due to the growth in one-person households. Thirty per cent of households in Polish cities are one-person, with the highest proportions in Warsaw and Łódź (38 and 35 per cent, respectively; GUS 2004a: 56). The rates of increase of one-person households between 1988 and 2002 range from 28 per cent for Katowice to 89 per cent for Wrocław. While the Czech figures are not directly comparable to the Polish ones (due to different time periods and different methodologies), they show similar trends, with a much stronger increase in the number of households compared to population growth, together with long-term decline in mean household sizes and growing numbers of small households. The trends evident in both the Polish and Czech data resemble those in other parts of Europe and are symptomatic of the second demographic transition (Ogden and Hall 2004: 101).

Table 6.5 Change in the numbers of population, households (total) and one-person households in the ten largest Polish and Czech cities, 1988–2002 and 1991–2001 (based on census data)

Poland 1988–2002	Change (in %) in number of		
	Population	Households (total)	1-person households
Warsaw	+2.0	+18.4	+72.4
Łódź	-7.6	+2.9	+40.3
Kraków	+1.6	+20.6	+84.4
Wrocław	+0.2	+20.6	+89.0
Poznań	-1.5	+13.3	+70.3
Gdańsk	-0.6	+15.4	+74.8
Szczecin	+1.2	+17.7	+78.8
Bydgoszcz	-1.1	+11.0	+53.6
Lublin	+5.1	+24.1	+87.9
Katowice	-11.4	-0.3	+28.3
Czech Republic 1991–2001			
Prague	-3.7	+0.1	+7.3
Brno	-3.1	+1.1	+8.1
Ostrava	-3.3	+5.0	+19.8
Plzeň	-4.4	+3.5	+18.8
Olomouc	+1.4	+3.3	+4.0
Liberec	-2.6	+7.2	+22.2
Hradec Králové	-2.7	+6.3	+21.7
Ústí n. L.	-1.0	+10.0	+33.1
Č. Budějovice	-1.9	+9.3	+22.9
Pardubice	-4.9	+6.1	+23.2

Note: * These numbers refer to so-called census households (*cenzové domácnosti*; see also Chapter 3, Figure 3.1).
Source: GUS, ČSÚ (several years), authors' work.

Housing Market Change: Quantitative and Qualitative Matters

The changes since 1989 have influenced all spheres of society, including housing. However, in both Poland and the Czech Republic, housing policy reforms were postponed for a long time and often accompanied by contradictions and inefficiencies. In the words of Donner (2006): 'In spite of governments' advocating the development of a true market-oriented system, the former pervading paternalism remains deeply rooted in society' (ibid.: 76).

Housing market transformation in the post-socialist period is a very broad topic and outside the scope of this volume. Reference has to be made to other work (Hegedüs and Tosics 1998, Pichler-Milanović 2001, Lowe and Tsenkova 2003, Lux 2003, Stanilov 2007a); only selected issues will be outlined here. First there is a discussion of the national housing stock, examining both the housing shortage and the problems of housing quality, before looking at urban housing markets, particularly their segmentation and the marginalization of public housing.

Housing stock, demand surplus and problems of maintenance Housing shortage has been a structural problem in Polish cities throughout the 20th century. The considerable damage during the Second World War exacerbated an already serious shortage of housing, a situation which continued under state socialism, where housing production failed to meet the great demand. The problem has not been solved in the period since the end of state socialism. In the early 2000s, the housing deficit was still estimated at between 1.3 and 1.5 million dwellings (Uchman and Adamski 2003: 134). In 2002, the entire housing stock in Poland amounted to 12.5 million dwellings, of which two thirds were in urban areas. By 2008, this number had grown to 13.2 million. Between the censuses of 1988 and 2002, the total housing stock increased by only 915,600 dwellings, or 9 per cent, in comparison with the 15 per cent growth recorded between the censuses of 1978 and 1988. Of the 12.5 million dwellings recorded in 2002, 11.8 million were actually occupied. At the same time, 13.3 million households were recorded in the census, considerably exceeding the number of dwellings (Table 6.6). Demand had grown much faster than supply, because average household size had fallen. In 2002, there were just 328 dwellings per 1,000 people, a number that increased to 337 dwellings in 2006 (data according to GUS). So, in spite of some progress, Poland's housing provision remains one of the poorest in Europe.

Table 6.6 Quantitative indicators of the housing markets in Poland, 1988/2002, and the Czech Republic, 1991/2001 (based on census data)

	Number of dwellings (in 000s*)	Dwellings per 1,000 population*	Number of households (in 000s**)	Mean household size**
Poland				
1988	10,717	289	11,970	3.10
2002	11,633	308	13,337	2.84
1988=100	*109*	*107*	*111*	*92*
Czech Republic				
1991	3,706	360	4,052	2.53
2001	3,829	372	4,271	2.38
1991=100	*103*	*103*	*105*	*94*

Note: * These numbers only refer to permanently inhabited dwellings. ** For the Czech Republic, these numbers refer to so-called census households (cenzové domácnosti; Chapter 3, Figure 3.1).
Source: GUS 2003, Bartoňová 2007; authors' work.

No less serious are problems related to the quality of the existing housing stock, since the physical conditions of many buildings, particularly in old built-up areas, deteriorated further during post-socialism. The number of dwellings in need of urgent repair is estimated to be between 600,000 and 700,000 (Uchman and Adamski 2003: 134, Billert 2004: 47). Many of them probably need to be demolished. As a result of the increase in the number of households, together with insufficient new building to replace the old housing stock, housing shortages, in terms of both quantity and quality, will certainly continue to be a major problem for Polish cities.

In addition, there is a shortage of building land for housing, particularly in the large cities. In 2003, the Law on Spatial Planning abolished plans from before 1995, but high costs, local disagreements and political constraints have delayed new local development plans. The lack of zoning plans has, in turn, delayed the construction of new housing. In 2005, only around 20 per cent of the country was covered by local development plans, and only 12 per cent of metropolitan areas (Śleszyński et al. 2007).

Just as in Poland, the Czech Republic saw a sharp decline in housing construction in the early years of the post-socialist transition – with numbers starting to rise again from the mid-1990s (Sýkora 2003: 53). The total number of dwellings in 2001 was 4.4 million, a 7 per cent increase compared with the figure in the 1991 census. Table 6.6 shows only permanently inhabited dwellings, because the available household numbers relate only to these dwellings. The increase in housing stock in the Czech Republic was almost the same as the increase in household numbers – 103 per cent increase in dwellings compared with 105 per cent in households but, as in Poland, household numbers considerably exceed the number of dwellings. Thus, in spite of an improved ratio of dwellings per 1,000 inhabitants in the inter-census period, the number of households without a separate dwelling (*nebydlící domácnosti*) grew from 16 to 18 per cent (Steinführer 2004a: 88). This suggests that, at least for 'starter' households, there was a deterioration of the overall housing situation during the first decade of the transition. Throughout the 2000s there was a further increase in new housing construction (with the exceptions of 2006 and, as a result of the global economic crisis, 2008).

Structural problems of urban housing markets   Denationalization, property transfers to the municipalities and, later, privatization of the housing stock (by restitution to former owners or their heirs, sale to sitting tenants or private investors) were the key processes in the residential sphere during post-socialism. A small segment of the nationalized housing stock was returned to previous owners or their descendants, but most of it was transferred to municipal ownership, with the aim of privatization in the future. Company and cooperative housing[2] was similarly affected in both countries. The large-scale privatization of different types of housing stock to sitting tenants has to be considered as the major trigger of post-socialist housing market change.

In Poland, improvement of housing conditions and reduction of the shortage of housing are gradually being achieved by the introduction of market mechanisms and the modification of national and local policies. The most important mechanism has been the mass privatization of cooperative, municipal and company flats. As a result, the ownership structure has changed. According to the census data of 2002, 55 per cent of the housing stock was in the private sector, compared with 44 per cent in 1988. By 2007, this share had risen to 63 per cent. By contrast, the share of municipal *(gmina)* housing decreased from 18 per cent in 1994 to 9 per cent by 2008. Cooperative housing made up 24 per cent in 2008, comprising both large, reformed state-socialist cooperatives as well as newly established 'ownership cooperatives' in former rental buildings. Long-term tenants who cannot or who do not want to buy, continue to pay regulated, that is non-market, rents while all newly-established contracts are based on market rents. At the same time, a special instrument has been developed to promote the construction of affordable (but not bottom-end) rental housing by non-profit social housing associations (*Towarzystwo Budownictwa Społecznego,* TBS); it aims at medium-income groups (more details in Uchman and Adamski 2003: 127–31). By the end of 2007, about 73,000 TBS dwellings existed (or 0.5 per cent of the entire housing stock; all data from ibid. and GUS). The income tax deduction introduced between 2002 and 2007, and the

---

2   In the state-socialist period, certain housing stock often belonged to and were rented out by certain companies. Cooperative housing was – apart from state-owned housing – the most common form of tenure in Poland and the Czech Republic until 1989.

new legislation and programmes targeted at lower-income social categories since 2006 are intended to bring about housing improvements for poorer groups.

In the Czech Republic, selling state, municipal or company-owned dwellings and the conversion of cooperative flats into private ownership were the most common forms of privatization. By the 2001 census, the rental sector – comprising both public and private rental dwellings – was still relatively strong (29 per cent of the entire housing stock compared with 39 per cent in 1991). It has continued to decrease since 2001, because of the continuing privatization of the municipal housing stock. Sitting tenants were the major beneficiaries of this process but private companies, estate agencies as well as starter households also acquired housing in this way. Unlike many other countries in transition, 'give-away' privatizations, that is the sale of public housing with a very large financial discount, was the exception and not the rule in the Czech Republic. However, the municipalities usually sell for a price considerably below market value: in Brno, for example, the price is reduced by 40 per cent in return for which the new owners must modernize the property. The decision about the privatization of the local public housing stock is made by the municipality (and the local parliaments) and, in the largest cities, this is deputized down to the individual districts. Thus, the speed and the form of the process differ from city to city, and sometimes even from neighbourhood to neighbourhood, and the selling price of otherwise similar dwellings has similarly varied. Meanwhile the bulk of municipal dwellings has been sold to the sitting tenants (Lux 2000: 10–11). By the end of 2007, the municipal housing stock in cities with 50,000 or more inhabitants had declined to 9 per cent, compared with 44 per cent in 1991. For example, Plzeň, saw a decrease from 33 to 6 per cent and Ostrava from 36 to 15 per cent. Brno and Ústí n.L. represent the two extreme cases among the second-order cities: in Brno the privatization of the municipal housing stock took place slowly and the municipal share is still relatively high (2007: 27 per cent, 1991: 37 per cent), the North Bohemian city of Ústí n.L. took the opposite path of selling almost all the rental flats in public ownership, which amounted to 53 per cent in 1991. In 2007, just 3 per cent were left in the public rental sector (all data according to Chlupová et al. 2009: 16). Former housing cooperatives have also been transformed, either privatized as jointly owned cooperatives or fully privatized and distributed to tenants (Sýkora 1996b).

The Czech housing market has changed only partially and slowly. As a result, and as in Poland, it can be characterized as somewhat fragmented. Tenants in public rental housing, who lived in their dwellings prior to 1990 (or their relatives, to whom they have passed on their right to live in the property – the so-called *dekret na byt*), enjoy strong tenants' (almost occupier-like) rights and, even up to 2010, regulated rents. A reform of the tenants' law has not occurred, so that one can speak of a deep deformation of the housing market, especially in cities (because the majority of people living in rural areas are owner-occupiers). Controlled rents are still a strong incentive to remain in the same dwelling for a long time, in spite of it probably being too large, rather than moving to a more suitable one. This is one of the reasons why the structure of flats does not correspond to the structure of households. Company housing has also largely been privatized (either sold to tenants or to entrepreneurs).

The structural problems of urban housing markets in Poland and the Czech Republic, in particular the segmentation and the postponement of housing reforms, meant that, in the first decade of the post-socialist transition, there were limited opportunities for newly-founded ('starter') households to find a dwelling either to rent or to buy. From about 2003 onwards, the position of these, as well as other low-income households, further deteriorated due to the sharp increase in housing prices and constructions costs on the eve of both countries' accession to the European Union in 2004. With respect to tenure, Poland and the Czech Republic are characterized by the increasing importance of owner-occupied housing, although there is still a significant, albeit declining, share of public rental housing. Private rental housing is negligible in both countries (Priemus and Mandič 2000) – as in most other countries in transition.[3] Cooperatives are a major segment in both countries and have their roots both in socialist tenure structures as well as in post-socialist privatization. As discussed below, almost all of these forms of tenure play a part in the inner cities of our four case studies, with the exception of the previously socialist cooperatives, which are more typical of the large housing estates.

Characteristics of the Four Case Studies

Łódź, Gdańsk, Brno and Ostrava: Short Introduction and Characteristics

Łódź and Gdańsk in Poland and Brno and Ostrava in the Czech Republic represent typical second-order centres of their countries. Although they differ in size – in 2005, Łódź had 768,000 inhabitants, Gdańsk 458,000, Brno 368,000 and Ostrava 310,000 inhabitants (see Tables 6.1 and 6.2 above) – they are important hubs of economic, administrative, commercial and educational activities. In the following, each city will be briefly introduced. Table 6.7 gives basic information on the four cities.

Łódź (Figure 6.3a) was, for a long time, the second largest city in Poland. Due to a significant population decline during the last two decades, it was overtaken by Kraków in 2007 (757,000 compared with 753,000 inhabitants). For centuries, Łódź had been a small, agricultural town with a limited local function. It developed as an industrial centre only from the 1820s onwards. As a result of the development of the textile industry, Łódź became one of the major economic hubs of textile production and distribution within the Russian Empire (until 1918) and the Second Polish Republic (from 1918 to 1939). The second half of the 19th

---

3   This, for example, is a salient difference compared to the situation in the neighbouring parts of eastern Germany, where both municipal and private rental housing still forms a major segment of urban housing markets.

and the beginning of the 20th century, often called Łódź 's 'Golden Age', led to an enormous growth of the city from 33,000 inhabitants in 1864 to 500,000 in 1911 (Riley et al. 1999: 22). By the end of the 19th century, 'Łódź was a huge monofunctional centre dominated by the textile industry, densely populated, with a chaotic spatial layout and a lack of tertiary services' (Liszewski 1997: 29). Until the Second World War, the city was characterized by an ethnically and culturally diverse population, whose largest groups were Jews, Poles and Germans. After the war, the Holocaust and forced resettlement of the remaining ethnic minorities across Poland and beyond, Łódź developed to become one of the major industrial centres of Poland, dominated by textile industry and engineering. But the city was already experiencing industrial and population decline before the beginning of the post-socialist transition in 1989. The main reason was the decline of the textile industry which had begun in the 1950s (Walker 1993: 1070–71).

Deindustrialization led to further economic decline, high unemployment and out-migration during the 1990s, as well as to considerable population losses (see Figures 6.8 and 6.9 below for more details). High mortality, suburbanization and out-migration mean further population losses are inevitable. However, as a result of housing shortages, population decline is not accompanied by housing vacancies. On the contrary, new, infill housing is being built. Demolitions are mainly a result of the poor physical state of the pre-Second World War housing stock. After the last administrative reform in Poland in 1999, Łódź has become the capital of a voivodship (administrative region at NUTS 2 level) and it is one of Poland's largest university centres (more than 118,000 students at several universities in 2006). During the last few years, the economic base of the city has improved. As a consequence, unemployment rates have decreased and some major investments have been made. Administratively, Łódź is divided into five boroughs (*dzielnice*): Śródmieście, Polesie, Widzew, Baluty and Górna and a further 70 districts (*jednostki urbanistyczne*).

Figure 6.3a Map with inner-city delimitations for Łódź
Source: Authors' work.

Figure 6.3b Map with inner-city delimitations for Gdańsk
Source: Authors' work.

Figure 6.3c Map with inner-city delimitations for Brno
Source: Authors' work.

Figure 6.3d Map with inner-city delimitations for Ostrava
Source: Authors' work.

Table 6.7 Basic information about the case-study cities

	Łódź	Gdańsk	Brno	Ostrava
Surface area (in km²)	294.4	266.0	230.2	214.2
Population (2008) (in 1,000)	747.2	455.6	370.6	307.8
GDP per head (in €)	10,410***	7,693*	9,951**	8,692**
Unemployment rate (2008)	6.8	2.6	6.1	8.4
Students at universities and secondary schools (in 1,000)	127.3	78.6	82.0	34.0

Note: * data for 2004; ** data for 2001; *** data for 2007 (most recent available data)
Source: Urban Audit, Municipality of Brno, USG 2009, USG/UMB 2009; Vaishar et al. 2009: 14.

Gdańsk (Figure 6.3b) is the major Polish port city and an important trade centre, with a population of 456,658 in 2006 and a history of more than 1,000 years. Located in northern Poland at the mouth of the river Vistula, it is the capital city of the Pomerania voivodship and forms the greater part of the Tricity agglomeration, which consists of Gdańsk, Gdynia and Sopot and has a total population of around 750,000 inhabitants. A former Hanseatic port (the Free City of Danzig) and the site of the first military operation of the Second World War in 1939, Gdańsk still bears the traces of a complex past. Cleansed of its German identity during the era of state socialism, it was heavily industrialized and portrayed as a working class centre, eventually becoming the cradle of the Solidarity movement *(Solidarność)* and a symbol of the Polish struggle for freedom and independence. At present, it is also a shrinking city with an ageing population, suffering from the negative effects of deindustrialization and considerable population loss. The administrative structure of Gdańsk consists of six basic divisions, called city boroughs *(dzielnice)*: Oliwa, Południe (South), Port, Śródmieście (Inner city), Wrzeszcz and Zachód (West) and is further divided into 27 districts *(jednostki urbanistyczne)*.[4] The earliest spatial development of the city proceeded from the historic city core *(Śródmieście Historyczne)* northwards, along the Vistula estuary, as well as alongside the main transportation axis, parallel to the coastline.

Brno (Figure 6.3c) is the second largest city of the Czech Republic, with almost 370,000 inhabitants in 2008 living on 230 km². Being the regional capital of South Moravia, it also serves as a seat for national administrative functions; for example, it hosts the Constitutional and Supreme Court of the Czech Republic. Administratively, Brno is divided into 29 urban districts *(městské části)*. Brno has been an important commercial, cultural and administrative

---

4 Neither the boroughs nor the districts have any executive role in the local government.

centre in Moravia since the Middle Ages. It was the capital of Moravia from the 17th century until 1946. Within the Habsburg Empire, Brno was one of the first cities to be industrialized (in the second half of the 18th century) and connected to the railway network (1839). The industrialization is reflected in the physical structure of the city, mainly in the former workers' quarters, where a complex mixture of manufacturing and housing can be found. The textile industry and engineering have for centuries played the most important role in the industrial sector. The extensive educational, administrative and cultural functions of Brno helped to establish a strong middle class that brought about the development of areas of villas and gardens. Today, Brno represents the most important trade-fair and exhibition centre in East Central Europe. It plays an important role as a centre for higher education, science and research (there are five public universities with a total of around 80,000 students). Since the mid-1990s, the surroundings of Brno have been characterized by intensive suburbanization. However, it is difficult to evaluate the population development in Brno since 1990, due to the existence of conflicting data about this development. While ČSÚ data report a decline of about 6 per cent between 1991 and 2006, there are indications that the real population development has stabilized, particularly after the 2000s, and that there was even some growth in recent years (ISEO data; Steinführer et al. 2010).

Ostrava (Figure 6.3d) is the third largest city in the Czech Republic. It was originally a small town (Moravská Ostrava) at the periphery of the Habsburg Empire, not far from the Moravian-Silesian border, where it crossed the Ostravice River, and the second largest settlement in the area (Slezská Ostrava). Administratively, Ostrava is divided into 23 urban districts *(městské obvody)*. Its large-scale urbanization started during the 19th century, when coal mining developed in the region, followed by iron metallurgy, production of railroad tracks and other heavy industries. Around the small town, a chaotic structure of mines, mining settlements, dumps, industrial enterprises and physical infrastructure developed during the 19th century. Ostrava was at the centre of this industrial area and the city grew rapidly in importance. Ostrava maintained its industrial role during the 20th century, when the city was called the 'steel' or 'black heart' of Czechoslovakia. After the Second World War, Ostrava's growth was strongly supported by the state socialist government. New satellite cities with between 80,000 and 100,000 inhabitants developed around the centre: Ostrava-Poruba from the 1950s, later Ostrava-Jih (South) in the 1970s and 1980s. Within the region, further larger cities for industrial workers came into being, and the whole region experienced a steady in-migration of mainly younger people, the majority of whom were workers with high salaries, because of their hazardous work (see Vaishar et al. 2009 for more details). After 1990, all the mines, together with the many factories of the heavy industry sector, were closed down (except the Nová Huť factory, which is now a part of the Arcelor Mittal Steel Company), resulting in high unemployment. Ostrava Technical University was relocated here in 1945 (from Příbram), and Ostrava University was established in 1991. Ostrava, because of its specific polycentric character (consisting of three parts, that is, Moravská Ostrava/Přívoz, Poruba and Ostrava-Jih), has a very specific form of suburbanization, with people moving to more distant towns and villages in the foothills of the near-by Beskids rather than to adjacent areas.

We conclude that there are *both comparable and contrasting characteristics of the case-study cities*. All of them have an industrial past. Łódź and Brno, for example, were called the 'Manchesters' of their countries (the 'Polish' and the 'Austrian' or 'Moravian Manchester', respectively). By the end of the 1980s, the textile industry was still the dominant industry in Łódź. Historically, Brno used to be a centre of textile industry and engineering, too, and it is also a well-known trade fair city. Ostrava was associated with mining, steel production and heavy industry for more than a century. Gdańsk, titled *Aurea Porta* (Golden Gate) of the Republic of Poland in the 17th century, played an important role as a harbour city and is still a centre of ship building. Moreover, all four cities host large universities and fulfil a number of further administrative and commercial functions that extend beyond their administrative borders.

But there are also significant differences between the cities that influence the development and current situation of their inner parts and city centres, respectively. The four cities experienced different paths of urban development in their recent history. Łódź and Ostrava only developed as cities from the mid-19th centuries. Brno and Gdańsk, on the other hand, are old medieval cities whose industrial areas developed from the late 18th and 19th century respectively, around and beyond the walled city of the time. This also led to a different structure of the urban body: while Brno and Gdańsk have a clearly defined (medieval) city centre,[5] Ostrava has a polycentric structure, and Łódź developed on a grid pattern where urban life was concentrated along the main axis, Piotrkowska Street, which was the location of the homes of the most important industrial families with their grand houses, and where most public and cultural institutions were situated.

The Inner Cities of Łódź, Gdańsk, Brno and Ostrava

(a) Łódź The inner city of Łódź *(Śródmieście)* was not destroyed during the Second World War and thus one of Poland's largest historic urban areas has been preserved. Characterized by its rectilinear street pattern, this area represents a major part of Łódź's cultural heritage (Kaczmarek 1997). According to the 2002 census, 66 per cent of the inner-city housing stock dates from before 1945, with 49 per cent of homes dating from before 1918. Some of the flats were originally very large (more than 100 m²), with generous room sizes and a high standard of amenities. However, a considerable proportion of

---

5   It is often argued that Gdańsk also has a second, functional centre situated in Wrzeszcz.

the dwellings in this area are today sub-standard housing. Before the Second World War, many homes in the inner city were in German or Jewish ownership. These dwellings became state-owned after the war and were assigned to Polish inhabitants. Many large dwellings were either divided up or given to more than one family. Some flats are still shared by two or more households, with a shared corridor or even a shared kitchen and bathroom.

Some renovation and change of use took place during the 1990s in the inner city (Riley et al. 1999). Many small enterprises replaced the large factories and, in some locations, housing was replaced by commercial and service activities. New shopping malls, such as Galeria Łódzka and Manufaktura, were established. The latter was created in the Poznański factory, formerly one of the largest textile companies in Łódź (Figure 6.4a). Renovation was restricted to historical sites and to selected buildings, which are now banks, hotels, offices and shops. To date, there has been little renovation of the old building stock and little refurbishment of housing (Figure 6.4b). Many of the new, post-privatization owner-occupiers do not have the money to renovate their flats and even less the stairwells, roofs or facades. The same is true for the municipality. As a result, many old houses are in a very bad state of repair and some are uninhabitable. Dilapidation and decay are the biggest problems faced by the old built-up areas today. Representatives of the municipality estimate that some 500 old houses in the city centre are uninhabitable. Yet, in spite of this, about 300 of these buildings are currently occupied. Poor condition and decay is the main reason for housing demolition. Between 1988 and 2002, 3,300 flats, or 9 per cent of the inner-city housing stock, were demolished.

In 2004/2005, the municipality undertook the revitalization project PROREVITA (Markowski and Stawasz 2007). This involved two small inner-city areas which were regarded as two of the least attractive residential neighbourhoods. Newly-built municipal rental housing (so-called TBS, see above) was developed here and will remain as rental housing. Recently, gentrification has occurred in individual streets or blocks, including former industrial sites, where upmarket loft housing is being created.

Despite recent in-migration, housing in the old areas of the inner city is still regarded as amongst the least attractive in the city. According to local experts, none of the large housing estates face problems as severe as those of the inner-city housing stock.

Figure 6.4a Inner-city Łódź. Renovated and reused textile factory
Source: Photo, A. Haase 2007.

**Figure 6.4b Inner-city Łódź. Refurbished and dilapidated housing adjacent
to each other**
Source: Photo, A. Haase 2007.

(b) Gdańsk The inner city of Gdańsk is not compact because of the city's location along the Baltic Sea coast. Thus, the pre-Second World War neighbourhoods were widely distributed in the central parts of the city, with a mix of residential and industrial areas. Besides the historical city centre, the central parts of Gdańsk are formed by the districts of Wrzeszcz, Oliwa and the shipyard area with its adjacent workers' settlements. Since Gdańsk was badly destroyed during the Second World War, all parts of the inner city today exhibit a mixture of pre-war and post-war housing stock, including prefabricated blocks. According to the 2002 census data, in ten out of the 27 districts of Gdańsk, more than 50 per cent of the housing stock was pre-war and in five it was 75 per cent or more. The districts with the highest proportions of pre-war housing stock, which are located north-west, north and east of the city core, show the specific character of Gdańsk as a seaport – its earliest development was on the Motława river and the Vistula estuary as well as alongside the main transport axis, parallel to the coastline, which linked Gdańsk to the neighbouring settlements and to the hinterland. Apart from their age, the inner-city areas are also characterized by a comparatively homogenous and densely built urban grid structure, in contrast to the outer districts, which consist mainly of newly-built socialist housing estates or individual housing.

Two inner-city neighbourhoods (Wrzeszcz Dolny and Nowy Port) are of specific interest for the empirical analyses in Chapters 8–10. Both of them have high proportions of pre-1945 housing and boast a rich history of former working-class estates, which have progressively deteriorated since the end of the Second World War. Two of the vital differences between the two districts are location and quality of the housing stock. Although both districts are identified as belonging to the inner city of Gdańsk, the location of Wrzeszcz Dolny is more central, while the more peripheral Nowy Port is regarded as less accessible and consequently, not as attractive (Figure 6.5a and b). The second difference is that despite the fact that 85 per cent of the buildings in Nowy Port were built before the Second World War, the demographic profile of the district is 'younger', due to a few high-rise blocks of flats which were built in the 1970s and which are inhabited by over 2,000 people, mainly young families with children. Both Wrzeszcz Dolny and Nowy Port nonetheless feature densely built-up pre-war housing structures with a distinctive inner-city quality and a potential for residential change, which makes them suitable for the purposes of our study.

Figure 6.5a Inner-city Gdańsk. Wrzeszcz Dolny
Source: Photo, A. Haase 2007.

Figure 6.5b Inner-city Gdańsk. Nowy Port
Source: Photo, A. Haase 2007.

(c) Brno Brno's inner city is built compactly around the medieval core. It developed in the course of industrialization during the 19th century and with inter-war building in the 20th century. Today, the inner city of Brno is a heterogeneous mixture of residential areas, services, including universities and hospitals, transport facilities, commercial space and old industrial brownfields, some of which are ready for new uses, including residential ones (Mikulík and Vaishar 1996, Kuča 2000). There has long been a spatial dichotomy in the inner city: the residential areas located west and north-west of the city centre are the most attractive places to live and offer up-market housing in tenement blocks and villas. The areas south and south-east of the city centre are, by contrast, associated with industry, working-class housing and environmental pollution. Even during the period of state socialism, these differences in the buildings, social structure and mental perceptions persisted and were reinforced in the post-socialist period (Steinführer 2003). The inner city of Brno is characterized by a relative lack of green spaces. There is just one large inner city park, Lužánky, which is close to residential areas with high-class pre-Second World War buildings, often in secessionist or functionalistic styles (Figure 6.6a). Although housing in the northern districts was neglected during the period of state socialism, these are still the most attractive residential areas in the inner city.

The old built-up areas south and south-east of the city centre are typical industrial districts with low- and medium-quality working-class housing (Figure 6.6b). The population density in these areas is still above average, as is the share of poorer households and Roma families. Many buildings are in a bad state of repair. They are close to the railway and to industry, so that environmental pollution is relatively high in these areas. Unsurprisingly, they are among the least attractive places to live in Brno.

The boundaries of the inner city area of Brno were delimited specifically for the work presented in this volume. Based on the division of Brno into 278 basic settlement units (ZSJ, *základní sídelní jednotky*) in the 2001 census, those ZSJ with more than 65 per cent of houses built before 1945, and where more than 65 per cent of dwellings were located in multi-storey buildings, were defined as inner city. These somewhat artificial boundaries were chosen in order to exclude both post-Second World War multi-storey residential buildings as well as detached and semi-detached houses from the inter-war period (Klusáček et al. 2007: 7). Thirty-three centrally located ZSJ met these requirements and are referred to as Brno's inner city in the following discussion.

Figures 6.6a Inner-city Brno. Secessionist building in Veveří
Source: Photo, P. Klusáček 2010.

Figure 6.6b Inner-city Brno. Workers' housing in Trnitá
Source: Photo, P. Klusáček 2010.

(d) Ostrava   Delimiting the inner city of Ostrava is particularly challenging since Ostrava is a polycentric city. For our analysis, we focused on the districts of Moravská Ostrava and Přívoz. Both districts started to develop in the second half of the 19th century. While today Moravská Ostrava represents something like the city centre of the entire Ostrava agglomeration and includes the cathedral, the market square with the city hall and a central shopping area, Přívoz is one of the oldest industrial settlements and it developed around the railway; it was largely populated by German inhabitants until 1945. Both areas show a mixed architecture: pre-1945 bourgeois tenement blocks and brick-built workers' settlements are adjacent to blocks built during the period of state socialism (Figure 6.7a). The territory is delimited by the railway in the north and west, the Ostravice River in the east, and the former Karolina coking plant, a large inner-city brownfield site awaiting redevelopment into a mixed residential and shopping centre, in the south.

In both Moravská Ostrava and Přívoz, the socio-economic structure of the residential population is mixed (Figure 6.7b). In some places, there is a concentration of Roma households, mainly in Přívoz. In Moravská Ostrava, near the cathedral and the market square, top-down gentrification has taken place, with Roma households being displaced to poor residential areas, and this area, particularly Stodolní Street, has become a centre of Ostrava's nightlife, where pubs and clubs are concentrated.

For the purpose of our empirical investigations, Ostrava's inner city was defined, somewhat arbitrarily, by geographical barriers. Its north-eastern border is formed by the Ostravice River, to the south-west by the railway to Frýdek Místek, the western by the railway to Přívoz-Sever and the eastern by Karolina brownfield site.

Figure 6.7a Inner-city Ostrava. Residential building in Moravská Ostrava
Source: Photo, A. Haase.

Figure 6.7b Inner-city Ostrava. Dilapidated house in Přívoz
Source: Photo, S. Martinát 2007.

Population and Housing Market Development in the Case-study Cities as Framework Conditions of Inner-city Residential Change

The following sections describe general trends of population and housing market development in the four case-study cities and provide a background for the in-depth analyses of the inner cities that are presented in Chapters 7–12.

Population Development

The trends in the *number of inhabitants* in the case-study cities from 1950 up to the present are shown on Figure 6.8. The city of Łódź experienced severe population decline as a result of the Second World War, but then saw steady population growth during the period of state socialism (Jakóbczyk-Gryszkiewicz 1997: 114–15). This was driven by high birth rates as well as by labour-related migration. The largest population was reached in 1984/85; since then, it has been declining. The apparent growth in 1988/1989 was a result of the enlargement of the city's boundaries. Population loss accelerated between the late 1980s and 2000s: between 1988 and 2006, the city lost almost 100,000 inhabitants. In 2007, the city had some 753,000 inhabitants.

Gdańsk also saw significant population growth after the end of the Second Word War. As a consequence of the war, including the Holocaust, flight and displacement, there were only about 120,000 inhabitants in the city in 1946, compared to about 256,000 in 1929. Between the 1950s and the mid 1980s, Gdańsk experienced continuous population growth as a result of industrialization and the resettlement of new inhabitants. Population growth stopped in 1985 and turned into a phase of stagnation prior to a small but steady process of population decline that started in 1995. From 1994 to 2007, the city lost about 7,400 inhabitants; so that in 2007, the city had some 456,000 inhabitants.

Brno experienced steady population growth from 1945 onwards, a trend which lasted until 1993, when the population peaked at 390,000. In-migration was the driving force for this population growth. Even in the 1980s, when natural population decline had set in, this was outweighed by net in-migration. However, by the early 1990s, there was no more net in-migration, and between 1994 and 2006, Brno's population decreased. The population register data for Brno indicate a loss of 22,000 inhabitants (or 6 per cent) between 1991 and 2006. More recent data, however, indicate stabilization and even some growth. In 2007, Brno had 368,000 inhabitants (or, according to the ISEO data set, almost 400,000).

Ostrava experienced considerable population growth after 1945 as a result of socialist industrialization and the construction of new housing estates, such as Poruba and Ostrava-Jih. With the exception of 1981–1983, when the city saw a slight population decrease, the growth lasted until 1990, by which time the city had 331,000 inhabitants. Since 1990, there has been continuous population decline, with the city's population falling by 22,400 from 1990 to 2006, mainly as a result of out-migration and, to a lesser extent, natural decline. In 2007, Ostrava had 308,000 inhabitants.

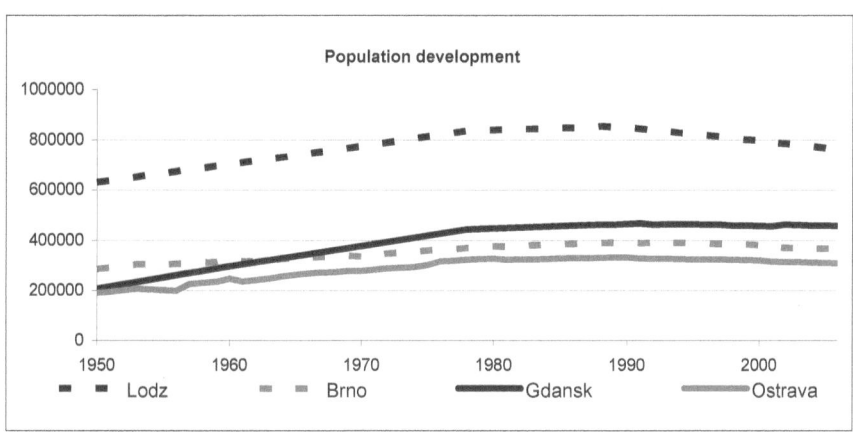

Figure 6.8 Population development in the four case-study cities 1950–2006
Source: GUS, ČSÚ (several years).

Łódź, Gdańsk, Brno and Ostrava have therefore experienced different paths of population development over recent decades. All four cities have, however, seen some population decline. Łódź, in particular, stands out, because it has experienced the greatest population decline of all Polish cities. Future projections from the Central Statistical Office assume a further significant decrease for Łódź (the projection for 2030 shows a 21 per cent decline from the 2002 figure of 785,000; GUS 2004b). For the other three cities, population projections are also negative, although there is, of course, a high degree of uncertainty, particularly with regard to migration. Projections for Brno and Ostrava estimate that population decline will continue to a less pronounced extent than in the past (Schneiderová et al. 2002, VCRR MU 2005).

The *causes of population decline* are similar in some respects in all four cities, but there are also differences. In Łódź, the initial cause of the decline was the excess of deaths over births, followed by, from 1999 onwards, increasing labour-related out-migration, interregional migration and suburbanization; in Gdańsk suburbanization played a more important role. As in many other second-order cities in Poland, one of the reasons for such suburbanization has been an unbalanced housing market, with high prices for flats and a limited availability of new housing within the city. As a result, the city of Gdańsk is losing people to the surrounding rural areas.

In Łódź, between 1988 and 2002, the general fertility rate declined dramatically (39.8 to 26.2 births per 1,000 women aged 15–49), so that now it has the lowest birth rate (and highest natural decline) of any of the large Polish cities (Jakóbczyk-Gryszkiewicz 1997: 115 and Figure 6.9). Much of the fertility decline has been a result of the postponement of childbearing, to which the difficult economic situation after 1989 contributed. Recent data, however, indicate some signs of a recovery in the birth rate. Out-migration has also contributed to population decline in Łódź. The numbers involved are lower than for natural decline, but are made up of both interregional migration and some suburbanization. From 1999 onwards, there has been net out-migration every year, including both long-distance migration and suburbanization. In-migration has remained low, which is emphasized by the fact that 66 per cent of inhabitants have lived in the city since their birth, a much higher proportion than in most other large Polish cities (Bierzyński and Węcławowicz 2008: 65; see also Kabisch et al. 2008: 39–45).

In Gdańsk, population losses are a result of both natural decline and out-migration, although their relative importance varies over time (Figure 6.9). The birth rate was lower than the death rate from 1996 until 2003 when it rose slightly, so that the two were almost in balance for both 2004 and 2005, a result of older women starting to have children later than expected, as well as the large cohort born in the 1970s and early 1980s having children. Moderate net out-migration has been continuous since 1995, because of the strong suburban outflow from Gdańsk to the neighbouring municipalities (the Gdańsk agglomeration is one of the largest national examples of a 'suburban ring'), as well as from net international out-migration, which has been in decline since 1990 (Grabkowska and Sagan 2008).

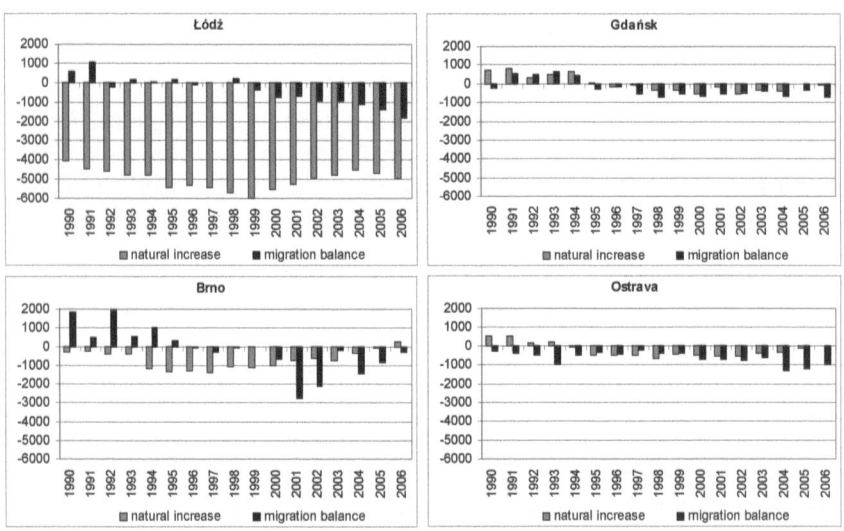

Figure 6.9 Natural population development and migration balance for the four case-study cities 1990–2006

Source: Grabkowska and Sagan 2008; Bierzyński and Węcławowicz 2008.

For Brno and Ostrava, both out-migration and natural decline have contributed to overall population decline. The number of people moving out of Brno doubled between 2000 and 2002 and then stabilized (at between 6,900 and 7,900) up to 2006. At the same time, the number of people moving into the city also doubled, but the numbers involved were smaller, so that there was a net out-migration between 2000 and 2006. Out-migration was mainly a result of growing suburbanization (Figure 6.9; see also Kabisch et al. 2008: 27–33). It was mainly the inner city that lost population, while the municipalities and districts surrounding Brno saw population increase. Municipal experts estimate a loss of at least 20,000 inhabitants to suburbia since 1990. In Ostrava, the number of out-migrants greatly exceeded the number of immigrants throughout the observed period (with a peak of 1,342 people in 2004). At the same time, from the end of the 1990s, suburbanization has been intensifying, with an annual out-migration of more than 1,000 people in 2006 and 2007 (Klusáček et al. 2007). Suburbanization is a demographically and socially selective phenomenon (Šnejdová 2006). Suburbanites in Brno and Ostrava are predominantly aged between 20 and 35. It is mainly families with higher incomes and qualifications who move to suburbia, seeking detached houses in rural surroundings. More recently, developers in the Czech Republic have changed their strategy and have begun to build homes for the middle class, so that suburban housing has become more affordable for a wider range of people. Despite the decline in the supply of cheap public finance in the last few years, suburbanization has not diminished. Both inner cities are therefore losing inhabitants and facilities and face spatial deconcentration. Besides its socio-spatial impacts, suburbanization also leads to growing costs of urban infrastructure, more commuters and traffic and increasing socio-spatial differentiation and fragmentation. But problems also arise in the suburban locations.

Both cities have been experiencing long-term natural population decline (Figure 6.9). The statistics show that the numbers of births and deaths were approximately the same at the beginning and at the end of the observation period, whereas between 1991 and 2005, there was an excess of deaths over births in Brno. In Ostrava, due to the generally younger population (a result of its specific history of post-Second World War urbanization), live births exceeded deaths up to 1993. In 2006, a slight natural increase in Ostrava was again recorded.

Housing Market Development

The quantity and quality of the housing stock as well as the various tenure arrangements have been discussed above, since they are crucial to how private households bring about residential change. The following section provides key information on the housing market situation for the four case-study cities, including the inner-city housing markets.

For the *Polish cities*, a comprehensive survey of the housing market situation is only possible on the basis of census data from 1988 and 2002. The housing stock of Łódź is much more diverse compared with other large cities in Poland because the city was not seriously damaged during the Second World War. In 2002, 15 per cent of the housing stock dated from before 1918, and a further 12 per cent dated from the inter-war period. However, in the inner city, two thirds of the housing stock dates from before 1945 (Bierzyński and Węcławowicz 2008: 73). In Gdańsk, only the historical city core was subject to massive wartime destruction. As a result, in 2002, 39 per cent of the housing in the city as a whole was pre-1945, while in the inner city, the proportion was 70 per cent.

Table 6.8 provides an inter-census comparison of housing market indicators for Łódź and Gdańsk. It shows that while in Łódź the housing stock remained almost unchanged, Gdańsk showed a considerable increase. Even more important, however, is the evidence of a substantial shortage of housing, revealed by the considerable difference between the total number of households and of available dwellings (2002: 36,000 more households than flats in Łódź and 20,000 more in Gdańsk). In Gdańsk, for example, in 1988 the average number of households per dwelling was 1.15, while in 2002 it had increased to 1.19 for the city and to 1.22 for the inner city (Grabkowska and Sagan 2008: 24). But there are also signs of an improvement of the housing situation during the post-socialist period: the size of dwellings, both in terms of room numbers and floor space, increased, while the number of persons sharing a dwelling decreased (as did the average household size).

As in other Polish cities, the housing market in Łódź and Gdańsk changed considerably after 1989. Over the past two decades, both the municipality and the cooperatives have been selling off their housing stock to sitting tenants for a very small percentage (around 10 per cent) of the real market price. As a result, there are many owner-occupied dwellings whose owners lack the financial means to renovate them, or even to maintain them. This is especially true for the inner city, where the old housing stock is in a far worse condition than in the newer estates.

The main indicator of changes in the local housing market in the post-socialist period is the considerably increased percentage of privately owned dwellings, occupied either by renters or owner-occupiers. In Łódź, the proportion of private dwellings rose from 13 per cent in 1988 to 28 per cent in 2002, while municipal housing stock declined from 42 to 24 per cent. The cooperative sector remained strong with 45 per cent (compared with 42 per cent in 1988; Table 6.8). However, compared to the city as a whole, Łódź's inner city had a much higher proportion of both private and municipal housing in 2002 (36 and 41 per cent, respectively) and a smaller proportion of cooperative dwellings (17 per cent; Bierzyński and Węcławowicz 2008: 75). In Gdańsk, the most common type of tenure in 2002 was cooperative ownership. With a share reaching almost 50 per cent, cooperative dwellings predominated over the private (25 per cent) and municipal sectors (23 per cent), as well as other types of tenure (Table 6.8). These are similar proportions to those in the 1988 census. However, there have been significant changes in the proportions of private and municipal dwellings, which demonstrates the dynamics of the processes of both restitution and privatization, even when the almost 15 per cent increase in the total number of dwellings in the inter-census period is taken into account (Grabkowska and Sagan 2008: 30–2).

Table 6.8 Indicators of the housing markets in Łódź and Gdańsk in 1988 and 2002 (census data)

	Łódź		Gdańsk	
	1988	2002	1988	2002
Housing stock:				
Total number of dwellings	313,764	331,668	138,249	164,229
Total number of permanently inhabited dwellings	n.a.	314,569	n.a.	157,609
Household characteristics:				
Total number of households	342,235	351,952	159,513	184,067
Total number of one-person households	88,792	124,601	32,568	56,925
Proportion of one-person households (in %)	25.9	35.4	20.4	30.9
Average household size	2.45	2.34	2.82	2.45
Housing quality indicators:				
Average number of rooms per dwelling	2.8	3.1	3.2	3.6
Average number of persons per dwelling	2.7	2.5	3.3	3.1
Usable floor space per dwelling (in m²)	45.7	52.1	50.7	64.9
Usable floor space per person (in m²)	17.1	21.1	15.5	20.3
Tenure structure:				
Proportion of private housing stock (in %)	13.4	28.4	8.1	25.3
Proportion of municipal* housing stock (in %)	42.5	23.6	41.7	23.9
Proportion of cooperative housing stock (in %)	42.3	44.8	45.5	48.9
Proportion of other tenure types (in %)	1.9	3.2	4.8	2.9

Note: * in 1988: state-owned housing stock.
Source: GUS (several years), Bierzyński and Węcławowicz 2008, Grabkowska and Sagan 2008, authors' work.

In the course of the privatization, new organizational forms to administer the housing stock developed, in particular associations of owner-occupiers (*wspólnota mieszkaniowa*), that is, people who have bought (and occupy) a home in a municipal building which still might have tenants (people who are renting). As a result, in a single building, there may now be at least three different types of tenure: firstly, long-term tenants, some of whom are sharing a flat with other households; secondly, owner-occupier newcomers who have bought their flat; and thirdly, tenant newcomers who are living in privately rented flats.

The municipalities have only a very limited influence on the housing market and its conduct and are only involved in small revitalization projects. In both Łódź and Gdańsk, new suburban housing provides a major pull factor for city residents to move. A specific characteristic of the newly-built housing stock, both within and outside of the cities in Poland, is that a considerable number of these new housing developments are guarded, fenced or gated (Smigiel 2009). Between 2004 and 2007, the housing markets in all the large Polish agglomerations boomed and prices rose significantly. In Łódź, for example, the price per square metre in

2004 was less than 3,000 Polish Złoty (PLN), but increased to almost 6,000 PLN in 2007 (Kabisch et al. 2008: 47). In the same period, the number of dwellings increased by only 4,000 (or 1 per cent; UMŁ 2008: 9).

Table 6.9 gives an overview of the local housing markets of the two Czech cities under investigation. Again, we still have to rely mainly on data from the last census in 2001. Quantitatively, the situation between 1991 and 2001 improved very little: there was a slight increase in the total number of flats (5,000 in Brno and 3,000 in Ostrava). As in the Polish cities, a shortage of housing is demonstrated by more households than permanently inhabited dwellings (14,000 and 8,000, respectively). Again, a trend similar to that in Łódź and Gdańsk is evident with regard to housing quality: the number of rooms per dwelling and the floor space both per dwelling and per person increased in the inter-census period.

Table 6.9 Indicators of the housing markets in Brno and Ostrava in 1991 and 2001 (census data)

	Brno		Ostrava	
	1991	2001	1991	2001
Housing stock:				
Total number of dwellings	158,555	165,366	132,806	135,912
Total number of permanently inhabited dwellings	151,671	151,724	125,969	128,388
Household characteristics:				
Total number of households	165,880	167,740	134,149	140,848
Total number of one-person households*	51,573	55,788	39,852	47,728
Proportion of one-person households (in %)*	31.1	33.3	29.7	33.9
Average household size*	2.34	2.24	2.45	2.25
Housing quality indicators:				
Average number of rooms per dwelling	2.3	2.7	2.4	2.6
Average number of persons per dwelling	2.4	2.6	2.4	2.5
Usable floor space per dwelling (in m²)	38.7	50.8	39.1	49.1
Usable floor space per person (in m²)	15.4	20.8	15.4	19.4
*Tenure structure**:*				
Proportion of privately owned buildings (in %)	73.3	76.5	59.2	62.7
Proportion of municipally-owned buildings (in %)	18.5	11.4	32.1	14.1
Proportion of buildings owned by cooperatives (in %)	7.3	5.5	8.3	7.5
Proportion of buildings with other tenure types (in %)	0.9	6.6	0.4	26

Note: * These numbers refer to so-called census households (*cenzové domácnosti*; see also Chapter 3, Figure 3.1). ** Building data had to be used because of the unavailability of complete flat property characteristics for the two cities for 1991. Therefore, a direct comparison with the two Polish cities as displayed in Table 6.8 is not possible.
Source: Klusáček et al. 2008 (based on ČSÚ data), authors' work.

Large property transfers from the municipality to the local inhabitants (and also to private companies) occurred in both Brno and Ostrava. However, the differences between the two cities are considerable (Table 6.9): While the share of residential buildings in municipal ownership in Brno decreased from 19 to 11 per cent between 1991 and 2001, in Ostrava it started from a much higher level and was halved (from 32 to 14 per cent). Again, the situation in the inner cities varies as a result of different historical development. In Brno's inner city in 1991, the share of municipal houses was 84 per cent. By 2001 it had declined to 46 per cent, while in Ostrava the decline was from 75 to 42 per cent. In the same period, the proportion of private houses increased from 10 to 35 per cent in Brno's inner city and from 14 to 25 per cent in Ostrava's inner city (Klusáček et al. 2008: 65–7). In the years that followed, these differences persisted, since in general, the municipality of Brno privatized its housing stock more reluctantly than any of the other major cities in the Czech Republic. In 2007, Brno still had 27 per cent[6] of municipally-owned housing, compared with 15 per cent in Ostrava.

As a consequence of the housing market transformation, rental housing in Czech cities has become increasingly marginalized in the period of post-socialism. In addition, there are very few privately-owned rental flats. Thus the owner-occupied sector is gradually becoming more important in the Czech Republic.

The whole country and, in particular, the two largest cities, Prague and Brno, experienced a property market boom after 2004, with rapidly increasing prices. In Brno, house prices increased more than threefold between 1998 and 2007, from 9,300 to 29,000 Czech Koruna (CZK) per square metre. A similar increase took place in Ostrava, where the average price for a flat grew by 350 per cent (from 410,000 to 1.4 million CZK). However, there was much more construction in Brno than in Ostrava: Brno saw an increase in building from 3.4 completed dwellings per 1,000 population in 2005 to 7.5 in 2007, while in Ostrava there was almost no change in the intensity of construction, with completed dwellings per 1,000 population increasing from 0.8 to 1.1 (all data from Vaishar et al. 2009: 36, 69–70). This discrepancy indicates the very different levels of prosperity of the two cities in this period, with Brno doing much better (which is also reflected in the different unemployment rates as shown in Table 6.7).

Summary

Łódź, Gdańsk, Brno and Ostrava are important second-order cities in East Central Europe. Post-socialist transition research so far has not had much to say about the fortunes and pathways of such cities. This introductory chapter on the four case-study cities has provided some basic information on population, household and housing market changes during the post-socialist period on the scale of the entire

---

6   The municipality of Brno itself states that the proportion for 2007 is 24 per cent (MMB 2008: 56).

city. It is obvious that, in all cities, a number of different and sometimes even contradictory processes occurred simultaneously. Yet, all of them experienced at least a short period of population decline, mainly due to out-migration. Particularly the rise in overall household numbers, which contradict long-term population trends, provides support for the hypothesis that second-order cities in East Central Europe also face the challenges of the second demographic transition. These demographic trends, however, meet particular housing markets which are characterized by an inherited imbalance between supply and demand, as well as by a number of persistent features of state socialism, in spite of all the transformations which occurred during the 1990s. The chapter also introduces the four inner cities of Łódź, Gdańsk, Brno and Ostrava. How the interplay of demographic and housing market changes, along with a number of other processes, affect residence and residents on the scale of the inner cities will be explored in the second part of the volume.

PART II
Empirical Investigations on and in Polish and Czech Inner Cities

Chapter 7

Old-new Diversity: Processes and Structures of Socio-Demographic Change in the Inner City

Annegret Haase, Adam Bierzyński, Maja Grabkowska, Petr Klusáček,
Stanislav Martinát, Zdeněk Uherek, Andreas Maas

Introduction

East Central European inner cities are still fighting against a long-prevailing prejudice. In contrast to their counterparts in Western Europe that have undergone processes of revitalization, regeneration and resurgence over the last few decades, they have traditionally been associated with deteriorating as well as physically and demographically ageing, old built-up areas in poor physical condition and in public ownership. In the western debates, the influx of new residents and distinct household types, as well as a diversification and rejuvenation of the inner-city population are highlighted in both the gentrification and the reurbanization literature, by contrast, it is widely agreed that East Central European inner cities continue to be inhabited by predominantly older and poorer people who either cannot afford better housing conditions or who were allocated their flats during the state-socialist period. Thus, it is the legacies of the previous system together with their persistence under new societal conditions that are stressed rather than change.

In our research on Polish and Czech inner cities, we certainly found evidence for the long-prevailing situation described above. But after a more thorough look at the statistics, and while listening to the answers of experts about recent socio-demographic changes within the inner city, we became increasingly confused. The picture that seemed to be clear at the beginning became more puzzling, more differentiated, more 'fluid'. Our scepticism was more in line with some other recent research. According to some scholars, there are – in apparent agreement with western patterns – more young and, partly, also more affluent people who are moving to the inner city (see Chapter 5).

Up to now, this striking news about socio-demographic change in East Central European inner cities has not received appropriate attention in the literature. Inner-city areas have not been the focus of urban research among East Central European scholars. If addressed at all, the centre of attention tended to be on the more extreme instances of socio-spatial change, such as luxury renovations, gentrification, the entry of western foreigners, physical dilapidation, ethnic concentration or the 'reinvention' of former Jewish quarters. While, from a western point of view, such tendencies might easily gain the label of gentrification, we do not want to start with this view for two reasons. Firstly, most East Central European scholars are cautious about applying this term and concept. Secondly, taking such a perspective immediately narrows the focus of investigation to processes of displacement, social upgrading and the like, and it might obstruct a wider view of socio-demographic changes which are better described as reurbanization of neighbourhoods after long-term decline.

Set against this background, this chapter challenges the story of population loss and ageing in the inner cities of Łódź, Gdańsk, Brno and Ostrava. We argue that the 'old story' of decline and ageing is no longer totally true, but also that we do not have a new, 'grand' story to tell. Rather, we present multiple, new, 'smaller' stories that partly overlap and whose long-term importance still has to be tested and has to be observed. We define socio-demographic change as the dimension of residential change that comprises the demographic and social characteristics of the residential population (that is their age, ethnicity, household structure and their spatial characteristics; see Chapters 2 and 3). Accordingly, socio-demographic change refers to the residents and it interacts with changes relating to the places of residence, that is housing market developments that were described in Chapter 6. Socio-demographic change is connected with increasing social and income inequality and the emergence of new groups at risk, such as long-term unemployed, large and one-parent families, people with little education and ethnic minorities, especially Roma, issues that cannot be discussed in detail in this chapter. The authors wish, however, to emphasize that they are aware of these connections.

The chapter is structured as follows: after this introduction, materials and methods used for the chapter are explained briefly. In the second section, the framework of analysis is presented: distinguishing between processes of socio-demographic differentiation and their results, that is persistent, newly emerged or rearranged socio-spatial structures, it identifies and explains the characteristics and underlying dynamics of this change for the four case-study cities. The third section expands on the process-related level, that is on differentiation. It reports on population decline and repopulation, ageing and rejuvenation, household change, gentrification and displacement, studentification and ethnic diversification. The fourth section looks at the results of differentiation and their spatial imprints – it analyzes patterns of segregation and fragmentation. The fifth and final section summarizes the results and makes some concluding remarks.

Materials and Methods

The chapter brings together the results of small-scale analyses which draw on census data from 1988 and 2002, for Polish cities, and from 1991 and 2001 for Czech cities, as well as annual population statistics (provided by the statistical offices and the municipalities) and a qualitative assessment of the relevant changes based on knowledge gained from interviews with experts as well as by area observation. For the inter-census periods (Poland: 1988–2002, Czech Republic: 1991–2001), we mainly use the census data obtained in the years mentioned above, while for the later period of time, that is after 2002 in Poland and after 2001 in the Czech Republic, we have to rely mainly on selected data from population registers and expert knowledge, which will be documented in the following analyses. The chapter operates on different spatial scales. It sets the focus on the inner city and smaller units like the neighbourhood, a block or even a single house; the entire city is used just as a 'mirror' for the inner city, mostly in a comparative manner. At this stage, we will mention just a few of the challenges that we face when investigating population development and its spatial consequences at different spatial scales.

- Our analysis has to come to terms with the contradictions between various statistical sources (see Chapter 2). In order to interpret the data appropriately, one needs to be familiar not only with the methodologies used to obtain the various data and, in our case, with different countries, but also with the current dynamics in the urban areas under investigation.
- We have to cope with the unavoidable gap between statistical data and urban reality. Official population data only reveal part of the story and may obscure the real processes going on in the cities. The 'real' population numbers are hardly detectable by the census or any other statistical data, due to the incomplete registration of urban dwellers (see Chapter 2).
- The assignment of an individual household to a certain type by statistics may differ from the way it sees itself. Therefore, research on urban population development is better approached with both quantitative and qualitative methods as demonstrated here.

Conceptualizing Inner-city Socio-demographic Change

When analyzing socio-demographic change in the four case-study inner cities, we distinguish between socio-spatial differentiation (in terms of a *process*) and the emerging new or rearranged *structures* of socio-spatial segregation (in terms of concentration, separation and fragmentation) as a result of the processes just mentioned; see also Rink 1997: 26).

We start from the position that the inner cities of our four case-study cities were already showing specific structures related to socio-demographic differentiation by the end of the 1980s, that is the beginning of our research period. Socio-spatial differences existed under socialism in different forms. On the one hand, socialist urban societies had 'inherited' patterns of separation from pre-socialist times. On the other, socialism developed its own specific forms of 'socialist segregation' either through deliberate processes, for example the concentration of the political élites and the assignment of certain privileged groups to prestigious housing estates or, as the unintended consequences of widespread 'ageing in place', due to low residential mobility during the state-socialist period. This led to a concentration of older people in inner-city areas and both younger and better-off households in the newly-built housing estates. In contradiction to the socialist idea of an egalitarian society, the state-regulated housing sector worked as a mechanism for the formation of significant socio-spatial differences during the socialist period. In the case of Polish cities, the level of socio-economic segregation increased, particularly during the 1980s, due to the rapidly increasing housing shortage and the largely corrupt methods of flat assignment. However, compared to western cities, the extent of residential segregation was much smaller. Since 1989, Polish and Czech inner cities – like their counterparts in the whole of East Central Europe – have undergone a new wave of socio-spatial differentiation.

Our research identified the following processes of socio-demographic differentiation in the four East Central European cities (see also Table 7.1):

- coexistence of population decline and repopulation,
- coexistence of ageing and rejuvenation,
- diversification of households (in terms of size, composition, stability),
- gentrification and displacement,
- studentification as a specific form of repopulation and rejuvenation and
- ethnic diversification.

These processes resulted in newly-emerged or rearranged structures of socio-spatial inequality in terms of:

- an increasing diversity of household types,
- spatial segregation (in terms of concentration and separation) of various social groups,
- spatial concentration of different age groups,
- forms of ethnic separation and exclusion and
- small-scale fragmentation of socio-demographic structures in the urban space.

In the following sections, we present firstly, the processes and secondly, existing or emerging structures as the result of socio-demographic change in the four case-study cities.

Table 7.1 Processes and structures (results) of socio-demographic change

Processes ...	... resulting structures
• coexistence of population decline and repopulation • coexistence of ageing and rejuvenation • diversification of households • gentrification and displacement • studentification as a specific form of repopulation and rejuvenation • ethnic diversification	• increased diversity of household types • spatial segregation (concentration and separation) • spatial concentration of different age groups • forms of ethnic separation and exclusion • small-scale fragmentation of socio-demographic structures in the urban space

Source: Authors' work.

Unpacking Differentiation: Detecting Processes of Socio-demographic Change in Four East Central European Inner Cities

Ambiguous Evidence of Population Decline and Ageing ...

A first examination of the population development in the case-study cities leads to the conclusion that population loss affects all four cities entirely and – to an even greater extent – their inner cities (Table 7.2). A second look, however, makes it obvious that there are differences as well as similarities between the cities. The 'story of decline' for both scales of investigation is striking, in particular for Łódź. Łódź grew from 762,000 to 854,300 inhabitants between 1970–1988, while at the same time the inner city lost 25 per cent of its inhabitants (a decline from 132,218 to 100,997). This process continued after the political turnaround: in absolute numbers, the population of the inner city decreased from 99,000 in 1990 to 79,000 in 2005: a fall of 20 per cent (Bierzyński and Węcławowicz 2008: 11). Consequently, inner-city decline is not only a post-socialist issue, however, the breakdown of state socialism accelerated population losses at the level of the entire city as well. The overall population declined by 8 per cent between 1988 and 2002 while at the same time the inner city lost 16 per cent of its population as a result of natural decrease, intra-urban and interregional out-migration.[1] The fact that the inner city still represents the district with the highest population and housing density in Łódź reflects the enormous densities during the 1970s–1990s (Dawson 1979: 381, Jakóbczyk-Gryszkiewicz 1997: 117–19).[2]

---

1   Information from an interview with a representative of the Łódź municipality (Łódź E2. E stands for expert).

2   Despite these losses, the inner city still shows the highest population density of any part of the city, which provides evidence for the density of housing in this part of the city

In Brno and Ostrava, too, the level of the decline of the inner city far exceeded that of the entire city (9 and 10 per cent compared with 3 per cent for both). In the case of Brno, these population losses could have already changed into a stabilization or even slight growth, as is suggested by some data for the years 2003–2007 (see Chapter 6 and Steinführer et al. 2010).[3] In Gdańsk, the lack of data for the inner city for 1988 hinders any thorough analysis of changes. Yet, according to existing census data for the whole city for 1988 and 2002 and the inner city for 2002, as well as local expert knowledge, it can be presumed that the spatial distribution of Gdańsk's population after 1988 has changed in line with the development identified for Łódź: while newly-built housing estates in the city peripheries have gained inhabitants, the inner city is still experiencing population losses (Grabkowska and Sagan 2008). The same is true for the two Czech cities where, according to local experts, suburbanization is now seen as the major challenge and even menace for inner-city population development and is likely to remain a major reason for inner-city population decline in the future.[4]

Apart from population decline, ageing turned out to be much more rapid at the level of the entire city than in the inner cities in all four cities during the inter-census period. This is because the inner cities were already 'older' at the beginning of the research period and the ongoing ageing process was thus less rapid during the inter-census period. Moreover, ageing in the inner cities has been attenuated by incipient processes of rejuvenation (see next section). Looked at through the 'residential lens', this means that, as a result of the legacies from state socialism, a considerable number of the flats in the inner cities are still inhabited by older (one-person) households who pay non-market rents. However, increasingly these older people are being replaced by their children and grandchildren as well as by other people, who are therefore making the inner city younger.

and for crowded housing conditions (Jakóbczyk-Gryszkiewicz 1997: 118–119).

3   According to the ISEO data (see Chapter 2), the population gain even amounts to 18,000 inhabitants between 2003 and 2007 (+5 per cent).

4   Information from an interview with representatives of the Ostrava municipality (Ostrava E6) and an association supporting the Old Ostrava (Ostrava E9).

Table 7.2 Population development and ageing in the case-study cities (entire city and inner city) for 1988/1991 and 2001/2002 (based on census data)

Population and Ageing	Łódź				Brno			
	Entire city		Inner city		Entire city		Inner city	
	1988	2002	1988	2002	1991	2001	1991	2001
Population (in 000s)	854.3	789.3	101.0	85.0	388.3	376.2	66.8	60.9
Population change	-7.6 %		-15.8 %		-3.1 %		-8.8 %	
Population 0–14 (in 000s)	161.3	97.8	19.2	11.4	76.4	54.0	12.1	9.3
Change in population 0–14	-39.4 %		-40.6 %		-29.3 %		-23.1 %	
Population 64+ (in 000s)	107.7	130.1	15.5	13.4	55.2	58.7	13.0	10.7
Change in population 64+	+20.8 %		-13.7 %		+6.3 %		-17.7 %	
Ageing index	66.8	133.0	81.0	117.5	72.3	108.7	107.4	114.5

Population and Ageing	Gdańsk				Ostrava			
	Entire city		Inner city		Entire city		Inner city	
	1988	2002	1988	2002	1991	2001	1991	2001
Population (in 000s)	464.3	461.3	n.a.	209.0	327.4	316.7	26.3	23.6
Population change	-0.6 %		n.a.		-3.2 %		-10.2 %	
Population 0–14 (in 000s)	104.2	66.7	n.a.	29.1	68.2	51.8	5.5	3.8
Change in population 0–14	-36.0 %		n.a.		-24.0 %		-30.9 %	
Population 64+ (in 000s)	41.9	62.9	n.a.	34.2	37.1	40.1	3.8	3.2
Change in population 64+	+50.1 %		n.a.		+8.1 %		-15.8 %	
Ageing index	40.2	94.3	n.a.	117.5	54.4	77.4	69.1	84.2

Note: 'Inner city' in the case of Łódź relates to the administrative district Śródmieście. In Brno and Ostrava, the term was generated analytically based on the administrative structure of Brno and Ostrava in 278 and 272 basic settlement units (BSU) in the 2001 census, respectively. In case of Brno, those BSU were taken into account where the proportion of houses built before 1945 was greater than 65 per cent and where the proportion of dwellings located in multi-storey buildings was also greater than 65 per cent. In case of Ostrava, the inner city was delimited mainly according to its geographical location and particular characteristics of the area. In Gdańsk, the inner city encompasses the historical city core and the adjoining urban units in which, at the time of the census in 2002, the share of residential buildings built before the end of the Second World War was equal to or higher than 50 per cent (altogether 10 out of the total of 27 urban units). The calculation of the household-related data for the Czech cities is based on the census households (*cenzové domácnosti*, Klusáček et al. 2007: 33).

Source: Bierzyński and Węcławowic 2008, Grabkowska and Sagan 2008 (based on GUS data), Klusáček et al. 2007 (based on ČSÚ data), authors' work.

In 1989 the four inner cities showed a comparatively 'older' profile than the total city, with a higher proportion of older people and a lower proportion of younger people. Łódź's inner city in the 1990s, for example, still displayed the most significant gender disproportion amongst all urban districts, a result of the large number of widows living alone (Jakóbczyk-Gryszkiewicz 1997: 120) and a concentration of female students in the student dormitories within the inner city.[5] But the drop in birth rates and out-migration of younger residents also contributed to ageing. The changes in age group distribution differ between the city as a whole and the inner city in Łódź, Brno and Ostrava. The proportion of under-15-year-olds declined in all three inner cities to a level that is comparable with that of the total city (Table 7.2). The high proportion of older people in the inner cities is due to their repopulation after the Second World War. In the case of Łódź these were repatriates from pre-war Poland's eastern borderlands and in-migrants from other Polish regions. In Brno, the abandoned housing stock formerly inhabited by the German population was quickly 'refilled' by Czech dwellers, due to the 'inherited' lack of housing resources in pre-war Czechoslovakia. The ageing index in Table 7.2 relates to the proportion of people aged 65 and over to the population younger than 15 years. Using Łódź as an example, it can be read as follows: in 1988, there were 67 people aged 65 years and over for every 100 aged 15 and under (and 81 in the inner city), by 2002 there were 133 (and 118 in the inner city). This means that ageing in the inner city started from a higher level at the beginning of the post-socialist transition, but it then increased more slowly than in the entire city. The situation in Brno and Ostrava is similar, the only difference being that, in both cities, the proportion of older residents in the inner city was still higher than in the entire city (as demonstrated by an ageing index of 115 versus 109 in 2001).

The majority of local experts interviewed in all four cities considered ageing to be a major challenge for future urban development. Some of them underlined the fact that, for example, Łódź and Brno belong to the 'oldest' large cities in Poland and the Czech Republic, respectively. These experts also emphasized the fact that ageing has a selective impact on urban space – it is perceived mainly as a problem of the inner city or the old built-up housing stock, but also as a future problem for large housing estates (Haase et al. 2009).[6] Interestingly, and in contrast to western cities, the interviewed local experts in the four cities perceived the phenomenon of 'living alone' as related to the old rather than to younger and middle-aged people.

---

5   This information stems from an interview with a representative of the Łódź municipality (Łódź E2). In 2002, the ageing index for the entire city of Łódź was higher than for the inner city but there was still a preponderance of female dwellers in the inner city (see Table 7.2).

6   Information from an interview with a representative of the Łódź municipality (Łódź E2) who also introduced the term *demographic waves,* that is a future replacement of population not only in the old building stock but also in prefabs built in the 1970s, whose residents are now in their 60s and 70s.

Thus, in all four cities under investigation, according to the census data up to 2001/2002, the inner cities were experiencing declines in terms of population numbers and household size but the decline in household numbers was significantly lower than that of total population.

... and of Repopulation and Rejuvenation

Although the overall population development in the four inner cities, at least up to 1991, was characterized by continuous decline and ageing, there are indications of a selective repopulation, where the prefix 're' relates probably more to a *qualitative change* than to a quantitative development.[7] When we speak of rejuvenation, we conceptualize it as a counter-process to ageing: it describes the increase of the proportion of younger age groups and the decrease of the mean age in an urban area. This process is exemplified here by Łódź, for which there is census data. From around 1988 Łódź's inner city has undergone a process of repopulation by smaller and younger as well as middle-aged households, the majority of which are childless households, a process that is only partly detectable in the statistics. This has led to a younger age profile or, at least, has countered the ageing process. The last census (2002) showed higher numbers of young people and lower numbers of older people in the inner city, compared with the city as a whole. Table 7.3 indicates that the in-migration to Łódź's inner city in the period 1989–2002 was driven mainly by younger age groups, especially by young individuals (students, apprentices and early-stage professionals), but also by a few younger families, as reflected in the higher percentage of children from 0–14 years, compared to the entire city. Up to 1988, in contrast, the in-migration of older age groups to both the entire city and the inner city was insignificant. The great majority of older inhabitants of the inner city either had lived here since birth or in-migrated before 1988. From 1989–2002, the inner city was characterized by an above-average in-migration of young residents: at the overall city level, just 15 per cent of all residents in their twenties had migrated to Łódź between 1989 and 2002, compared with 33 per cent of all residents aged 20–29 who had migrated to the inner city in the same period. Looking at the age group 30–39 (early-stage professionals), the shares are 10 per cent for the whole city and 11 per cent for the inner city. Taking into account that this is only the registered in-migration and what has been previously said about the influx of unregistered residents, we must conclude that the ageing process of the inner city is not as advanced as it appears from the official statistics and that the scale of depopulation is also overestimated. The local experts were ambivalent about whether the inner city was experiencing rejuvenation.

---

7 With this definition, we deliberately go beyond the usage of *repopulation* in the British debate as a term describing simply the influx of new residents into an urban area after a period of decline and population losses (Lever 1993, Ogden and Hall 2000, Burton 2003).

Thus, what can be observed now could be interpreted as the beginning of a 'resurgence' of the inner city that will continue over the next few years. The importance of in-migration is also indicated by the fact that, in the inner city, the proportion of those who have lived in Łódź since birth is the lowest amongst all five city districts (Bierzyński and Węcławowicz 2008: 65).

Table 7.3 Łódź: in-migrants up to 1988 and 1988–2002 by age group

Age class	Łódź entire city				Łódź inner city			
	up to 1988		1989–2002		up to 1988		1989–2002	
	Total	%	Total	%	Total	%	Total	%
0–14	248	0.2	3,151	3.2	19	0.2	399	3.5
15–19	1,738	3.2	3,010	5.6	207	3.3	687	10.6
20–29	7,667	6.0	19,380	15.3	795	4.5	5,389	33.0
30–39	15,772	17.2	8,963	9.8	1,902	19.3	1,060	10.8
40–49	35,905	26.9	3,561	2.7	3,823	26.4	387	2.7
50–59	43,413	36.6	1,657	1.4	3,484	34.2	209	2.0
60–64	19,986	54.4	466	1.3	1,542	53.0	56	2.0
>64	79,201	60.9	1,761	1.4	8,467	63.2	168	1.2

Note: The percentages relate to the total numbers of single age groups in Łódź.
Source: Bierzyński and Węcławowicz 2008: 67–68.

However, it is important to remember that no data are available on population change for any of the four cities more recently than that of the last censuses that took place at the beginning of the decade. But, as a result of the qualitative research, it can be assumed that population decline has probably been slowed down by in-migration. This assumption was strengthened by a number of experts in the four cities, especially for areas such as Wrzeszcz Dolny in Gdańsk or Veveří in Brno. In Brno, the repopulation hypothesis is also supported by the fact that the municipal sewage plant, which was constructed for 400,000 inhabitants, is used to full capacity; this does not correspond to the official population data.[8] In the case of Gdańsk, local experts confirmed that there is a trend towards the repopulation of inner-city areas by younger professionals, including families with younger children, for whom it is more attractive to live centrally than in the suburbs.[9] To what extent outflows are

---

8   Information presented at the workshop *Strengthening the city* in Brno in January 2009.

9   Information from an interview with a representative of the Gdańsk municipality (Gdańsk E2).

'balanced' by inflows cannot be stated exactly – here, we have indeed to wait for the next census or further migration data from small-scale surveys.

The qualitative research at least increased our awareness that there is a counter-trend which is not only bringing population back into the inner cities but which is also changing the age and household composition of the inner-city population. In terms of the age structure, repopulation contributes to rejuvenation. However, the process is less visible in, or detectable by, the statistics than in reality, since many of the young newcomers are not registered.

The corresponding evaluations made by the interviewed inner-city residents showed that they view the process as normal.

As a preliminary conclusion from the material presented here, one can identify a *juxtaposition of population decline and ageing* as powerful and ongoing prevailing trends of urban and particularly inner-city development on the one hand but *simultaneous processes of in-migration, decelerated ageing or even rejuvenation* at a modest level on the other. Interestingly, these processes and their results, especially today's coexistence of old and young residents and their distinct housing needs and preferences, are only partly perceived by local experts. Local experts assume that the inner city will become increasingly attractive for younger (and better-off) households, but that this is contingent on renovation and improvement of housing standards. Current inner-city housing demographies are therefore believed to be more static than they really are. Furthermore, the residential mobility of younger inhabitants is under-estimated (see also Haase et al. 2009).[10]

Household Change: Increasing Diversity and Heterogeneity

In all four case-study cities, population trends in the 1990s were similar to those of many other second-order cities across Europe, as well as being in line with general demographic trends (Ogden and Hall 2000, Buzar et al. 2005). Population decline was accompanied by a much smaller decrease in household numbers,[11] which was mainly driven by a sharp increase in the number of one-person households (Table 7.4). Consequently, mean household sizes diminished.

The following discussion of household transformation in the four cities is divided into two sections. In the first, we deal with the inter-census period from 1988 to 2002 in Poland and from 1991 to 2001 in the Czech Republic. In the second, we discuss trends since the last censuses.

The inter-census period can be characterized as follows (Table 7.4): while the absolute number of inhabitants has declined in the four cities, the absolute number of households has increased, due to downsizing. In the case of Łódź, however, the tremendous losses at the inhabitants level could only be balanced out at the

---

10 The information stems from interviews with representatives of the Łódź municipality (Łódź E2) and two real estate agencies (Łódź E1 and E6).

11 For the definition of households in the Polish and Czech statistics see Chapter 3.

household level with respect to the entire city (+3 per cent), while in the inner city, the number of households also decreased (-9 per cent), in spite of a significantly growing number of one-person households (so that 40 per cent were one-person households in the inner city by 2002 compared with 16 per cent in the city as a whole; see also Bierzyński and Węcławowicz 2008: 31). Simultaneously, the share of 3+ households was decreasing at the entire and inner-city level, from 45 to 36 per cent and from 39 to 34 per cent, respectively. To put it differently: while in 2002 in inner-city Łódź, almost 40 per cent of all households were one-person households, only one third of them were made up of three or more people.

In all four cities, and at both spatial scales, households became smaller on average. In Brno's inner city, the decrease in mean size of households was less prominent than in the other cities, because of the concentration of a large Roma population, characterized by a higher share of families with many children and larger households. The trends in smaller territorial units with a predominantly Czech population were similar to those observed in Łódź (Klusáček et al. 2007). The same holds true for the decrease of 3+ households at the scale of the inner city, which was also much smaller than at the entire city level. The situation was comparable in Ostrava. At the entire city level, the population loss was 'balanced' by a growth in households, while the inner city showed a net decrease in households during the inter-census period, despite more smaller households, which underlines the impact of population loss. The mean size of households decreased from 1991–2001 in the whole of Ostrava, as in the other three cities but was significantly lower in the inner city (Table 7.4). While the number of one-person households increased, the number of 3+ households decreased. The inner city, however, showed a somewhat different picture. While it conforms with the overall city trend, the increase in one-person households was much less significant compared with that of the entire city (changes range from 5–20 per cent for the overall cities) again, probably due to the presence of the Roma population with their generally larger households. Gdańsk saw, like Łódź, an increase in one-person households (changes range from 40–75 per cent for the inner cities) and a decrease in 3+ households. But the share of one-person households in the inner city – although in both cases it was higher than for the entire city – differed (Łódź: 40 per cent; Gdańsk: 33 per cent). Therefore, the mean household size in the inner city of Gdańsk in 2002 was the same as that for the entire city, while it is significantly lower in Łódź. For the inner city of Gdańsk, the lack of small-scale data for the 1988 census rules out any evaluation of change, but on-site research and experts' information indicate that the trends are not too different from the other three cities and, especially, from Łódź.

Table 7.4 Household development in the case-study cities (entire city and inner city) for 1988/1991 and 2001/2002 (based on census data)

Household development	Łódź				Brno			
	Entire city		Inner city		Entire city		Inner city	
	1988	2002	1988	2002	1991	2001	1991	2001
Households (in 000s)	342.2	352.0	41.1	37.5	166.0	167.7	31.0	29.0
Change in households	+2.9 %		-8.8 %		+1.0 %		-6.5 %	
One-person households (in 000s)	88.8	124.6	13.0	15.1	51.6	55.8	11.3	11.2
Change in one-person households	+40.3 %		+16.2 %		+8.1 %		-0.9 %	
3+ households (in 000s)	81.6	73.0	16.2	12.6	65.3	59.7	9.8	8.8
Change in 3+ households	-10.6 %		-5.6 %		-8.6 %		-10.2 %	
Mean household size	2.45	2.34	2.33	2.14	2.34	2.24	2.15	2.10

Household development	Gdańsk				Ostrava			
	Entire city		Inner city		Entire city		Inner city	
	1988	2002	1988	2002	1991	2001	1991	2001
Households (in 000s)	159.5	184.1	n.a.	83.5	134.1	140.8	11.5	11.1
Change in households	+15.4 %		n.a.		+5.0 %		-3.4 %	
One-person households (in 000s)	32.6	57.0	n.a.	27.8	39.8	47.7	4.1	4.3
Change in one-person households	+74.8 %		n.a.		+19.8 %		+4.9 %	
3+ households (in 000s)	88.4	79.4	n.a.	33.9	57.4	50.6	4.3	3.5
Change in 3+ households	-10.2 %		n.a.		-11.8 %		-18.6 %	
Mean household size	2.82	2.45	n.a.	2.47	2.44	2.25	2.29	2.13

Source: Grabkowska and Sagan (based on GUS data), Klusáček et al. 2007 (based on ČSÚ data), Bierzyński and Węcławowicz 2008, authors' work.

When investigating household transformation in terms of their type or composition, we have to be clear that the definitions in the Polish and Czech censuses provide a specific, data-gathering-related understanding of, for example, family households and non-family households. In some respects the census definitions conflict with the understanding of household and family in this volume (see Chapter 3). Another incongruity may occur between the census data on household types and the self-perception of the resident households themselves.

While the Polish census distinguishes only between family households (including marriages and cohabitations with and without children as well as lone parents) and non-family households (one- and multi-person households; see also Bierzyński and Węcławowicz 2008: 29, 32, 40), the Czech census distinguishes four categories: (1) complete family households including married and cohabiting couples with or without children, (2) non-complete family households including one-parent households and single empty-nesters, (3) multi-person non-family households and (4) individual households including one-person households (Klusáček et al. 2007: 33). There is, at least in the Polish census data, information about whether family households represent marriages or cohabitations. However, in the Polish and Czech data it is unspecified as to whether the family households are younger or older, how many children they have or whether one-person households are younger or older, live alone or in other arrangements, for example, flat shares. As a result the data can only be used to a limited extent to detect shifts towards alternative ways of living, compared with the core family. However, they provide some basic information that serves as a starting point for the qualitative analyses presented in Chapter 9.

Tables 7.5 and 7.6 provide the census data for these typologies for the years 1988 and 2002 (Poland) and 1991 and 2001 (Czech Republic) as well as for changes between these years. The data show the following:

- The proportion of family households decreased, whereas the percentage of non-family households increased, especially in the two Polish cities.
- The decrease in families was mainly due to the decrease in families formed by couples with children. By contrast, the proportion of couples without children, whether married or cohabiting, and lone-parent families increased.
- Cohabitation played only a minor role among couple households, at least up to the early 2000s.
- The proportion of lone parents ('incomplete families' in the Czech census) saw a significant increase during the inter-census period in all four cities.
- Compared with the entire city, the share of non-family households within the four inner cities was already higher in 1988/1991. During the inter-census period, the inner cities saw a further decrease in family households and an increase in non-family households. In 2001/2002, their share was between 37 per cent (Gdańsk) and 44 per cent (Łódź), reaching levels that are similar to Western European cities.
- Comparing the four cases, there are some differences. While the two Czech cities (2001) and Łódź (2002) showed higher proportions of non-family households, the inner city of Gdańsk had a lower share in 2002. Here, the most striking feature is the sharp decrease in family households, especially of married couples without children, whereas the proportion of married couples with children decreased less dramatically as did the proportion of lone-parent households. For Gdańsk, no conclusion can be drawn, due to the lack of data for the inner city in 1988 (see above). However, when looking only at the 2002 data, it appears that out of all four cities, the increase in both total household numbers and numbers of non-family households was greatest in Gdańsk.

Was there an increasing diversity and heterogeneity of households in the inter-census period? From the data the following can be concluded: *Firstly*, there was a decline of what the census defines as family households, irrespective of whether the couples are married or have children. The decline in family households is mainly a result of a decrease in households with children. The relative importance of childless couples, no matter what their age, increased in all four cities. To put it differently, the decrease in couple and family households coincides with an increasing importance of one-person and multi-person non-family households. *Secondly*, the data show a growing importance of lone-parent households, particularly in the inner city where they have become 'normal' among households with children. *Thirdly*, there are obvious commonalities between the four cities at the entire and inner-city level. At the same time, there are also differences relating to both the national context and the local specifics of the four case-study cities. *Fourthly*, the Polish and Czech census data do not, however, allow a clear distinction between different household types, as we distinguished them in Chapter 3. It is impossible to differentiate 'multi-person non-family households' into flat shares of students and young professionals and housing arrangements established by unrelated older (impoverished) persons sharing a flat for economic reasons – let alone the issue of the subjective meaning attributed to such arrangements by the residents themselves.

Our consultations with local experts and interviews with inner-city residents provided evidence that the statistics reveal only a part of the complex process of household transformation. This holds true also for the period after the last censuses in Poland (2002) and the Czech Republic (2001). Knowledge about the role of silent transformations of the inner-city residential structure is (still) limited among local experts (see also Haase et al. 2009). There is some understanding of the change of housing preferences and the growing interest among young people in living in the inner city. But it is generally assumed that young (pre-family) households still prefer to purchase a home on the urban periphery or in suburbia, because the inner city is seen as an unsuitable place for families to live and to bring up children. In the experts' view the inner cities are not only abandoned by families but also by young couples who wish to found a family The reasons given relate to the push factors of the inner city (no playgrounds, noise and air pollution, lack of safety) and pull factors of single-family housing in suburbia (spacious housing, back gardens, safe playgrounds). Only a few voices mentioned the inner city when talking about good places to live with a family, mostly in Ostrava and Gdańsk.[12]

---

12 The information stems from interviews with a representative from a real estate agency in Łódź inner city (Łódź E6).

Table 7.5 Household types 1988 and 2002 in Łódź and Gdańsk

Household type	Łódź entire city			Łódź inner city			Gdańsk entire city			Gdańsk inner city
	1988	2002	% change	1988	2002	% change	1988	2002	% change	2002
Households in total	342,235	351,952	2.8	41,117	37,506	-8.8	159,513	184,067	15.4	83,519
Non-family households** (% of total number of households)	28.5	37.9	37.0	35.0	43.6	13.7	22.4	33.3	71.3	36.7
Family households* (% of total number of households)	71.5	62.1	-10.8	65.0	56.4	-20.9	77.6	66.7	-0.8	63.3
Families										
Married couples (in %)	78.3	70.2	-20.3	75.3	63.3	-33.9	80.2	73.6	-9.2	69.5
- without children (in %)	27.9	26.9	-14.3	28.6	22.6	-40.5	22.0	24.6	10.7	24.4
- with children (in %)	50.4	43.3	-23.6	46.8	42.7	-29.8	58.2	48.9	-16.8	45.1
Cohabiting couples (in %)	n.a.	3.1	-	n.a.	4.3	-	n.a.	2.8	-	3.1
One-parent households (in %)	21.7	26.7	9.8	24.7	32.4	5.0	19.8	23.6	17.8	27.4

Note: * family households = married or cohabiting couples with and without children, lone parents. ** non-family households = one-person households, multi-person non-family households.

Source: Bierzyński and Węcławowicz 2008, Grabkowska and Sagan 2008, authors' work.

Table 7.6 Household types 1991 and 2001 in Brno and Ostrava

Household type	Brno entire city		Brno inner city			Ostrava entire city			Ostrava inner city		
	1991	% change	1991	2001	% change	1991	2001	% change	1991	2001	% change
Households total	165,880	1.1	30,996	28,994	-6.5	134,149	140,848	5.0	11,499	11,128	-3.2
Non-family households** (% of total number of households)	31.5	16.2	37.2	42.9	7.9	30.0	35.8	25.4	36.1	41.5	11.3
Family households* (% of total number of households)	68.5	-5.8	62.8	57.1	-15.0	70.0	64.2	-3.7	63.9	58.5	-11.4
Family households											
Complete family (in %)	80.7	-12.1	73.8	67.9	-21.8	84.0	76.8	-11.9	80.1	71.4	-21.0
- without children (in %)	37.4	5.4	38.8	37.6	-17.6	36.6	40.7	7.0	34.9	36.5	-7.4
- with children (in %)	43.3	-27.3	35.0	30.2	-26.6	47.3	36.1	-26.6	45.2	35.0	-31.5
Incomplete family (one-parent households) (in %)	19.3	20.7	26.2	32.1	4.4	16.0	23.2	39.2	19.9	28.6	27.0

Note: * family households = complete family households (married or cohabiting couples with and without children) and incomplete family households (lone parents). ** non-family households = one-person households, multi-person non-family households.
Source: Klusáček et al. 2007 (based on ČSÚ data), authors' work.

Younger households without children, such as students and young professionals, are not considered as a client group in the housing market. The concentration of student flat shares in the inner city is not mentioned by the majority of the interviewed experts. However, incipient processes of an influx of students into the inner-city housing market, with the exception of dormitories, as described (below) by a real estate representative from Gdańsk, bring about new demand groups for particular housing market segments (Chapter 6). Immigration from abroad plays no major role in either (inner) city, with the exception of the Roma populations in Brno and Ostrava, who to a large extent moved in from Slovakia after 1990 and who occupy the eastern (Brno) and northern (Ostrava) parts of the inner city, where housing conditions are particularly poor.

Since the next censuses in 2011 will also provide limited information about household change – independently of which part of the city is considered – additional qualitative research is and will be indispensable for gaining a more comprehensive picture of how households are changing. Change includes not only an increasing heterogeneity but also a growing instability of household and living arrangements. Since census data provide insights only for a moment in time every ten years, they are not able to detect the increasing instability of household arrangements. Examples of household careers that underwent changes over a time span comparable to an inter-census period will be presented in Chapter 9.

Gentrification and Displacement

There has been much discussion about 'post-socialist gentrification'. The debate emerged rapidly after the systemic change in 1989/90 and became a fashionable topic for both Western and East Central European scholars. But the more case studies were carried out, the more doubts have arisen about exactly which processes of up-market residential developments and related socio-economic change were taking place in East Central European cities (Standl and Krupickaite 2004, Sýkora 2005). While acknowledging the fact that there are developments of this type in most of the big cities, scepticism is appropriate and justified, mainly with respect to their scope, character and consequences for the urban space, which makes it difficult to (simply) apply 'western' models in many cases. Gentrification in its classical, 'western' understanding – as a possible counter-process not only in social but also in demographic terms – is not a large-scale issue in any of the four cities. Some cases of spot gentrification, restricted to single blocks and streets, can be detected (see Sýkora 2005). For Łódź's inner city, Marcińczak (2007: 74, 77–8) describes this phenomenon. He uses the terms 'pocket' and 'façade' gentrification and argues that '[…] it seems reasonable to assume that the inner city is witnessing the process of "pocket" or "façade" gentrification rather than the fully-fledged phenomenon leading to social and physical revitalization of whole districts [as it] takes place in Western Europe and North America' (Marcińczak 2007: 74). So, upgrading and the related displacement of resident tenants will almost certainly remain restricted to smaller areas or single investment projects. In the near future, the resettlement (or displacement) of workers residing in architecturally valuable but neglected historical building stock opposite the former Poznański textile factory that was transformed into a shopping centre by means of investment-led and state-supported renovation, is planned by the municipality in cooperation with private investors – it is assumed that the displaced residents will be assigned new flats in other districts of the city (Uchwała... 2004, Łódź E2). The municipality plans to cooperate closely with the investors. The same is planned for another complex of pre-Second World War workers' housing south of the city centre (Księży Młyn and the former Scheibler's textile factory). Both are good examples of the fact that, in post-socialist cities, often municipal weakness or *laissez-faire* politics and private engagement and primacy are loaded with conflicts of interests, particularly in areas with competing developmental needs (see also Tsenkova 2009: 303).

Figure 7.1 Nowa Street in Łódź
Source: Photo, A. Haase 2008.

In some residential streets in Łódź, for example, new middle-class, owner-occupied housing has been erected following the demolition of old dilapidated housing stock and the resettlement of the lower-class residents who rented the flats here (Figure 7.1). In Ostrava, deliberate displacement and restructuring of an inner-city high-value area (Stodolní Street) occurred in the early 2000s (see also Lux

2004: 46). Before regeneration, the area was one of the most neglected in the inner city, populated by poor people and with a high proportion of Roma. These were forced to leave before the area was renovated and revitalized as a restaurant and bar area, which nowadays is one of the centres of nightlife in Ostrava. According to local experts, nobody really knows where the displaced Roma inhabitants of Stodolní Street have gone to – they are assumed to have settled in areas where relatives live – particularly in the north of the inner city (Přívoz-Sever; see below and Figure 7.2). Spot gentrification is not as apparent in Brno, however, there are plans to revitalize centrally located inner-city areas close to the main station which today are populated by a large proportion of Roma and socially disadvantaged Czech inhabitants. In both cities, the presence of Roma is assessed by local experts as a fundamental obstacle to any gentrification. In Brno, there are also a few investor-led projects that target a gentrification-like development by selling single inner-city condominiums to more affluent owners. Local experts, however, deny the existence of 'area gentrification' and do not expect anything like it in the near future (Mulíček and Seidenglanz 2004: 37). *In-situ* upgrading by new owner-occupiers (often former tenants) is another, though under-researched, issue in all four cities (see also Chapter 6).

Figure 7.2 Stodolní Street in Ostrava
Source: Photo, S. Martinát 2007.

Studentification: Signs of the Discovery of the Inner City by Students

Students were not typical inner-city residents in any of the four cities until the collapse of state socialism. After 1989, however, their preferred place of residence started to change. Although most students still live in student dormitories, this share has decreased over recent years. As in Western European cities, students increasingly use flat sharing and subtenancy as alternative forms of housing. The reasons are, however, pragmatic and relate to comparatively lower housing costs than in the dormitories as well as proximity to the place of education (see in more detail Chapter 8).[13] A 'flat-share lifestyle' such as is typical in countries like Germany or Austria (Steinführer and Haase 2009b) has not really developed (yet) in any of the four case-study cities, although there are some examples where flat shares have become meeting points for students life (Grabkowska 2007: 141–2). Students are looking for living space in different parts of the city including the inner city, but also in the large housing estates.[14] In our four case-study cities, there are some inner-city areas which have seen a rising influx of students during the past few years and which are highly appreciated as housing locations among students, for example, Wrzeszcz Dolny in Gdańsk (Buzar and Grabkowska 2006) and Veveří in Brno.[15]

The importance of students as new residents of the inner city (outside of dormitories) is less visible in the statistics than in reality. Even among the residents of the affected areas, the phenomenon is only partly recognized, as our interviews with residents show.

The presence of students in the inner city is a strong argument for reconsidering the magnitude of inner-city depopulation with alternative means, rather than with official statistical data alone. According to our interviewees, only some of them are registered at the place of residence in the city where they study. This is true particularly for those who do not live in the student dormitories but share a flat with others. Only some of them have a legal contract with the flat or house owner. The others have just an oral agreement which is, however, not seen as a problem that makes their housing situation unstable or otherwise precarious. Whether or not a student has a legal contract for his or her room or flat does not, as a rule, correspond with whether he or she is registered (see above). In most cases, students with legal contracts are not registered either.

---

13 This information stems from interviews with Polish and Czech students in 2008 (see Chapters 8 and 9).

14 According to the information of local experts and our own observations (interviews with residents), the housing estates are – due to moderate housing costs and good transport connections – preferred places for student flat shares.

15 See also one of the online platforms where students' housing is supported by listing offers and requests for flat or room share within big university cities in Poland and the Czech Republic, for example, Gdańsk see Stancja. Ogłoszenia wynajmu. Gdańsk and for Brno the platform for student housing of the Masaryk University.

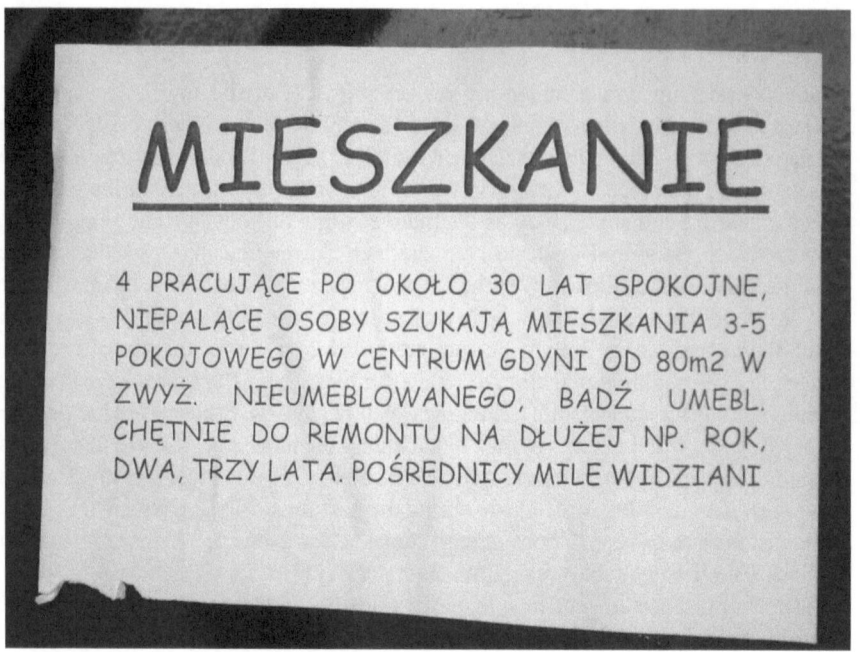

Figure 7.3 Advertisement of young professionals looking for a flat

Note: Remark: In this advertisement, four young professionals are looking for a flat in the inner city for a limited period of time (1–3 years). They are ready to accept both a furnished or unfurnished flat and to do renovation work themselves.

Source: Photo, A. Steinführer 2007.

The in-migration of students contributes to repopulation and rejuvenation but also transforms the educational structure of the inner-city population. It shapes a new image of the inner city as an attractive place to live for young people who are in a transitory phase of their life, a phase where most of them have not yet decided how to plan their future life, for example, whether to have a family or not. We can conclude that we found some evidence of a significant number of (partly) unregistered newcomers living in the inner city for a limited period. However, to speak of studentification, a process discussed in recent Western European literature mainly in relation to gentrification and inner-city upgrading (Smith and Holt 2007, Steinführer and Haase 2009b), seems to be premature for our four cities. Processes that might represent the beginning of studentification could be observed in all four cities, without being a mass phenomenon. Future research is necessary to follow the development and to assess whether studentification is becoming a new characteristic of the inner cities of Gdańsk, Łódź, Brno and Ostrava.

Ethnic Diversification: Newly-emerging Diversity

In Polish and Czech cities, in contrast to western cities, ethnic diversity and immigration played a marginal role during the era of state socialism. An exception was the Roma population in Czech cities (Navrátil et al. 2003). After 1945 most Polish cities had lost their ethnic heterogeneity as a result of the Holocaust and ethnic expulsions. This is especially true for the two Polish case-study cities, Łódź (inhabited before the Second World War by 60 per cent Poles, 31 per cent Jews and 9 per cent Germans) and Gdańsk (95 per cent Germans, 4 per cent Poles). Questions about nationality and ethnic status were not included in the census during the period of state socialism; as a result the 1988 census lacks these data. According to the census conducted in 2002, there were only 1,975 inhabitants (0.25 per cent of the population) in Łódź who declared a nationality other than Polish (Bierzyński and Węcławowicz 2008). The situation is similar in Gdańsk. Here, the census data for 2002 show the extent to which Gdańsk, in terms of the ethnic origin of its inhabitants, is almost a homogenous city: 96 per cent of Gdańsk's population declared Polish nationality, while only 0.4 per cent acknowledged non-Polish nationality. The remaining 3.5 per cent who did not declare a nationality might have the ambivalent identity of the Kashubians, that is members of the local ethnic group settled in and around the Gdańsk region (Grabkowska and Sagan 2008).

In the two Czech cities, the proportion of ethnic non-Czech population is higher than in the Polish cities – it adds up to 5 per cent in Brno and 8 per cent in Ostrava in 2001. For the inter-census period 1991–2001, different trends are visible for the two cities: while the proportion in Brno increased from 3 to 5 per cent, it decreased only very slightly in Ostrava, from 8.5 to 8 per cent. In both cities, however, the inner city exceeds the total average – in Brno, it was 5 per cent in 1991 and increased to 6 per cent in 2001. In Ostrava, it decreased from 11 per cent in 1991 to 9 per cent in 2001. The latter trend might be due to rising unemployment in Ostrava that made the city less attractive for foreign workers who had come before 1989 for jobs in industry (Klusáček et al. 2007: 60–61). In both cities, the largest groups of migrants, Ukrainians and Vietnamese, do not concentrate in the inner city (Uherek 2009).

As for the ethnic structure of the non-Czech population, the share of Slovaks and those who do not declare their ethnic affiliation is comparatively high in both cities. Both groups might relate to the Roma population, since a considerable number recently migrated from Slovakia and Roma are likely not to declare their ethnic affiliation to avoid discrimination. Only a small number of the Roma who live in Brno and Ostrava declared their ethnic affiliation in 2001 (Table 7.7). There is a lack of knowledge about how many Roma live in the Czech Republic altogether (Langhamrová and Fiala 2003). The two Czech case-study cities are no exception here. While, for example, only 658 Roma lived in Ostrava according to the last census, experts' estimations vary from 20,000 to 40,000 (see, for example, Lux 2004). The numbers of all other ethnic groups is negligible. Roma in Brno live mainly in the southern and eastern parts of the inner city (Trnitá, Zábrdovice) while in Ostrava, those of them who stayed within the inner city live mainly in its northern part, close to the main railway station (in the district Přívoz; see also Klusáček et al. 2007: 58–60).

Table 7.7 Declared nationality (národnost) of Brno and Ostrava inhabitants in 1991 and in 2001 (census data)

Declared nationality	1991				2001			
	Brno		Inner city		Brno		Inner city	
	Total	%	Total	%	Total	%	Total	%
Czech	375,892	96.8	63,704	95.3	356,473	94.8	57,327	94.2
Non-Czech of which:	12,404	3.2	3,109	4.7	19,699	5.2	3,546	5.8
Slovak	7,137	1.8	1,359	2.0	5,795	1.5	986	1.6
Roma	1,497	0.4	890	1.3	374	0.1	206	0.3
Other (including Polish, German, Ukrainian)	3,770	1.0	860	1.3	13,530	3.8	2,354	4.1
Declared nationality	1991				2001			
	Ostrava		Inner city		Ostrava		Inner city	
	Total	%	Total	%	Total	%	Total	%
Czech	299,381	91.5	23,464	89.3	292,148	92.2	21,535	91.1
Non-Czech of which:	27,990	8.5	2,799	10.7	24,596	7.8	2,105	8.9
Slovak	17,508	5.3	1,273	4.8	11,016	3.5	700	3.0
Roma	1,593	0.5	373	1.4	658	0.2	114	0.5
Other (including Polish, German, Ukrainian)	8,889	3.0	1,153	4.9	11,922	4.4	1,291	6.0

Source: Klusáček et al. 2007.

Both Poland and the Czech Republic are countries where emigration has played a role during the last two centuries, although to a much greater extent in Poland (Drbohlav 2000). Whereas in Poland the 1980s was a decade of (labour-related) out-migration (mainly to the US and Western Europe) for approximately 1 to 1.3 million people, in Czechoslovakia up until 1989, no more than 10,000 people a year left the country for work and other reasons. After the breakdown of state socialism, however, both countries have become, 'almost unnoticed', increasingly attractive for immigration – a new phenomenon for societies that were more used to coping with out-migration rather than to integrating and getting along with foreigners. Immigration has been rising in both countries during recent years, but up to now has not impacted much on the urban demographic landscape in quantitative terms (Drbohlav 2000, Kotowska 2001, Uherek 2003). Immigration has different dimensions. The most important for our purpose are labour-related, seasonal immigration (highly qualified and low-qualified), concentrated in large cities like the four case study cities (Drbohlav 2000, Okólski 2000: 157) and newly evolving processes of re-migration and Polish immigration from abroad, that is the influx of persons of Polish origin both from the US and Western Europe (for re-migration see Iglicka 2002, Biernath 2006: 206) as well as people of Polish origin who have been repatriated from Kazakhstan (Okólski 2000).

Foreign immigrants have not played a major role in Polish cities up to the present, except for Warsaw and some other large cities (Drbohlav and Čermák 1998: 100–101). For Łódź and Gdańsk, so far they are not important. In Łódź in 2002 2,006 people (0.25 per cent of the population) declared that they had come to Łódź from abroad since the last census; in Gdańsk the number was 1,845 (0.4 per cent of the population). In Brno and Ostrava, the share of immigrants has increased recently. But while in Brno it rose steadily from 1996 to 2006 (from 2 to 4.5 per cent), in Ostrava it has oscillated between 2–3 per cent up to 2007 (Klusáček et al. 2007). For both Brno and Ostrava, local experts expect an increase in foreigners in the next few years. While in Brno immigrants with higher education are most typical, in Ostrava immigrants tend to be less well-educated and look for low-skilled jobs. According to the ISEO estimations (2008),[16] approximately one quarter of all foreigners in Brno live in the inner city. In Ostrava, by contrast, the city district most populated by foreigners is Ostrava South (large housing estates), which is home to every second foreigner.

Socio-spatial Separation, Concentration and Fragmentation

In the last section, the processes of socio-demographic differentiation in the four cities were discussed. Now the resulting patterns of socio-spatial separation, concentration and fragmentation in the urban space are examined.

Separation is conceptualized here as the physical distance of groups, based on place of residence or a spatial pattern, which sorts population groups into various neighborhood contexts and shapes the living environment at the neighborhood level. It brings about physical *concentrations* of particular residential groups. In this way, both separation and concentration interact on a small scale. They describe spatial patterns that are produced by underlying economic and social processes that structure the housing market, in combination with existing characteristics such as social class, position in lifecycle, ethnic background or religion. Together, they form patterns of what is usually termed socio-spatial or residential segregation and they

---

16   See also www.mvcr.cz/clanek/statistiky-pocty-obyvatel-v-obcich.aspx?q=Y2hud W09MQ%3d%3d.

reflect social differentiation in the urban space (Massey and Denton 1998: 282, Hamnett 2001: 164, Schäfers 2001: 302). Social distance becomes manifest in spatial distance. Because they are a result of differentiation, separation and concentration lead to stronger differences (Dangschat 2000: 209, Häußermann and Siebel 2004: 140). *Fragmentation* is defined as the representation of social, demographic, ethnic differences within the urban space and results in mosaic-like or 'splintered' patterns of socio-demographic structures (see also Buzar et al. 2007). While on the level of the whole city and larger urban areas, fragmentation seemingly 'balances' the extent of residential segregation, at the district or neighbourhood level, it is a representation of distance within a smaller urban area.

There are two main reasons why we do not use the term segregation as a (static) pattern. On the one hand, we have not found large-scale patterns of segregation in any of our four cities. Socio-demographic differentiations in space occur mostly on a small scale and are fragmented. Furthermore, there are differences between the observed processes – that is why we use terms such as separation and concentration. On the other hand, we share the caution of some East Central European scholars who wonder whether socio-spatial differentiation in post-socialist cities should be approached by the concept of segregation or if one should at least speak of 'post-socialist segregation' in order to emphasize that there is a difference, compared to what we know from Western European and U.S. cities (Sýkora 2009a). Post-socialist segregation, which is a matter of controversy among East Central European scholars, refers both to patterns of differentiation that have existed from the period of state socialism and to those that have developed only within the course of the transition (ibid. and Szelenyi 1983). It includes a wide range of processes – up-market developments as well as the deterioration of areas and can be found in all parts of the cities. Meanwhile, there is much evidence for post-socialist segregation in East Central Europe, but again, with a clear focus on the capital cities (Kährik 2002, Standl and Krupickaite 2004, Ruoppila 2006, Sýkora 2009a: 1–2). For both the Polish and the Czech cases, one has to emphasize that patterns of segregation changed only slowly during the 1990s but more rapidly from the early 2000s onwards, triggered, in particular, by large-scale suburbanization.

Social Segregation and Ethnic Exclusion

Until 1989, typical 'socialist' patterns of spatial inequality could be found that partly reflected pre-socialist differences and that were partly induced by the reality of socialist housing policy (see Chapter 6 and above). In the case of Łódź, for example, an analysis based on the 1988 census data provides clear evidence that the late socialist city showed a significant level of spatial separation depending on, for instance, the educational and professional status of the population (Węcławowicz 1993).

Since the beginning of the new millennium, Polish and Czech inner cities have undergone a new wave of socio-spatial differentiation – the gap between better-off and poorer residents has widened. Changes after 2000 relate to both a deepening of existing differences and newly-evolving differentiation as a result of the transition. Both dimensions are issues in the scholarly debate. Scholars report on increasing concentrations of poverty and of socially deprived residential groups in parts of the inner city (for Brno: Burjanek 1997, for Łódź: Markowski and Stawasz 2007: 60–66). Some even speak of tendencies towards social and spatial exclusion (Jałowiecki and Łukowski 2007, Sýkora 2007, Jarosz 2008), the separation of poor and rich (Szczepański and Ślęzak-Tazbir 2007, 2008) and the growth of an urban underclass in large industrial cities (Grotowska-Leder 2002 who uses the example of Łódź), spatial polarization (Sagan 2000c: 70–71, Węcławowicz 2002b: 74–9, Marcińczak 2007: 65) or even of 'Polish favelas' (Jędrzejko 2008). These processes are identified in most cases for old, dilapidated built-up inner-city areas; old building stock is often closely related to decline, impoverishment and ageing. The inner city of Łódź has, for decades, been a case in point for this phenomenon. It was mentioned by Hamilton and Burnett (1979: 283–4); later, scholarly work described a story of ongoing decline, impoverishment and crime in the pre-Second World War housing areas (Kaczmarek 1997, Warzywoda-Kruszyńska and Grotowska-Leder 1997, Wolaniuk 1997). At the same time, several local experts saw the inner city as one of the major challenges for urban revitalization.

The discussion about socio-spatial separation and its consequences emerged soon after the systemic change (see, for example, Sýkora 1993). Apart from the debate on the concentration of poverty and exclusion, the issue of the concentration of wealth has also become a leading topic among urban scholars. According to recent studies, there are – seemingly in accordance with western patterns – more young and, partly, also more affluent people who are moving to the inner city (Parysek 2005, Sýkora 2005, Buzar and Grabkowska 2006; see also Chapter 10).

Set against this background, our case studies lead to the following question: do we observe tendencies of socio-spatial segregation and exclusion, too? Although our analysis is not exhaustive in this respect, the following points can be made. There are indications that existing socio-demographic differences in space have survived and partly deepened after 1989 and especially after the turn of the millennium. This is true for both the built environment and the social context which, apart from the social dimension of separation, also leads to a 'symbolic differentiation' of residential districts by residents and persons from outside.[17]

The cities of Brno, Łódź and Gdańsk may serve here as examples. Brno's inner city continues to be characterized by the pre-socialist dualism between better-off residential areas in the west and poorer ones in the east of the city (Steinführer 2004, Klusáček et al. 2007: 8). In the eastern parts of the inner city, there are overrepresented groups, including older people living alone, (lone-parent) Roma families with many children and the unemployed (see also Table 7.8 below). Poor quality housing stock is usually the home to poor (or excluded) people. Unattractive housing areas close to industrial areas and derilict land persisted in the transition

---

17   See the studies of Steinführer (2002, 2004b) who reports on continuity and ruptures in districts with good and bad images in Brno during the 1990s.

period. In Łódź, some old built-up areas in bad condition are inhabited mainly by lower status residents and generally associated with these groups, while generous housing stock from the 1920s and 1930s is currently becoming increasingly more attractive for wealthier residents (authors' consultations with local experts;[18] see also Bujwicka and Michalska-Żyła 2007: 277, Markowski and Stawasz 2007: 60–66). Areas of pauperization, inhabited to a great degree by residents on low incomes, are concentrated mainly in the inner city (see also the developing discourse on urban poverty and exclusion in Poland: Węcławowicz 2003: 91–117, Jędrzejko 2008). Among the local experts interviewed, many associated the old built-up areas with a concentration of poor inhabitants. This is also true for a least some parts of the inner cities of the other three case-study cities. The following assessments by local experts in Gdańsk underline this claim.

In some places, specific demographic structures (older people living alone, families with several children) overlap with bad housing conditions (Marcińczak 2007: 77: 'enclaves of poverty', Markowski and Stawasz 2007: 69, Bierzyński and Węcławowicz 2008: 38) and low social status, as well as high unemployment (Marcińczak 2007: 73–4). Here, one could also speak of a 'consolidation' or 'perpetuation' of existing socio-demographic structures during and also through the post-socialist transition. In both cases, the concentration of certain socio-demographic groups (older people living alone, lone mothers, families with many children, Roma in Brno) is typical for this kind of socio-spatial separation and is closely linked to impoverishment and exclusion (Marcińczak 2007, Szczepański and Ślęzak-Tazbir 2007: 317–18).

In Gdańsk, the hierarchy of more or less attractive residential areas is determined by the topography: the most prestigious areas are situated in the 'upper terrace' peripheries (here topography plays an important role, since almost all housing developments before 1989 were concentrated in the 'lower terrace'), so that the moraine hills represented a barrier which only disappeared with the liberalization of the real estate market and the explosion of private building activities for housing in the 1990s. However, because of the poor (public) transport links to the 'upper terrace', the 'lower terrace waterfronts' (for example the districts Zaspa and Przymorze) have recently again become more attractive and popular, even if they are dominated by prefabricated socialist housing stock. Least attractive locations are situated close to areas that lie between the water and the hilly west, that is near to and south from, the inner city. Here, concentrations of dilapidated areas inhabited by an older and low income population, such as Dolne Miasto, Letnica or Nowy Port, formed through the socialist period and were 'inherited' by post-socialist Gdańsk.[19] In the two Czech cities, it is different residential groups who either live in poor housing conditions or who have problems finding

---

18   Information from an interview with a representative from a real estate agency in Łódź inner city (Łódź E1).

19   Information from an interview with representatives from the Gdańsk Development Office in October 2007 (Gdańsk E2).

appropriate housing. While the former is true mainly for older people and low-income households and Roma, the latter applies primarily to young ('starter') households and young families (see also Lux 2004: 33–6). Areas where Roma are concentrated are sometimes called 'Bronx', which refers to the Manhattan district inhabited mainly by Afro-Americans. This is generally true for inner Brno but it was also mentioned by local experts in Ostrava when they spoke about Roma areas in the inner part of this city (for example by a representative of the housing department of the municipality: '[...] this is Ostrava's Bronx').[20]

To speak about ethnic exclusion is only relevant for the two Czech cities and there mainly with respect to the Roma population that are concentrated in particular parts of the city as a whole, as well as parts of the inner city. The exclusion of Roma is a very complex phenomenon. The most striking relationship is between the segregation of Roma because of their ethnicity and their *de facto* social exclusion as an impoverished residential group (see also Lux 2004). In other words: Many urban neighbourhoods with a population of lower socio-economic status often coincide with areas in which Roma are overrepresented (Sýkora 2009a: 10).

At a general level, the Roma population is the group most affected by social exclusion in the Czech Republic, not only in urban environments. This was true for the state-socialist period and continues to be true today. There are both enduring concentrations of Roma who underwent further deprivations after 1989 and new concentrations of poverty in Czech cities (ibid.: 11). The inner cities form one focus of the concentration of Roma (this is the case in both Brno and Ostrava); others are the 'worst' parts of large housing estates and dilapidated areas at the periphery of a particular city (this applies to Ostrava). As far as age-specific segregation is concerned, the areas where Roma concentrate are especially young areas, because of the presence of a high percentage of families with many children and the large proportion of children altogether. For both cities, statistical data show that the ageing index is the lowest and differs significantly from other locations in those areas where Roma are concentrated (Klusáček et al. 2007). At the same time, these areas show comparatively high percentages of households with five or more people, which also indicate the existence of larger families. In both cities there are particular areas where this applies, for example Bratislavská Street and the Trnitá district in Brno, as well as the area around Stodolní Street (until its renovation in the early 2000s) and the Přívoz-Sever district in Ostrava. This means that the negative image of such areas is due first to the fact that Roma live there and second that these areas are poor and dilapidated where nobody wants to live and, furthermore, supposed to be dangerous. In some cases, the segregation of Roma is even actively supported by particular activities of municipal politics (Sýkora 2009a: 11). The deliberate displacement of Roma from Stodolní Street in Ostrava (discussed earlier) might serve as an example from our case-study cities in this context.

---

20   This information stems from an interview with a representative from the Ostrava municipality (Ostrava E6); the term was also mentioned by other local experts in Ostrava.

Age-specific Differences

Age-specific concentration was a typical pattern in all four cities during the state-socialist period, when older people increasingly concentrated in the inner city and younger households in the large housing estates in the outer parts of the city. When speaking of 'age-specific concentration', we refer to the concentration of different age groups and related types of households and living arrangements, for example, the concentration of children (and family households) or of older age groups. When looking at our case-study cities, it is clear that concentration of age classes is an issue in all four of them. But the prevailing picture of differentiation has partly changed since 1989. Here we focus on the example of Łódź where, in 2002, in the inner city there was a concentration of children and adolescents (0–17) in the northern and south-eastern part of the case-study area, due to the above-average share of larger families in these areas. This demonstrates an overlap of two dimensions of concentration: the social and age-specific (Marcińczak 2007: 73–4, Bierzyński and Węcławowicz 2008: 18, 24, 38). There is an even higher separation of the post-working age group (64+); concentrations occuring in the north-eastern part of the inner city, which coincides with a concentration of one-person households – older widows – in this area (Bierzyński and Węcławowicz 2008: 18, 28, 35).[21]

Inner-city Small-scale Fragmentation

Socio-spatial differences exist on various spatial scales. After 1989, the processes of socio-spatial differentiation, described above, have led to a mosaic-like or 'splintered' structure of these differences. In all four of our case-study cities, we find evidence for this phenomenon. To put it differently: We can observe the '[…] simultaneous presence of a given social group in both socially heterogeneous as well as socially homogeneous places' (Sýkora 2009a: 3). Evidence for this small-scale fragmentation, which also contributes to exclusion, will be given here for the two Czech case-study cities Brno and Ostrava. When we move down the spatial hierarchy within the inner city and look at selected small-scale areas, we can detect considerable differences between the areas where Roma are concentrated and areas that are populated mainly by Czechs. Tables 7.8 and 7.9 show these differences – the focus is on selected indicators for the census years 1991 and 2001. In the case of Brno, the areas Bratislavská and Trnitá are areas where a high proportion of Roma live, in Ostrava it is the area Přívoz-Sever. In these areas, the ageing index shows, for the period 1991 to 2001, a process of rejuvenation, probably due to the in-migration of Roma families after 1989 and the presence of younger families with a comparatively large number of children. In contrast, some of the areas which are mainly inhabited by a Czech population (Gorkého and Konečného náměstí in

---

21   This relationship was also confirmed by local experts from the municipality (Łódź E2, E5).

Brno and Lázně, Ostrava-střed and Radnice in Ostrava; see Tables 7.8 and 7.9) underwent processes of ageing that were much more extensive than the average values for the inner city indicate. For example, the percentages of 5+ households in the areas where Roma concentrate exceeds the average value for both inner cities and those of the areas inhabited by non-Roma. In line with the above-average presence of larger households, there was also a decrease in the proportion of one-person households during the inter-census period in Bratislavská and Přívoz-Sever, which contrasts with the general trend (see above). The socio-demographic specifics of the areas are accompanied by socio-economic characteristics such as above-average percentages of inhabitants with basic education and below-average percentages of those with higher education, and especially university degrees, as well as high rates of unemployment. The same is true for the housing conditions (see also Chapter 6 of this volume), which are characterized by a high average number of persons per flat and few square metres per person, features that relate to extreme housing density and overcrowding. The data, hence, support the thesis of socio-demographic (plus socio-economic and housing quality) fragmentation within the inner city. Fragmentation in this context represents social and ethnic distance that is manifested by spatial distance at a small-scale level. Differences become clearer when a smaller spatial scale is chosen. Consequently, fragmentation reflects heterogeneity between small units of the inner city which, at a smaller scale, represents concentration and distance alike.

The Future: Increasing Separation or 'New' Heterogeneity?

When cross-referencing the evidence of post-socialist socio-spatial differentiation in the four cities under investigation, we arrive at two assumptions. Either there is a growing separation of different socio-demographic (and socio-economic) groups within the inner-city space, which is reflected by the strengthening and cementing of patterns of concentration, separation and fragmentation of particular residential groups. Or there is a 'new diversity' that is also reinforced by small-scale fragmentation that results in different socio-demographic (and socio-economic) groups living in the same area and in a way counteracts separation.

Table 7.8 Selected socio-demographic data for inner-city neighbourhoods of Brno in 1991 and 2001

Brno	Inner City		Bratislavská		Gorkého		Konečného náměstí		Trnitá	
	1991	2001	1991	2001	1991	2001	1991	2001	1991	2001
Ageing index	107.4	114.5	73.0	41.6	87.9	108.5	108.9	161.3	90.9	76.6
Census households by size										
one-person (in %)	36.5	38.7	43.8	40.5	30.3	36.3	31.9	34.7	36.5	39.6
5+-person (in %)	3.8	3.2	7.8	9.0	4.8	3.1	3.3	2.6	5.1	4.1
Education and unemployment										
Basic education (in %)	28.6	20.8	43.5	40.9	24.8	16.9	22.8	15.4	40.4	28.1
University education (in %)	14.9	17.5	6.4	6.9	19.0	21.8	20.2	23.2	5.7	8.0
Unemployed of total population (in %)	n. a.	7.2	n. a.	17.2	n. a.	5.5	n. a.	4.5	n. a.	9.1
Housing										
Average number of people per habitable room	1.2	1.2	1.4	1.5	1.2	1.1	1.2	1.1	1.3	1.3
Average number of square metres of habitable area per capita	16.7	18.8	14.6	13.6	17.8	20.6	18.7	21.0	15.4	16.4

Source: Klusáček et al. 2007 (based on ČSÚ data), authors' work.

Table 7.9 Selected socio-demographic data for inner-city neighbourhoods of Ostrava in 1991 and 2001

Ostrava	Inner City		Lázně		Ostrava-střed		Přívoz-Sever		Radnice	
	1991	2001	1991	2001	1991	2001	1991	2001	1991	2001
Ageing index	68.4	84.9	94.2	158.4	21.2	50.6	138.4	40.6	88.1	120.4
Census households by size										
one-person (in %)	35.8	39.0	31.2	32.7	28.5	35.7	49.9	43.9	26.3	33.5
5+-person (in %)	4.1	2.8	2.5	1.4	5.5	1.2	5.6	8.7	4.9	2.9
Education and unemployment										
Basic education (in %)	32.3	23.4	22.3	15.2	26.2	18.7	57.2	49.1	26.9	18.5
University education (in %)	13.3	15.4	24.4	26.9	14.5	14.3	2.7	2.8	22.6	24.5
Unemployed of total population (in %)	n. a.	9.8	n. a.	5.4	n. a.	6.8	n. a.	18.7	n. a.	7.0
Housing										
Average number of people per habitable room	1.1	1.1	0.9	0.9	1.1	1.0	1.2	1.5	1.0	0.9
Average number of square metres of habitable area per capita	15.6	18.9	20.1	22.5	n. a.	15.9	16.7	13.4	19.8	23.4

Source: Klusáček et al. 2007 (based on ČSÚ data), authors' work.

There is no single or simple answer to this dilemma. It is probably dependent on the perspective of the research and the researchers themselves. In our case, we claim that there is simultaneity of different processes of socio-spatial differentiation that leads to a *juxtaposition of a variety of developments*, namely:

- a *consolidation of existing patterns of separation* (the concentration of older people and poorer families within the inner city and the continuing concentration of Roma in parts of the inner city with a bad housing stock and reputation),
- the *emergence of new socio-spatial differentiations* (the deliberate displacement of Roma from Ostrava's Stodolní Street and of poorer households from highly regarded inner-city areas in Łódź for replacement by up-market developments),
- an *increasing heterogeneity* caused by the influx of new residential groups (first and foremost young people in education or early-stage professionals – see Chapter 9 – or better-off households that encourage small-scale gentrification, also Roma that strengthen enclaves of ethnic segregation and furthermore contribute to a rejuvenation of the inner-city population) and
- the *increasing small-scale fragmentation* (the existing differentiation of socio-demographic and socio-economic groups within the inner city in 1989 has been deepened and diversified by the in-migration of mainly younger residents and also better-off households).

The question is whether the future will bring more separation or maintenance of diversity. Critical readers might argue that the diversity found could merely represent a transitory stage of development that is always followed by clearer patterns of separation. Unfortunately, we cannot give a final answer to this question. After cross-referencing the opinions of the experts, however, it becomes obvious that most of the interviewed locals, whether experts, stakeholders and residents, see the inner city as a place that might, in the future, become increasingly attractive for younger (and better-off) inhabitants – as in Western European cities. The fact that future changes of the residential population of the inner city could be effected mainly by younger and more affluent people and could subsequently aggravate separation is widely accepted. Therefore, 'pockets' of poverty in both cities (inhabited, in Brno and Ostrava, to a large extent by Roma; in both cities, local experts speak of 'the Bronx') were identified by local experts as a problem that will either progressively disappear or should even 'be removed' from the inner city. This is also true for the two Polish cities, here with respect to poorer households. In this regard, interviews with a variety of local experts made it obvious that up-market development processes such as gentrification are seen by the municipalities and housing market experts as appropriate means to shape 'a good future' for the inner city. Revitalization projects (both municipal ones as well as *in-situ* upgrading in the course of the large-scale privatization of the housing stock in both cities), are therefore planned with the aim of bringing about not only a physical renovation but also a change of residents. Less euphemistically, one could say that, in all our four cases, gentrification is accepted and appreciated as a form of inner-city change, although it is accompanied by a resettlement as in Łódź (or displacement) of workers' families in favour of transforming their homes to high-class housing. In Brno and Ostrava, the replacement of Roma from the inner city is seen, by and large, as an improvement of the housing and living quality of this area. Apart from the 'Roma issue', gentrification is assessed as a strategy for smaller areas (streets or blocks) but not as an instrument at a larger scale. Due to a lack of public and third-party funds, the municipalities depend on the activities of private investors. Consequently, the needs of investors currently dominate the agendas of inner-city change. The displacement of the poor is therefore accepted and perceived as a 'normal' (or even 'natural') process.

From the authors' standpoint, the judgement about the fortunes of the four inner-city areas is far from unanimous. On the one hand, we observed particular residential groups moving out of and into the inner city. On the other, we observed different residential groups co-residing, for example, newcomers with long-term dwellers and different socio-demographic and socio-economic groups within the same neighbourhood. We also observed real, and were told about further planned, displacement of poorer inhabitants in favour of new up-market developments. The results could be interpreted as an 'old-new diversity'. This term includes both the heterogeneity of the residential population as we find it at the moment in all four inner cities and the fact that this heterogeneity relates to both persistent and changing patterns, inertia and new developments, continuity and change. What is more, this 'old-new diversity' might be a transitory stage or it could describe a longer-term state that could undergo further differentiation in the future. As a result, the future of the inner cities of Łódź, Gdańsk, Brno and Ostrava need not be 'either decline and ageing' or 'revitalization and gentrification', but could consist of a mosaic made up of heterogeneous groups that live more or less separated lives. Spatial separation and small-scale fragmentation do not form opposites in this mosaic; their appearance depends on different household and housing-market driven processes that might at a particular point of time strengthen separation while maybe leading, at a later stage, to a stronger fragmentation. Fragmentation may also lead to exclusion; it can refer to very small-scale units such as a street or even a block. At the moment, all four of the inner cities under analysis are 'at the edge' or, as Petsimeris (2005: 255) claimed, at the point where they are undergoing a bifurcation in their trajectories with simultaneous downward and upward changes.

Conclusion

The years after 1989 brought fundamental changes to Polish and Czech inner cities. Summarizing the most striking results of the research presented in this chapter: *first,* they show an 'old-new diversity' encompassing both remarkable change and inertia. It needs a comprehensive view to detect both changes and persistencies. Often, change is more visible and more to the fore. But the current situation of Polish and Czech inner cities cannot be understood without their persistent patterns in terms of both the built and social environment. Therefore, any change or increasing diversity has to be understood against the background of persisting structures. *Second*, inner-city restructuring 20 years after the beginning of the post-socialist transition is still characterized by the consequences of systemic change; these though are increasingly interspersed with other global or long-term trends such as demographic change or European integration. So we show that there are different dimensions of underlying dynamics which generate a complex, diversifying and fragmenting picture of the four inner cities which were the subjects of our analyses.

In order to examine the idea of 'old-new diversity' as a common characteristic of all four inner cities, Figure 7.4 and Table 7.10 provide a comparative view on the processes and results of socio-demographic and socio-spatial differentiation in the four case-study cities. The table does not say much about scope and scale; it just represents an overarching assessment drawing on all empirical evidence gathered during the three years of the research project. It is clear that our four case studies challenge the prevailing story of population decline and ageing as the predominant processes that characterize East Central European inner cities. Although population decline and ageing continue to be important processes, they are increasingly overlapping with other processes and appearances that indicate a change of the residential population. This is also true for the stories told about inner cities before the fall of state socialism – reality at that time was most probably much more a result of adaptation in place than residential mobility and thus also characterized by a certain degree of change. Today, there is not 'the one or grand new story', for example that of gentrification, there are many new stories and many of them are still in an incipient stage and hard to detect by statistical analyses (see Figure 7.4). This 'old-new diversity' is being created by prevailing structures and inertia as well as by newly-emerging structures and a variety of rearrangements. While 'old' refers to the continuing processes of population decline and ageing, processes that were predominant in pre-1989 times, 'new' relates to recent phenomena such as new in-migration, rejuvenation and a wider range of socio-spatial differentiation, although these processes often are not easy to disentangle from each other. Residential change occurs at various scales; it might be more or less visible – usually, processes of in- and out-migration draw more attention than in-place rearrangements when household compositions change in one and the same flat. That means that it does not happen only by residential mobility but also by adaptation in place. Moreover, we found various dimensions of dynamics (spatial, temporal). This makes patterns of change diverse and hard to grasp by standard methods or approaches of urban research.

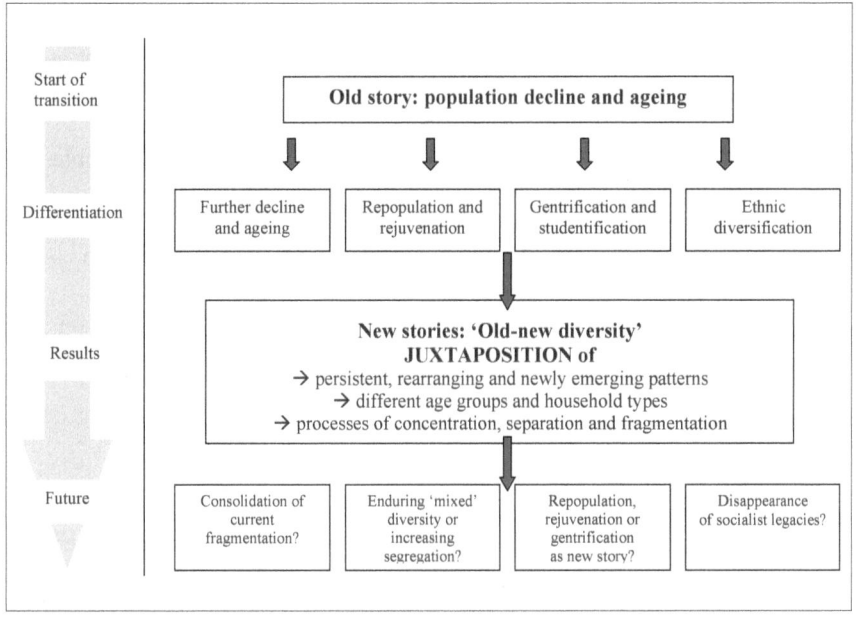

Figure 7.4 Systematization of inner-city socio-demographic change
Source: Authors' work.

Socio-demographic change in Polish and Czech inner cities is driven by diverse population groups in terms of age, professional background, income and household type. It is not only younger in-migrants who search for a transitory place to live, it is also families with dependent and adult children, as well as empty-nester couples who have chosen the inner city as the place to live. Since we cannot back these assumptions by numbers, we want to stress that we relate our conclusions explicitly to the observed phenomena and qualitative information. Seen from the viewpoint of our model of residential change, the detected in-migration leads – see Chapter 2, especially Figure 2.1 – not only to a diversification of socio-demographic patterns of the residential population, it also diversifies what residents need and want from the inner city as a place of residence. In-migration therefore not only changes the socio-demographic structure of the inner-city residential population; it also alters existing housing careers and preferences (Chapters 8 and 11). The new 'amalgam' or, less euphemistically – the 'old-new diversity', relates to the legacies of pre-socialism, socialism and post-socialism alike and is re-configured continuously. In other words: when we claimed the socialist development was more dynamic, we have to recognize the current development as also being dynamic. So, the mosaic of processes identified in our research will not be frozen but rather will lead to new patterns (Figure 7.4).

When comparing the processes of socio-demographic change and patterns of differentiation in the four case-study cities listed in Table 7.10, it becomes obvious that:

- only in Łódź and Ostrava does population decline represent a problem; in Brno and Gdańsk it is much more moderate or even debatable,
- population decline is more relevant for the four inner cities,
- the simultaneousness of ageing and rejuvenation was observed in all four inner cities,
- household changes show similar characteristics in all four inner cities; the only exception is the Roma population in Brno and Ostrava (different to the two Polish cities),
- the in-migration and presence of students represents an increasing impact factor at least in three out of the four cities (the evidence for this process in Ostrava is weak),
- there is evidence for small-scale upgrading processes or spot gentrification in all four inner cities while direct evidence of (planned) displacement was reported only for Łódź and Ostrava where low income population groups were/are affected (in the Czech case many Roma),
- the perpetuation of poverty in particular places of the inner city is a wide-spread phenomenon (there is evidence in three out of four of the inner cities) while new pockets of poverty are mainly related to places of ethnic segregation of Roma in the Czech cities Brno and Ostrava,
- processes of small-scale upgrading, perpetuation of poverty enclaves as well as the formation of new concentrations of poverty represent evidence for an increasing socio-spatial differentiation as well as processes of segregation (or separation) and
- there is a rising fragmentation, too, which makes it difficult to identify larger-scale segregation patterns.

Table 7.10 Processes of socio-demographic change and patterns of differentiation in the four case-study inner cities (1988–2009)

Process/pattern	Łódź	Gdańsk	Brno	Ostrava
(A) Processes of socio-demographic and socio-spatial change				
Population decline	++	+	+	+
Ageing	+	+	+	+
Repopulation (recent in-migration)	+	+	+	+
Rejuvenation	+	+	+***	+***
Decrease in household numbers	+	n.a.	+	+
Downsizing of households	+	n.a.	x*	+
Increase in one-person households	+	+	x*	+
Decrease in 3+ households	+	+	+	+
Decrease in core-family households**	+	+	+	+
Increase in one-parent households	+	+	+	+
Influx of students/ 'studentification'	+	+	+	x
Spot or 'façade' gentrification	+	+	+	+
Displacement (or similar processes)	+	0	x*	+*
(B) Patterns of socio-demographic/socio-spatial differentiation				
Increased diversity in household types	+	+	+	+
'Perpetuation' of poverty concentration	+	+	x*	+*
New 'pockets' of poverty/exclusion	0	0	+*	+*
New 'pockets' of wealth	+	+	+	+
Socio-spatial concentration/separation	+	+	+	+
Age-specific concentration	x	0	+*	+*
Ethnic segregation and exclusion	0	0	++	++
Socio-spatial fragmentation	+	+	+	+

Note: ++ strong evidence; + evidence; x little or ambiguous evidence; 0 no evidence/ process or phenomenon does not occur; * (mainly) related to Roma population; ** married couple with at least one dependent child; *** relates to both the presence of Roma and the influx of younger households
Source: Authors' work.

Cross-referencing the situation in the four inner cities, the following conclusions can be made:

- Although decline and ageing are still prevailing patterns, the importance of other processes such as repopulation, rejuvenation, up-market developments and concentration of poverty – both in quantitative and qualitative terms – has increased after the last censuses. These processes are different in their scope and scale; their simultaneousness, however, pushes forward socio-spatial differentiation and fragmentation.

- During the last two decades, all four inner cities have undergone processes of change of their residential population. Set against this background, they are, on the one hand, real *zones in transition* as defined at the beginning of the 20th century by the Chicago School. On the other, they are, at the same time, real *zones of inertia*, too, because of the presence of established, that is pre-transition structures and arrangements that further shape their socio-demographic picture, for example, municipal housing in the inner city inhabited predominantly by older people.

- We observe a *rising heterogeneity of the residential population in terms of household types and living arrangements* which is also reflected in housing arrangements, an issue which is expanded on in more detail in Chapters 8–10. Whether this rising heterogeneity is due rather to a differentiation of (existing) household types and living arrangements or the evolution of new ones cannot be decided at this stage and needs further research. We underline, however, that both persistent and new forms of socio-demographic differentiation are coexisting and influencing the current patterns of residential structures in our four inner cities.

- Since the temporalities of change and persistencies are different, we are, most probably, currently witnessing a *transitory period of the coexistence of different residential groups in close neighbourhood.* The rising fluidity of living and housing arrangements, residential biographies and the housing market will contribute to a dynamic change of the current coexistence of persisting, rearranging and newly-established patterns.

- Diversity is closely related to a deepening of existing or inherited as well as newly-emerging patterns of socio-spatial concentration, separation and fragmentation within the urban space. Separation in relation to exclusion becomes clearest with respect to the concentration of poverty and – in the two Czech cities – of the Roma population.

- *Repopulation and rejuvenation* have brought about new processes of spot gentrification and early-stage studentification that bring new residential groups and their preferences and lifestyles into the inner city. This, however, always gets more associated with the displacement of poorer residential groups from high-quality inner-city locations.

The dynamics of socio-demographic change is different for the 1990s, which we can assess by means of the census data and the last decade for which we have to rely on qualitative research. Qualitative research produced evidence for the assumption that the socio-demographic changes observed at an early stage of the inter-census period became stronger and more important after 2001 and 2002, respectively. However, any forecasts about future development trends are difficult to make; therefore there are question marks in the bottom boxes of Figure 7.4. The existing heterogeneity could develop in different directions which leads to new questions for further research: Will we see a consolidation of current patterns? Will the observed diversity on a small scale continue or will socio-spatial differences and segregation increase? Will the observed incipient processes of rejuvenation and repopulation become the 'grand story' of the future and will persistencies such as ageing and depopulation disappear?

Chapter 8

Households as Actors I:
Housing Careers and Housing Arrangements

Annett Steinführer, Annegret Haase, Maja Grabkowska

Introduction

The conceptual approach of this volume, as described in Chapter 2, considers private households, with their residential location decisions, mobility behaviour and dwelling adaptations, as key players in residential change. In this vein, the following chapter will highlight the role and behaviour of households in the inner-city housing markets of Polish and Czech second-order cities, with a particular focus on housing careers and the actors' housing and living arrangements. In contrast with Chapter 7, the research is based exclusively on interviews and considers the period after the turn of the millennium, that is beyond the time frame captured by the censuses in 2001 and 2002, respectively.

With this chapter, we seek to answer questions about the types of households found in the inner cities of our case studies (in qualitative terms), the circumstances which led them to their current place of residence, and the residential patterns which have evolved.

After this introduction, the results are presented in four sections. First, we shed light on the local housing markets through the eyes of our interviewees and in terms of the tension between new market opportunities and restrictions, on the one hand, and state-socialist legacies, on the other. Second, we group the housing and living arrangements of the respondents before considering, in particular, student households as new actors in East Central European housing markets. We then present two selected housing careers in more detail. Finally, the findings are summed up and conclusions drawn.

Materials and Methods

The findings presented in this chapter and in Chapter 9 draw on a set of qualitative interviews with inner-city residents and, to some extent, with local experts as well (see also Chapter 2). Sixty two interviews with inner-city residents are the major data source.[1] They were carried out by the authors in the inner cities of

---

1   In Chapter 2 we mentioned 69 interviews, but not all of them covered all aspects discussed in Chapters 8 and 9. The interviews will be denoted as 'E' for experts and 'HH' for

Łódź, Gdańsk, Brno and Ostrava in 2007 and 2008.[2] The sample includes a mix of household types that was purposively created using a snowball system. Our main focus was on non-traditional household types, such as singles, non-married couples without children or unrelated adults sharing a dwelling (approximately 50 per cent of each inner-city sample). The remaining interviews were conducted with families with children and with older, long-term residents. Table 8.1 summarizes the major socio-demographic characteristics of these interviewees. It groups the households according to the typology presented in Chapter 3 and into households living on their own or with others and having children or not. Yet, these categories capture just a certain feature of their daily life – their living and housing arrangements might vary considerably. One-person households, for example, comprise a broad variety of living arrangements, such as singles, widows/widowers or divorcees. They might be in different age groups and live on their own, have a partner elsewhere (living apart together, LAT) or share a flat with others. The actual living and housing arrangements of the interviewees will be described later on in this chapter.

Since the contacts were made via a snowball system, there are, of course, limits to such attempts, as well as imbalances in the structure of the individual samples of each inner city (for example, with regard to gender). However, as with qualitative research in general, this chapter, as well as Chapter 9, does not intend to describe 'representative' patterns of residential change but to understand, embed and explain real-world behaviour and its consequences for inner-city residential change.

The semi-structured interviews focused on the residents' housing biography, their current housing situation and everyday life, their subjective evaluations of the residential environment, housing preferences, as well as future plans. They were based on guidelines which were similar in the four cities, but adapted to local specifics and thus not completely identical.

The interviews were analyzed by employing hermeneutical interpretation and contrasting techniques. After an examination of each individual interview, a generalized and comparative analysis was carried out to identify overarching commonalities and differences.

---

households (that is residents), respectively, and supplemented by the respective interview number and the city (for example, Brno HH1). For Gdańsk, additionally, the inner-city area (Nowy Port and Wrzeszcz Dolny) is referred to by either adding 'N' and 'W', respectively (for example, Gdańsk HH N2). All the names used are pseudonyms. The quotations are cited without interruptions and 'errs'; pauses are indicated by ... and omissions by [...]. All interviews in this and the following chapter were conducted in Polish and Czech, respectively, and transcribed. The quotations were translated into English for this volume.

2   The interviews in Ostrava were conducted and analyzed by Kateřina Sidiropulu Janků (Sidiropulu Janků 2008).

Table 8.1 Socio-demographic characteristics of the interviewees per case-study city

	Łódź (PL)	Gdańsk (PL)	Brno (CZ)	Ostrava (CZ)	SUM
Number of interviews with inner-city residents	16	15	15	16	**62**
Number of interviewees*	21	22	17	18	78
- male	8	11	6	6	31
- female	13	11	11	12	47
Household types					
- one-person household (single, widow/er, divorcee, living apart together)	3	4	5	2	14
- cohabiting couple	5	3	3	2	13
- family household with dependent child/ren (one or two parents, patchwork family)	2	4	4	11	21
- family household with adult children	5	1	-	-	6
- unrelated persons living together (including singles and couples living in flat share)	1	3	3	1	8
Age groups of the interviewees					
- 20–29 years	10	11	6	2	29
- 30–39 years	4	8	9	11	32
- 40–49 years	1	2	1	3	7
- 50–59 years	3	1	1	1	6
- 60 years and older	3	-	-	1	4
Housing tenure of the interviewees					
- owner-occupiers	8	13	6	13	40
- renters (including sublease, legal or illegal)	13	9	11	5	38

Note: * Some interviews were carried out with several, sometimes even all, household members. Therefore, the number of interviews is not identical with the number of interviewees.
Source: Authors' work.

New Opportunities and Old legacies in Local Housing Markets

Flat-hunting households always face certain conditions and restrictions that are not only shaped by their socioeconomic status, types and amount of capital (economic, social and cultural) or stage in the life cycle, but also by the specific local housing market. As briefly described in Chapter 6, housing reforms in Poland and the Czech Republic went half-way down the transition path, thus creating new opportunities and restrictions by keeping the position and the status of certain market segments more or less untouched. These fragmented local housing markets influence residential preferences and location decisions alike. In this section, we will describe the framework conditions that the households face from their perspective. Thus, it is not so much the 'hard' facts about the housing markets of Łódź, Gdańsk, Brno and Ostrava that will be presented but rather the interpretations made by inner-city residents which arose out of their own process of dealing with both personal and structural opportunities and restrictions.

Despite the 'free market' rhetoric, not all segments of the housing markets in Poland and the Czech Republic are equally accessible to all consumers with a similar social status. To put it differently: denationalization, partial privatization of the old housing stock to both investors and sitting tenants, as well as the development of a large-scale post-1990 housing segment of flats, single homes and semi-detached houses, led to a somewhat chaotic pattern of tenure options that no blueprint had foreseen when the transformation of the housing sector began in the early 1990s. As already explained in Chapter 6, in both Poland and the Czech Republic there are still segments of the housing market occupied by sitting tenants who have been living there for decades and whose specific rental arrangements are protected and guaranteed by law. Although these segments have shrunk in the course of the post-socialist transition – particularly due to mass privatizations to the tenants themselves but also to private investors – they still exist to varying degrees in the individual cities and continue to affect individual housing decisions. However, the segment of private rental housing (be it legal or illegal) as well as cooperative and owner-occupied housing have become more important and have contributed to a diversification of both local housing markets and individual housing careers.

Therefore, there are persistent differences between newcomers in the housing market ('starter' households) and 'sitting' residents, that is households which already occupied a dwelling before 1990. Despite the two decades which have since elapsed, this is still a basic criterion which makes a difference because of the housing reforms of the transition period – or, to put it more precisely: the housing reforms which did not take place. Therefore, the methods of finding a flat vary between these two groups and with respect to different housing market segments.[3] Table 8.2 provides a schematic overview.

---

3   The widely used practice of staying in the parental flat and sharing it with parents and, maybe, even further family members, will be discussed below.

Table 8.2 Finding a flat – household strategies according to tenure

	'Sitting' households (occupied a dwelling already before 1990)	**'Starter' households** (came into being during the post-socialist transition and started flat hunting then)
Tenancy in municipal housing stock	• allocation • 'inheritance' • dwelling exchange • 'purchase'/bribery	• 'inheritance' • 'purchase' • allocation • legal or illegal renting (at market conditions) • renting with purchase option • dwelling exchange
Tenancy in privatized (former municipal or cooperative) housing stock*	–	• legal or illegal renting (at market conditions)
Owner-occupation	• inheritance • purchase from municipality during privatization (below market price)	• inheritance • purchase from municipality during privatization (at or below market price) • purchase from estate agency (at market conditions)
Cooperative housing	• acquisition over time • inheritance	• acquisition during privatization • inheritance

Note: * Other private rental dwellings are such a marginal segment in both countries that they are not considered here. Even for these, date of occupancy (before versus after 1990) also makes a difference.
Source: Authors' work.

Due to the sampling procedure described above, most of our interviewees are 'starter' households. This implies that when they began flat hunting, they were faced with housing markets in transition which, however, were and still are characterized by some persistent patterns of how to find a flat or to sublet it. Equally important, they usually had limited economic resources and therefore had to adjust their personal wants and desires in terms of housing because of the excess of housing demand. Inheritance, purchase and renting turned out to be major strategies in order to enter the housing markets. In the following, we will briefly characterize each flat-finding strategy from the interviewees' perspective.

Inheritance

Inheritance – legal or semi-legal – is a strategy which has been adopted in several housing market segments. It was already an important practice during state socialism and it continues to be so. Inheritance means both the legal transfer of a particular dwelling to someone else and also the *de facto* or legal right to use a certain flat. In the Czech Republic, this right *(dekret na byt)* is often transferred from grandparents to grandchildren. In Poland, while it is legal only for children to inherit the tenancy from their parents, in practice, cross-generation transfers are also quite common. This is why the term *inheritance* is also used for rental dwellings – and it is the way that the residents themselves view this practice.

> This flat used to belong to my uncle, I got it after he died. That's why I am here, actually. I did not make a conscious decision to live in this part of the city, at this place. That was just a coincidence. (Łódź HH11)

> The people who live in these buildings are mainly retired people, pensioners or young people who also take over flats from their parents or grandparents […]. (Gdańsk E7)

Inherited flats are a good option to begin with for a household entering the housing market for the first time. And it is certainly not by accident that it is usually the grandparents' dwellings which are taken over; the parents still use another flat. This cross-generation relationship with housing was also very common during state socialism. A good example of how such persistent strategies survive under new structural conditions is provided by Lenka, a 22-year-old student. She lives with her partner in a flat share in inner-city Brno but is registered at her grandmother's flat in another part of Brno, where she has no wish to live at all. However, both she and her family regard this as an option for her future housing career:

> Lenka: Actually, I am still officially living somewhere else, because I am registered as a resident at my granny's place on the Cejl [street], so we won't have any problems over there. […]
>
> Interviewer: And if I am allowed to ask a somewhat indiscreet question: if you are registered with your granny, you will have a chance to get that flat later on, won't you?

Lenka: Yes, that's exactly why daddy wanted me to register there. [...] Well, it was as simple as that, my dad made up his mind and told me to register there, so I did. (Brno HH5)

Inheritance (or 'quasi-inheritance'[4]) is thus a way by which young inner-city dwellers replace relatives in a particular flat and keep it in family ownership (or 'quasi-ownership'). Replacement also occurs through the second strategy to be described, that is purchase.

Purchase

Buying a flat is the second major option for flat-hunting households. In Poland and the Czech Republic, the large-scale privatization process of the municipal housing stock occurred less dramatically and more slowly than in other countries of Eastern Europe, which today can be characterized as '"super" owner-occupied nations' (Lowe 2003: XIII). However, housing privatization has to be regarded as a main driver of housing market dynamism, as it offers new opportunities to buy, sell or rent dwellings. Again, we will shed light particularly on the practices related to the former municipal housing stock, since this was the major type of housing we found in the inner parts of the four case-study cities.

Privatization was, in the first instance, intended for sitting tenants. Where there was a lack of interest, the flats were sold off to other households or to private companies. Many long-term (and, thus, older) residents bought their dwellings – not always for themselves, often for the next generation or the one after. This phenomenon is also observed by local experts, as the following quote from Gdańsk shows.

[...] because for Gdańsk it's a widely known fact that a great number of students move into the city, so their parents very often have to make a choice, if their child did not get a room in a students' hostel, they very often bought a flat if they could, possibly a two-room flat, one-room if they could not, for their children to live in while they are studying. They assumed that after the five years of studies and so, they could easily sell the flat, not knowing if their child would stay in Gdańsk or not. (Gdańsk E8)

That flat was bought by my parents and by my grandfather, when I was a teenager. (Łódź HH15)

In both the Polish and the Czech cities under investigation, there are several cases where not all dwellings in a building were sold to sitting tenants, particularly

---

4   This term echoes the notion of 'quasi-property' in housing markets under socialism and refers to the strong emotional and material ties to rental flats by their long-term dwellers (Šmídová 1996 and Chapter 12).

if the tenants were very old, as a result of their lack of economic capital or the absence of relatives interested in purchasing the flat. Sometimes, these remaining tenants continue to occupy their flats, enjoying legal protection and paying so-called regulated, that is non-market rents (see also Chapters 6 and 12). Yet, they are always threatened with displacement or being 'sold out'. One interviewee in Brno describes the logic of such a process. In this case, the building as a whole had been sold off to a private investor, due to the lack of interest by the sitting tenants in buying their dwellings.

> And there is that guy who has enough money to buy this house from the municipality, including the tenants so to speak, who live here on lease. And afterwards he starts evicting them, somehow, and those who manage can move out, so he will have those flats at his disposal, and then he offers them to others at market price. [...] That's how we met him, too. [...] There are people here who expect some sort of compensation, if you ask me, or something. They are in that sort of situation, they signed an agreement that they will move out [...]. (Brno HH3)

Indeed, because of these fragmented tenure patterns, even in one single building, a number of conflicts can arise, for example, with respect to reconstruction efforts, the entry of new owner-occupiers, or different feelings of responsibility for the property bought or rented. But such conflicts between different groups of residents do not only run along the lines of 'sitting' versus 'starter' households or long-term protected tenants versus newcomer owner-occupiers. There is also a difference between long-term residents who were invited to buy their rental flats at subsidized prices, on the one hand, and new residents who bought their flat at market price from a real estate agency or the municipality, on the other. The following quotation illustrates the views of the latter type of resident. The interviewee claims that the new owner-occupiers and former tenants still need to learn to be owners.

> That was a municipal flat, and maintenance was provided by the municipality or the real estate firm that was working on behalf of the municipality. That's why tenants did not have to take care of anything. So they never learned to take care. On the other hand, I used to live [...] in a cooperative flat. We had to keep order there, had to do different sorts of maintenance work ourselves. [...] But these people from the municipal flat did not have to do anything, and the big problem now is, let's put it this way, even if they have got a little used to it by now, you still have to convince them that nothing is going to happen on its own, that it's their property now [...]. The municipality sold those flats to them, by and by, so that's how they came to be their property [...]. (Ostrava HH4)

As discussed in Chapter 6, the privatization of the municipal housing stock is not yet complete in any of the four cities. Thus, housing market dynamism in this specific segment will certainly continue to be a characteristic of second-order cities in Poland and the Czech Republic.

Renting

Renting relates to both public (municipal) and privately owned dwellings. Finding a municipal rental flat used to be a constant problem during the transition period. At first this was a result of the inherited housing shortage and the sharp decline in new public investments as well as to the factual 'closure' of the municipal (formerly: state-owned) sector, which privileged sitting tenants and their relatives, for instance via dwelling exchange or the opportunity to inherit the right to use the flat. Later on, as a consequence of the privatization process, the availability of municipal flats declined so a shortage of rental housing in that segment developed. Waiting lists for municipal flats – a legacy from state socialism and one of the main options for acquiring a flat before 1990, as our interviewees from 'sitting' households also confirmed – still exist in all four of our cities. Yet, none of our 'starter' households relied on this option, which always means long periods of waiting.

However, as was the case during state socialism, there are other ways to enter the public rental sector, by, for example, buying the right to use the flat from a sitting tenant (be it a family member or not) or reconstruction at one's own expense. Thus, all sorts of capital – economic, cultural and social – are invested in order to find the first starter flat, which then makes any subsequent steps in the housing market possible.

> So we were all looking [for a flat] like mad, H. [boyfriend] was also searching the internet for days on end and waiting for something to show up, and he was phoning everywhere and then we came here, and since we were the first we were lucky, so we were the ones to get the flat. Otherwise there would have been so many interested in this flat. (Brno HH5)

> I was really keen on finding a flat and there were many such flats on offer in downtown Brno. So we would get something, or we would get a rental agreement for one of those flats, on condition we pay for the renovation. [...] Also, there were really many of them, there were so many of those flats here in the city centre. [...] I know that many of my friends or people of my age also came here because that was their only real chance to get a flat. (Brno HH11)

A couple of interviewees also entered the rental segment (in one of the ways listed above) in order to become owner-occupiers in the short or long run. They engaged, for example, a real estate agent to provide them with a dwelling that was already approved by the municipality for privatization.

Yet, while all the aforementioned strategies refer to more or less legal options, most rental arrangements we found were at least in some sense illegal. Oral contracts only and rents paid in cash were reported fairly often.

> Interviewer 1:   Okay, so who of you is actually renting the flat, what about the rental agreement?
>
> Interviewee 1:   We have an agreement among ourselves. Just an oral agreement. It has been like that for two years now, and everything is fine.
>
> Interviewee 2:   Mrs I. [official tenant] trusted us and handed us the keys after our first instalment. We liked the flat, and she put so much faith in us that we did not have to have an agreement.
>
> Interviewer 2:   And who of you is registered here as a temporary resident?
>
> Interviewee 1:   No, that's all among ourselves only. Just as an agreement among private people.
>
> Interviewer 1:   So everyone is registered at their parental homes?
>
> Interviewee 1–3:   Yes, at our parental homes. (Łódź HH13)

This illegality also implies uncertainty and a rather precarious position for the tenants (and, to some extent, also for the landlord). But illegal renting-out is a wide-spread, well-known and thus quasi-legitimate strategy in all of our case study cities. Practically all our interviewees know about illegal tenancy – some are actors in this game themselves, all the others have at least heard about it. Talking about such a practice in her residential building, one interviewee in Brno states:

> [...] that none of us is much interested in that, because everybody would sublet this flat even if it is state property.[5] And you can see this at the beginning of every university term when the students stick their names on the nameplates and letterboxes, and a fortnight later, the landlord will remove their names and tell them they just cannot do that, because if anybody should ask they should say they don't live there, they are only watering the flowers or something like that. (Brno HH12)

Our research showed, moreover, that there are different attitudes towards inherited, purchased and rented space. Although the flats found are often regarded as temporary dwellings, the efforts to adapt the place to personal wants and needs range from doing just the absolutely essential things to creating a home (see also Chapter 10). The variety of tenure forms also reflects the different financial backgrounds of young urban dwellers: often, their dependence or independence

---

5   This is a remarkable linguistic example of how strong the institutional legacies from state socialism still are in the Czech housing market. Although the flats were already transferred in 1991 from state to municipal ownership, some interviewees today still speak of 'state' flats (similarly, Steinführer 2004a: 86). However, among our interviewees, this term was only used by residents from Brno (and not in Ostrava).

from their parents' funding is decisive, as well as whether they have any (regular) disposable income or not (see also Grabkowska 2007: 142–3). In a couple of cases, the new flat was taken over through 'non-market activities' (Buzar and Grabkowska 2006: 172): either the young in-migrant moved into or remained in the flat of relatives, or he made an informal agreement on rental payment with the owner of the flat.

In all four inner cities under investigation, we thus find evidence for a new housing market dynamism in which inherited residential patterns overlap with new practices in a complex way. Residential change is brought about not least by the large-scale privatization of the municipal housing stock to various investors (sitting tenants, 'starter' households, but also private investors who then sell single dwellings to private households). As a rule, once these dwellings and buildings are in private hands, reconstruction efforts emerge which are partly subsidized by the municipalities (see the example in Chapter 12). These efforts, in turn, change the physical appearance of parts of the inner city.

Suburbanization, the major urban transformation process in all of the four cities, is an additional trigger of this 'silent' repopulation of the inner cities. In several cases, our interviewees mentioned that their landlord had left the inner city for suburbia but wanted to keep the inner-city flat for a variety of reasons (such as additional and tax-free income, or children, for whom this dwelling was to be kept in family ownership or quasi-ownership).

> When I started studying, at that time all students were living in the housing estates on the outskirts [...] now I think they are moving downtown. Well, that's because quite a few people who used to live downtown are building or have built homes elsewhere, they are now moving out of their downtown flats. (Brno HH1)

Although the newcomers are not evident in the official statistics (Chapter 7), the residents themselves observe this new phenomenon and reflect upon it. They describe both the inner cities and their houses as characterized by a juxtaposition of 'sitting' and 'starter' households, of old and new residents (in terms of age), of privileged long-term tenants with comparatively low rents and newcomers paying market rents. Based on such observations, the interviewed experts also draw implications for the future of the inner cities.

> In the old district of Nowy Port, that's where the elderly live mainly. Maybe that will change in ten or twenty years, because as far as we can see, those elderly people, some of whom actually own those flats, sign them over to the young, in most cases to their grandchildren. (Gdańsk E5)

Whether newcomers or sitting tenants, short- or long-term dwellers, the inner cities of Łódź, Gdańsk, Brno and Ostrava are characterized by a great variety of housing and living arrangements. These will be described more systematically in the following section.

Diversifying Housing and Living Arrangements

Since the selective sample aimed, in the first instance, at including non-traditional household members, we met a number of newcomers to the inner city who had been living there for a relatively short period of time and were usually not registered there. These were mainly younger people in education but also early-stage professionals. Their household types ranged from singles and cohabiting couples to flat sharers and even more complicated, multi-household arrangements (for example, a flat owned and inhabited by a single as well as by a cohabiting student couple as unregistered subtenants). This complexity is not detectable in either of the data sources usually referred to – even if there is some information about 'non-family, multi-person households', for example, the census data for Łódź, shows certain concentrations of such households within the inner city (Bierzyński and Węcławowicz 2008: 32, 40). From such data sources, however, it is impossible to identify clearly arrangements such as flat shares of students or young professionals, since these non-family households could also be established by unrelated older (and probably impoverished) persons sharing a flat for economic reasons, let alone the issue of the subjective meaning attributed to such arrangements by the residents themselves.

Table 8.3 lists all the household types, living and housing arrangements that we found during our interviews with inner-city residents in Łódź, Gdańsk, Brno and Ostrava. Starting from the original table introduced in Chapter 3, where our general conceptualization of household types, living and housing arrangements can be found, it shows the variety of household types in the samples of the four case-study cities. It has to be pointed out that research on households always faces the problem of inevitable differences between external (researchers') perspectives and the self-perception of the households themselves. This is, for example, true when a person from Brno in a LAT (living apart together) relationship is classified as a one-person household but perceives herself as living as a couple in two flats, or when a flat share in Łódź consists of a cohabiting couple and a single who jointly perceive themselves as pragmatic co-residents, or when a flat in Ostrava is inhabited by a single man who owns it and a cohabiting couple who are subtenants. Table 8.3 takes the researchers' viewpoint but, in its differentiation of households, living and housing arrangements, also tries to do justice to subjective perspectives.

Table 8.3 Household types, living and housing arrangements in the case studies*

Basic household type	Living arrangements	Housing arrangements
1. One-person household	Single	Living alone Flat share
	Widow/er	Living alone
	Divorcee	Living alone
	Living apart together (two persons of opposite sex with intimate relationship), not married	Two persons, each living separately
2. Cohabiting couple**	Married/non-married couple of opposite sex, without children	One common residence Flat/house share *Subtenancy*
	Married empty-nester couple	One common residence
3. Family household with dependent child/ren	Married/non-married couple of opposite sex with at least one child	One common residence
	Patchwork family/step family with at least one child	One common residence
	One parent with at least one child	One common residence
	Divorcee with at least one child	With former partner: two separate dwellings (with child/ren who commute)
4. Other types of family household	Married/non-married couple with at least one adult child	One common residence
	Three-generation household	One common residence
	Two-generation household of grandparent(s) and grandchild/ren	One common residence
5. Unrelated persons living together	Single	Flat share
	Non-married couple	Flat share Subtenancy
	Living apart together (two persons of opposite sex with intimate relationship), not married	(Separate) flat share (Separate) subtenancy

Note: * Based on the general typology introduced in Chapter 3 (Table 3.1); non-italics: evidence from Łódź, Gdańsk, Brno and Ostrava; in italics: not in the general table. ** In Poland, for both married and non-married couples, the expression 'partner relationship' is used. Whereas married couples are also called marriages, non-married couples are termed 'consensual relationships' or 'cohabitation' (Okólski 2006).
Source: Authors' work.

From Table 8.3, the following conclusions can be drawn:

- It clearly shows that there is a large variety of types for each category. In this vein, we can justifiably speak about a diversity of household types, living and housing arrangements in all of the four case-study cities.
- It is impossible to equate household type, housing and living arrangements. Knowing one does not yield clearly drawn implications for one of the other categories. As a rule, one household type can live in a variety of living and housing arrangements. This is especially true for one-person households and flat sharers. When we look at one-person households, we find them in four different living arrangements – as singles, divorced, widowed persons or as LAT – as well as in different housing arrangements. Living alone occurs in one flat, in a flat share or even in two flats (in the case of LAT).
- Not only non-traditional household types, but also family households are subject to diversification. Among our interviewees, we found a number of other living arrangements in two-generation relationships between parents and children (irrespective of age). Apart from core families consisting of a married couple with at least one dependent child, we also found non-married couples with one or more dependent children, patchwork families with at least one dependent child, married and non-married couples with adult children as well as single parents and three-generation family households. This diversity of modern families today also indicates that there is an increasing acceptance of family household arrangements beyond the pattern of a married couple with one child or several children.
- Not only social and demographic changes contribute to this diversity. There are also persistent household types, living and housing arrangements that have endured since the period of state socialism. Among them are, for example, families with adult children, three-generation arrangements or 'cross-generation' households, that is grandparents with grandchildren. Yet, in spite of their similar form, their subjective meaning today might be different. Taking the example of the cross-generation household, it was already pointed out above that besides a close grandparent-grandchild relationship, such an arrangement can also be grounded on rational deliberations about future tenure options. Flat sharing of parents and adult children is another typical household type which is caused primarily by a lack of low-price housing and the fact that many young people remain in their parents' flat until they have finished their education. In contrast to many Western cities, subtenancy also plays a role as a housing arrangement, including, for example, cohabiting couples that share a flat with other households. Flat-share arrangements do not only include the younger type mentioned earlier, but also cases where different one- and multi-person households were forced to share a flat in the past and still live in such arrangements. This applies mainly to middle-aged and older age groups, mainly (widowed) one-person households and empty-nester couples.

• Some living and housing arrangements listed in the original Table 3.1 were not found among our interviewees (for example, same-sex unions). This certainly has to do with the selective sampling. However, we can also question the somewhat 'metropolitan' bias in the original table as well as the typology's appropriateness with regard to second-order cities and to national contexts where such unions are both less common and not widely accepted, as, for example, in Poland.

Table 8.3 does not say anything about the stability or permanence of the listed types and arrangements. But according to our empirical research, it is also the limited stability of many living and related housing arrangements that contributes to the diversity described. Thus, temporality, as a factor influencing housing biographies and living arrangements, also has to be considered.

Therefore, also with respect to housing and living arrangements, we can speak of an 'old-new diversity': some arrangements existed during state socialism, others have emerged (in a more than negligible quantity) only during the last few years (for example, cohabiting couples and flat shares). 'Old-new' diversity, moreover, refers to the coexistence of long-term housing arrangements and transitory units. It has to be emphasized that both persistent and new forms of socio-demographic differentiation coexist and influence the current patterns of residential structures in our four inner cities.

Throughout our research, one specific – and in Polish and Czech cities new – group of actors bringing about inner-city residential change emerged, namely students. The following section looks specifically at this group.

Student Households as New Actors in Local Housing Markets

As already discussed in Chapter 7, there are signs of a 'studentification' in the four investigated cities. While this process is not at all surprising from a 'western' perspective, it is indeed a new phenomenon in East Central Europe. During state socialism and well beyond it, students used to live in halls of residence (provided that they were allocated a place there), remained in their parental home (and sometimes had to commute over considerable distances) or lived as subtenants in flats of relatives or private landlords across the cities and their hinterland. In both Poland and the Czech Republic, however, the costs for the halls of residence increased significantly during the transition period. Moreover, in the Czech Republic, the state-financed housing allowance for students was reformed in 2005: as a result students now receive direct subsidies for all types of housing, whereas previously, housing allowances were paid by the state to the halls of residence, rather than on an individual basis. This change in the legal framework can be considered as an important milestone for students becoming real actors in local housing markets. What is more, the number of students has grown considerably in all four cities in the past ten years or so. In Brno, for example, the number of students at Masaryk University alone, which is the largest institution of higher education in the city, increased threefold between 1996 and 2007 (from 13,000 to 38,000; data provided by Masaryk University administration as of December 2007). In Gdańsk, the number of students grew from some 42,000 in 1998 to almost 79,000 in 2008 (USG 1999 and 2009).

From our interviews, we know that students look for private rental flats in different parts of the cities under investigation and in different kinds of housing stock. They are willing to accept a variety of options. For students coming from places outside their university town, a typical progression is to start in a hall of residence and then to move to flats in different parts of the city. The inner city is regarded as advantageous, because of its location and, moreover, as no more expensive than the communal housing in the dormitories.

> Interviewer: Do many students rent private lodgings these days?
> Student: Yes, I've noticed that more and more of them do because the prices are becoming equalized, and, you know, the convenience of living in a private lodging is far greater than in a hall of residence. It is more comfortable, so the students decide to rent, especially as the cost of living in a hall of residence keeps going up […]. (Gdańsk HH N5)

> It's the freshmen who first of all live in the students' hostels because they don't yet know where to go and what to do. And they choose to live there because they like to socialize, because there they can be together. On the other hand, I know that they mostly take these flats, well, because they can afford them much more easily. (Brno HH11)

At the same time, the landlords (and 'quasi-landlords') are also aware that students provide them with the opportunity to improve their income. 'What I think, yes, if you ask me, a lot of people had that idea, there are a lot of students here, and they would like to make some money off them' (Brno HH5).

Studentification leads not only to overall residential change in the inner city, by transforming its socio-demographic and socio-economic structure, its amenities, meeting places, shops and pubs. It is also to be found on the micro-scale of the individual household. The typical arrangement of a student flat share in our cities is two persons per room. In a number of cases, these are couples sharing a flat with others, while sometimes it is more complicated, with one couple being the main tenants and other persons being their subtenants. But also unrelated others share one room located in a larger flat where quite a few others may live. Some Czech interviewees deliberately refer to their housing arrangement as 'student households' (for example, Brno HH1, HH2). What is more, they relate it to the notion of 'typical' or even 'traditional', thus to the idea of a well established type of housing arrangement.

> If there is more than one family unit or something, don't know what to call that [...]. That's what you would call, well, flat sharing, a typical students' flat. Such a students' household. (Brno HH2)

> I am here because of my studies, because I took my degree here and I am working here, and I have been living in different types of flats. At first, me and a friend were sharing a flat, as many do while they are studying [...]. (Łódź HH11)

> Well, I think that here in Brno among students this is rather traditional, but it is a bit different as we are not that crowded like most other students, also due to the fact that this flat is rather cheap in comparison with others. So it's rather comfortable and we two [she and her partner] simply see it as home. (Brno HH5)

These student arrangements are thus considered 'normal' – although historically, they are indeed a new phenomenon appearing in the second decade of the post-socialist transition. At first glance, the formal structure of these new housing arrangements (usually shared rooms) very much resembles the long-term tradition of student housing in halls of residence (where single rooms did not exist). But the students underline the freedom from control and latitude in choosing how and with whom to live, as well as the proximity to the city centre and the university, which they consider as major advantages in comparison to most student dormitories. The transitory character of their living and housing arrangements was of no concern to the interviewees; they looked at it pragmatically and through the lens of the current priorities in their professional and private lives. How transitory such arrangements might be and what this new mobility means for housing biographies will now be shown by the example of one Polish and one Czech interviewee.

Housing Careers in Transformation: Two Examples

From a Western European perspective, young in-migration to inner-city districts also leads to an increase in non-traditional living and housing arrangements and, at the same time, to increasing housing mobility and numbers of short-term housing arrangements and, therefore, a large turnover rate. Among our interviewees we also identified a small segment of highly mobile inner-city dwellers that do not fit into the prevailing pattern of housing biographies in East Central Europe. A Czech survey in 2001 found that almost half the respondents (43 per cent) had changed their housing location just once between 1960 and 2001, while 29 per cent had never moved to another home or town. Only 11 per cent of the sample had experienced a housing relocation three or more times so far in their lifetime (Sunega et al. 2002: 26). Due to the chronic housing shortage during state socialism, most young people had to remain in their parents' flats until marriage and sometimes even after they were married and parents themselves. Housing occupancy was therefore very stable, often amounting to decades of non-mobility. However, this does not mean that household compositions did not change several times, or that the way the home was used was not adapted time and again (Steinführer 2004b: 267–96). The following two examples are therefore very different to these traditional housing careers.

Petr is about 30 years old and has already had quite a few different residential locations in inner-city Brno. His brief but very mobile housing career resembles in its basic pattern those of other young interviewees: originally from a small town in Bohemia, he came to Brno to study during the 1990s. He first lived in halls of residence, which he changed twice.

> Quite a while ago, when I was still a freshman, I was living in students' hostels in different parts of Brno, maybe for a year and a half. [...] I moved out of the hostels because I didn't want to live there anymore, because I didn't find living there exactly fun. [...] Next time I got to live in a students' hostel, three of us were sharing one room, and I didn't know the others. [...] Afterwards, I somehow got to live in the Mánesky [another students' hall], there we had to share the showers and loos, that was not exactly to my liking either [...] and at that time, I had already got to know some people because I had been there some time, so I knew someone to move out and share a flat with [...]. (Brno HH3)

Leaving the halls of residence was, in Petr's case, not linked to the desire to live on his own but rather to his wish to continue to live communally, but in a qualitatively different way. Figure 8.1 shows that, after having left the halls of residence, Petr's housing behaviour turned out to be no less mobile. Before moving to his current home – an owner-occupied dwelling in the eastern parts of the inner city, which he shares with his girlfriend – he lived in six more locations, most of them in the outer parts of inner-city Brno.

All his accommodation before the current one were rented and generally in very poor physical condition ('the water supply was often interrupted, and the electricity did not exactly work well either'), with a large number of temporary inhabitants and a considerable residential turnover. Most of these arrangements were based on contracts (there was always 'simply a signed sheet of paper'), but it is not clear whether or not these were legal arrangements – however, from the interviewee's perspective, this did not matter at all.

Not only Petr's housing location but also his household composition was subject to change several times, partly because of his changing living arrangements. With his former partner, he lived in two different housing arrangements: once as a cohabiting couple on their own and once in a flat share.

---

1. Small town (Bohemia): *parental home*
2. Brno Komárov (peri-rural district at the urban edge): *shared rental room in halls of residence*
3. Brno Královo Pole (outer part of the inner city): *shared rental room in halls of residence*
4. Brno Stránice (between inner city and urban edge): *shared rental room in halls of residence*
5. Brno Zábrdovice (inner city): *shared rental room, 10–12 persons in derelict shelter*
6. Brno Střed (inner city): *4–5 persons, 2 bedrooms, renting*
7. Brno Štýřice (outer part of the inner city): *cohabiting couple (2 persons), 2 bedrooms, renting*
8. Brno Židenice (outer part of the inner city): *3 cohabiting couples, 3 bedrooms, renting*
9. Brno Královo Pole (outer part of the inner city): *4 unrelated persons, 2 bedrooms, renting*
10. Brno Zábrdovice (inner city): *subtenancy in friends' dwelling (3 persons, 3 bedrooms)*
11. Brno Zábrdovice (inner city): *cohabiting couple (2 persons), 2 bedrooms, owner-occupation*

Figure 8.1 Housing career of Petr (about 30 years old) from Brno
Source: Semi-structured interview in December 2007.

At the time of the interview, Petr had been cohabitating for three months in what had previously been a municipal flat which he had bought on the free market. Price and location were major factors that brought them there – as well as the expectation that the area was revitalizing which would impact on the economic value of their dwelling and improve its future price. From Petr's perspective, the privatization of the municipal housing stock is a major trigger for this process.

> All the flats here are being sold and all the people are moving away, so this district will just look completely different in ten years' time. I still believe that this [neighbourhood] will just improve. [...] Because they are turning everything into private property, Brno, the City of Brno is selling all these houses, and, let's put it this way, speculators will snatch them all up, and they will then sell everything to people who want to live in a flat of their own and who will be willing to invest money in their property. (Brno HH3)

While Petr and his partner are themselves actors in this process, they consider their current flat as transitory ('this flat is just for the time being'). Nevertheless, they invested money in buying and in refurbishing the flat – regardless of the fact that their housing careers are certainly far from being 'finished'.

Natalia, from Gdańsk Nowy Port, represents a similar example. At the time of our research, she was 31 years old. Natalia is a Gdańsk native and had moved seven times before the time of the interview, mostly within the city but also to the hinterland and back. At the moment, she lives in a patchwork family with her second husband who brought his child from a former relationship into the marriage. Natalia lived – apart from her parents' residence – in the home of relatives and, then, in her own home during her first marriage at a place outside Gdańsk. Afterwards she returned to her parents' home and then moved again to a separate flat when she entered the second relationship, first as cohabitation and then within a marriage and a patchwork family (Figure 8.2).

---

1. Gdańsk: *parents' home*
2. House in the countryside outside Gdańsk: *parents' home*
3. Gdańsk Suchanino (large housing estate): *relatives' home*
4. Gdańsk Morena (large housing estate): *parents' home*
5. Pruszcz Gdański (small town neighbouring Gdańsk): 3 bedrooms, 2 persons (*marriage 1*)
6. Gdańsk Morena (large housing estate): *parents' home*
7. Gdańsk Przymorze (large housing estate): 2 persons (*cohabitation/marriage 2*)
8. Gdańsk Nowy Port (inner city): 2 bedrooms, 3 persons (*patchwork family*)

Figure 8.2 Housing career of 31-year-old Natalia from Gdańsk
Source: Semi-structured interview in September 2007.

Asked about the reasons for her numerous relocations, Natalia simply replied that they were related to various changes in her personal life. Having lived for years either in large housing estates in Gdańsk or in suburban locations, she deliberately looked for a flat in an old tenement house (*kamienica*). Her first choice had been Wrzeszcz Górny, a prestigious neighbourhood in inner-city Gdańsk, but this was impossible because of the housing costs there.

> We deliberately looked for an old building and for a flat on the ground floor. And, furthermore, the size of the flat was important for us, mainly in relation to the costs. I do not hesitate to confess that we wanted something in the old part of Wrzeszcz, but this area has such high prices that, in order to be able to live around the Jaskowa Dolina street, one would have to win the lottery and even that would be not enough [...]. (Gdańsk HH N1)

Natalia claims to be very content with her current dwelling, and she does not intend to move out in the immediate future. Her high level of satisfaction with the flat is a result of the fact that it fulfils several of her household's needs: it is located in a pre-Second World War building which is in line with her preferences. Its size and the additional working space in the basement allow her to work as a furniture restorer, which is both her job and leisure activity combined. Its layout provides space and intimacy for all the three members of her patchwork family, as well as their pets. The flat's good accessibility eases the organization and coordination of the daily routines of the different household members. And after Natalia's stepson moves out (he is planning to study in another city), the couple plan to rearrange the flat: 'When the young one leaves, then we will clear out his room and move our bedroom there, while the current bedroom will be transformed into an office and occasional guest room' (Gdańsk HH N1).

In the conceptual section of this volume, we claimed that residential change occurs not only through housing mobility but also because of *in-situ* change. The housing biographies of Petr and Natalia indeed provide evidence for both options and how they are combined in one and the same household. While Petr has so far solved changing housing and household needs through mobility, Natalia's relocations were conditioned by her changing household conditions – but once in a place where she feels at home, she prefers to adapt her dwelling rather than changing it again, at least for the time being.

We do not want to claim that these post-socialist housing biographies – characterized by high mobility and a fluidity of living and housing arrangements – are 'representative' for newcomers to the inner city, let alone inner-city residents in general (for further examples see Buzar and Grabkowska 2006, Steinführer et al. 2010). Such a hypothesis cannot be proven on the basis of qualitative research. Even among our interviewees the examples of Petr and Natalia are rather extreme with regard to their housing mobility. Nevertheless, they provide evidence that the housing careers of young urban dwellers in East Central Europe today are characterized by a relatively long transitory phase between leaving the parental home and some sort of settling down. This is not to suggest that founding a family or sharing a flat when cohabiting necessarily implies the end of a housing career (or even means there will be no further changes of partner). But the housing biographies of most of our interviewees are in sharp contrast to those of their parents. Our research seems to suggest that it is becoming both more common in Polish and Czech cities and – which is important to emphasize – also accepted to go through different stages of transitory living and housing. Yet, it also needs to be pointed out that transitory dwellings and flexible housing arrangements are not only the result of voluntary choice. Rather, transitory rental arrangements in the inner city are a result of the high housing costs in the private rental market and the owner-occupied sector which allow only a limited access to housing for newcomers. The inner-city in-migrants adapt to this situation by not only sharing flats, but also even single rooms. In addition, informal arrangements without legal contracts, which are characterized by great uncertainty and a lack of reliability (in judicial terms), are frequent – although our interviewees do not see this as a major problem.

There is, therefore, a strong interplay of housing and living arrangements on the one hand, and the related housing biographies, on the other. Most of our interviewees are newcomers to the inner city and often, though not always, have deliberately chosen housing there (see also Chapter 9). These households have the potential to stabilize the residential function of the inner city structurally, since they create new residential patterns of inner-city housing and living which might attract other people to live there. However, diversifying housing biographies are by no means restricted to younger age groups. Among our interviewees, we also found middle-aged households, mainly families with dependent children, or early empty-nesters, who have deliberately moved to the inner city in the rental or owner-occupied sectors and who intend to remain there. These households have, in most cases, a medium or higher level of disposable income and represent higher educational groups.

In summary, residential differentiation not only occurs in relation to the behaviour of particular residential groups, living and housing arrangements but generally in the interaction of individual life courses of various household types and their members ('agency') and the given housing market conditions ('structure'), leading to a variety of post-socialist housing careers and inner-city residential patterns.

Summary and Concluding Remarks

Inner-city housing markets in post-socialist cities represent structures of opportunities and restrictions. In this chapter, we have paid particular attention to newcomer households that came into being after the transformation of the housing sector had been well underway. Consequently, these households were faced with legacies from the old system and their complex overlapping with newly developed structures. We found a close interplay between households as actors and the urban environment. With their housing choice, households react not only to given settings, such as the housing market, but also to the conditions of their own professional and private lives and to their alterations. They make decisions based on preferences and constraints. In doing so, they shape housing arrangements and biographies that impact back on the urban space and inner-city residential patterns. Thus, we come back to the logics of the two sides of residential change and the close interlinkage between residents and their residence, as discussed in Chapter 2. Nevertheless, it would be short-sighted just to claim that the evolving post-socialist residential patterns simply coexist with persisting quasi-socialist ones. Such an interpretation loses sight of the far-reaching transformation, particularly of property structures, during the transition. Although dwellings in Polish and Czech inner cities today are still subject to inheritance (and 'quasi-inheritance'), to exchange and illegal renting out, these strategies exist within new societal frameworks. To put it differently: capitalizing on flats today means something different than it did before 1989, because of the much greater importance attached to private property.

Our research has focused particularly on a certain socio-demographic group: younger households in a transitory life phase which, as housing mobility research has demonstrated, are known to be much more mobile than, for example, people in older age groups. However, we argue that this housing behaviour is indeed something new in Polish and Czech cities. These newcomers to the inner cities differ in their housing and living arrangements, but non-traditional households are a significant group. Their quantitative impact cannot be assessed on the basis of existing data bases or our research but, qualitatively, they matter. Along with this process, students have become new actors in inner-city housing markets, bringing about residential change by appropriating space partly left empty by recent suburbanites. They employ different strategies to find a starting place where they can live and create a number of different housing and living arrangements.

In this chapter, we have discussed the Polish and Czech examples mainly in terms of their similarities, without exploring deeper for differences because, on the basis of qualitative evidence, this would imply fairly hypothetical statements. Moreover, we have argued from the logics of our interviewees. We are aware that this perspective also has its limitations. Perhaps the decisive question is not whether a single household decides to stay or to leave the inner city. It is rather to what extent and how inner-city neighbourhoods undergo changes through this in-migration, either because the newcomer households remain, even in spite of changing internal arrangements, or because they function as door-openers that draw further newcomers into the neighbourhood after they leave for elsewhere. Housing careers and housing arrangements in flux thus contribute to diversifying pathways of inner-city neighbourhoods.

Chapter 9

Households as Actors II:
Attitudes towards Living in the Inner City

Katrin Grossmann, Annegret Haase, Annett Steinführer,
Maja Grabkowska, Adam Bierzyński

Introduction

In Chapter 8, we concluded that new and more dynamic housing careers are found both among students and early-stage professionals compared with those of their parents' generation. As a result of new opportunities in local housing markets, the inner city provides a place where new households and people with new living arrangements are able to settle.

This chapter examines the role of place as seen through the eyes of the residents. What attitudes do residents have towards the inner city as a place to live? Is it their place of choice? Are inner-city residents here because they are attracted to the inner city? If so, what exactly attracts them? Or is it a 'negative' decision, evolving from individual and structural constraints? Can we assume that the households which live here or have recently moved here are also likely to stay in the inner city in the future?

This chapter presents a typology of attitudes of inner-city residents towards housing and living in the inner cities of Łódź, Brno and Gdańsk. The objective is to assess the attitudes that inner-city dwellers have towards their residential location, how they came here and to what extent they feel attached to the place. In order to answer these questions, we examined the attitudes of our interviewees towards the inner city as a place of residence, their mode of residential decision-making and their criteria for evaluating the inner city. With the help of this knowledge, we can say something about the attractiveness of the inner city for different groups of residents, as well as something about preferences, opportunities and constraints.

After describing this typology of attitudes, we reflect on a phenomenon that was apparent in a number of interviews: the transitory character of the current housing arrangements. Many interviewees see themselves as temporary residents of the inner city, assuming further relocation along with changes in their personal circumstances. We identify 'transitory urbanites' as important actors in current inner-city change in Polish and Czech second-order cities. We will discuss what this phenomenon means for the residential function of the inner city today and what it might imply for the future.

Material and Methods

The interview sample was introduced in Chapter 8. The interviews were analyzed by employing hermeneutical interpretation and contrasting techniques. After an examination of each individual interview, a generalized and comparative analysis was carried out, in order to identify overarching commonalities and differences. Based on these analyses, types of attitudes held by interviewees towards inner-city housing and living were identified. The typology is a result of an interpretation of the interviews, based on inductively identified features and dimensions (see also Kluge 2000) in order to construct types of interviewees.

The types are conceptualized as clusters (of cases) that have common characteristics and can be differentiated from other clusters. Types are thus a combination of characteristics (see also Kluge and Kelle 1999: 75–8, following the early definitions by Lazarsfeld 1937, and Barton 1955). The typology was created in a series of stages (Figure 9.1).

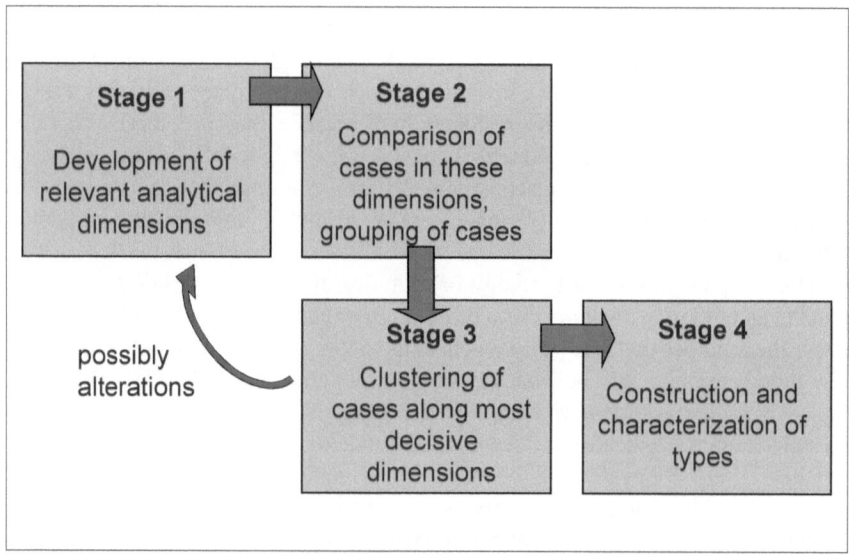

Figure 9.1 Typology building: A stepwise procedure
Source: Kluge (2000).

First, we defined the relevant analytical dimensions concerning the relationship between inhabitants and their residential environment. This was developed inductively via a common coding system, where we identified features important for a structuration of attitudes towards the residential location. These were:

- *life cycle:* household type, life cycle phase, educational and professional situation,
- *city-mindedness:* preference for or aversion to living in the inner city, advantages and disadvantages,
- *type of building stock:* preference for brick buildings (or other types of housing) or indifference,
- *voluntariness* of living in the inner city in general and the current residence in particular,
- *tenure:* renter, owner, subtenant – legal or illegal, and
- *residential environment:* importance of (access to) green space.

Second, we compared the segments of interviews involving these categories and identified similarities and differences across the cases. Groups of cases were then constructed. *Third,* cases with similar characteristics were clustered along the most decisive dimensions to form types. *Finally,* the types were further differentiated and characterized.

Attitudes Towards Inner-city Housing and Living: A Typology

Figure 9.2 summarizes the types of attitudes towards the inner city. The typology was created along two dimensions. The first dimension is the mode of the residential location decision that brought the interviewees to the inner city, in other words, whether people moved there intentionally or by chance. While a purposeful location decision for the inner city seems to be a recent phenomenon, structural and personal circumstances have long influenced residential location decisions. People were allocated administratively to a dwelling in the state-socialist period, they 'inherited' a dwelling from relatives or moved to the inner city because a specific dwelling was available and there were no reasonable alternatives. This 'by chance' mode of housing location decisions might thus be specific for post-socialist cities or, more precisely, for location decisions made under conditions of severe housing shortage. Accordingly, this mode of location decision relates more to structural than to individual restrictions.

The second dimension of the typology is the criteria interviewees use to judge the inner city as a residential environment. We found two categories of evaluation criteria that produced different attitudes: emotional and rational criteria. Emotional criteria are, for instance, the expression of a special liking for architectural features or a feeling of dislike for the housing environment. Rational criteria can be the proximity to places relevant for the daily life of interviewees or cost-benefit considerations. Certainly, in many interviews these criteria appeared side by side. In each individual case, we explored which criteria played a more prominent role in the narratives. If both emotional and rational criteria appeared and neither dominated, we found that resulting attitudes were again different enough to form a separate type ('mixed criteria'). Altogether, seven types of attitudes of residents are distinguished (Figure 9.2).

Mode of residential location decision / Criteria for the evaluation of the inner city	Intentional choice	By chance	
Emotional criteria	*Type 1 a)* **Aesthetic urbanite**	*Type 1 d)* **Urbanite by chance**	*Type 2 a)* **Frustrated inner-city resident**
Mixed criteria	*Type 1 b)* **General urbanite** *(balance between aesthetic and functional preferences)*	*Type 2 b)* **Ambivalent inner-city resident**	
Rational criteria	*Type 1 b)* **Functionalist urbanite**	*Type 2 c)* **Pragmatic inner-city resident**	

Figure 9.2   Attitudes toward inner-city housing (typology)
Source: Authors' work.

Within these seven types, there is a group of convinced inner-city residents that we call urbanites (types 1a, 1b, 1c and 1d) and a group of less convinced inner-city residents (type 2a, 2b, 2c). In the following, each type is described briefly. Each description starts with a case that is prototypical followed by some information on other cases and how the type varies.

Type 1a) The Aesthetic Urbanite

> These [city centre] flats are for fanatics like us.
>
> (Gdańsk HH W3)

The aesthetic urbanite moved intentionally to the inner city; an inner-city location was a priority in making the decision for the current residential location. This preference is mostly emotionally motivated. Interviewees state, sometimes enthusiastically, that they have a special liking and preference for the physical (but also social and cultural) features of the inner city. In the evaluation of the inner city residential location, the aspects of atmosphere, building aesthetics or the milieu are most important. Other criteria rank second.

Table 9.1 Representatives of the type 'aesthetic urbanite' and their characteristics

City	Number of interviews	Characteristics of the interviewee and the household
Łódź	3	· Łódź HH 2: Mother of family with adult child, the parents in their early 50s, both professionals, lives in a brick building · Łódź HH 8: Family with adult child, parents in their late 40s and early 50s, both professionals, lives in a brick building · Łódź HH 12: Patchwork family, parents in their late 30s and early 40s, three dependent children, lives in a brick building
Gdańsk	2	· Gdańsk HH N1: Patchwork family, parents in their early 30s and late 40s, both professionals, one dependent adult child (18 years, student), lives in pre-1945 brick building · Gdańsk HH W3: Married childless couple in their early 30s, both professionals, live in a pre-1945 brick building
Brno	2	· Brno HH 7: Single in his late 20s in owner-occupation, with one flat-mate (*prototype described in the paragraph below*) · Brno HH 16: One-person household, female and in her 40s living apart together with her partner in Prague, professional with Austrian background

Source: Authors' work.

Tomáš (Brno HH7) is the prototype of the convinced inner-city resident with such strong emotional bonds towards this residential environment that he cannot imagine living elsewhere. He is 28 years old and works as a designer. He personally experienced some alternatives – a privately owned home and a block of flats[1] in a large housing estate *[panelák]* – and particularly rejects large housing estates for psychological reasons: 'I could not put up with the location.' Inner city means to him a critical mass of people who are similar to himself, no anonymity, as well as a wide range of cultural opportunities, pubs and clubs. In the flat, he loves the original details of the interior: 'There were some basic things […] not done. For example,

---

1 By *block of flats* we mean prefabricated buildings typical for socialist mass-housing construction from the 1960s to the 1980s. We distinguish between single blocks (in Polish *blok, blok z wielkiej płyty* and *panelák, panelový dům* in Czech) and large housing estates *(sídliště* and *blokowisko* in Czech or *wielkie osiedle mieszkaniowe* in Polish).

there were some four layers of lino on the floor. When I had taken them all off, there was beautiful parquet flooring underneath [...] which needed to be sanded only.'

Living in the inner city also has a social dimension for him. After having described the responsibilities he has as a flat owner, he was asked whether this was due to the particularity of the property. He answered:

> I rather felt as if, don't know, in a psychological sense there is some difference between a prefab [panel] house or a housing estate and here. It is that all people know each other, basically. [...] Even if this here is in the city centre [ve středu města], it seems to me as if it were in the countryside.

For all interviewees of this type, it is the combination of physical and social characteristics that make the special atmosphere of inner-city living. Maria, originally from Austria, now living in Brno, calls this 'the spirit of it [the inner city]'. She finds it in the beautifully refurbished residential environment and lively street cafes that, in her view, resemble the inner city of Vienna. For some, the spirit is mostly in their own flat or building, for example, for Marian, a father in a patchwork family, who deliberately chose to buy a flat in a dilapidated brick building in Łódź's inner city: 'Quite an old house, but it is beautiful, there is some stucco work inside and parquet flooring. A rather untended staircase, but beautiful ... It was also very important that it is in the city centre' (Łódź HH 12). An interviewee in inner-city Gdansk pointed out the small-town character of the old neighbourhoods and a social environment where social networks can be established easily (Gdańsk HH N1).

What serve as arguments for the inner city, in turn, serve as arguments against other places, especially against the large housing estates, but also against suburban living. Negative factors are anonymity and 'concrete architecture'; once again, emotionally justified criteria: 'I would not go back to that Manhattan [the central area of multi-storey blocks]. It was freezing there. I don't know what it is like now, but people lived anonymously there, with all that concrete around' (Łódź HH 8).

Another negative factor is the low ceiling height of flats in blocks, which creates, according to the interviewees, a feeling of 'restriction': 'For me, *bloki* are just so cramped. The ceiling comes right down on your head. I just don't like that cramped feeling' (Łódź HH 2).

Type 1b) General Urbanite

> The parquet flooring is a convenience of this flat, and the brick walls too. ... Of course, the locality is also a convenience.
>
> (Łódź HH 10)

General urbanites moved intentionally to the inner city and appreciate its aesthetic and functional advantages. The intentional decision to move to the inner city is thus grounded in emotional and rational criteria, which are both equally important for the evaluation of the inner city as a residential environment.

Table 9.2 Representatives of the type 'general urbanite' and their characteristics

City	Number of interviews	Characteristics of the interviewee and the household
Łódź	2	· Łódź HH 10: One-person household, male, professional in his 40s, lives in a tenement building (brick) from the 1950s · Łódź HH 15: Professional in his late 20s, cohabits with his partner who is also a professional, he lives in a block of flats
Gdańsk	6	· Gdańsk HH W4: Professional in his early 40s cohabiting with his partner, a student in her mid-20s, lives in a pre-1945 brick building · Gdańsk HH W5: One-person household, male, professional in his late 30s, lives in a pre-1945 brick building · Gdańsk HH W6: Married couple in their 40s, both professionals, one dependent child, lives in a pre-1945 brick building *(prototype described in the paragraph below)* · Gdańsk HH W7: One-parent household, female, professional in her early 50s, two dependent children (one of them is 18 years), lives in a pre-1945 brick building · Gdańsk HH W8: Married couple in their early 30s, both professionals, one dependent child, lives in a pre-1945 brick building · Gdańsk HH W9: Student flatshare, two females, three males, all in their mid-20s, live in a tenement block (brick) from the 1950s
Brno	1	· Brno HH 10: Woman on maternity leave in her early 30s, core family with two children, lives in a pre-1945 brick building

Source: Authors' work.

Wawrzyniec (Gdańsk HH W6) is in his 40s and lives in a 140 m² flat with his wife and teenage daughter. They moved there from a flat in one of the new (post-1989) housing estates located on the so-called upper terrace (see Chapter 6). The main reason for this change of dwelling was that a large part of the family's monthly budget was consumed by the costs of daily commuting by car. Since Wawrzyniec and his wife both work in Wrzeszcz, it became their 'natural' destination. However, it took them almost a year to find the 'right' flat. Their choice was conditioned not only by such parameters as size and layout but also a certain 'feel' of the old architecture: '[...] we liked this flat. It has got a couple of advantages, I would say, both the looks and the layout of the rooms, it is quite an elegant flat.' Yet, despite Wawrzyniec's liking for the pre-Second World War tenement houses in Wrzeszcz, it is the central location and proximity to shopping and other facilities which he regards as the main advantages of inner-city living. 'There is one fundamental advantage, that is good access to public transport, you can easily get to your work and other places, for example, for shopping or for entertainment, it is simply much easier from here and also less expensive.'

General urbanites appreciate both the qualities of their place of residence and those of the inner-city location ('closeness'):

> [...] there is no doubt, the layout of this flat is a big advantage, the way it has been designed. Another advantage is the parquet and the brick walls. ... Of course, the locality is yet another advantage. ... Everthing's close by. It's close to the railway station. Once I used to work as a teacher in Warsaw, so it was just a five or ten minutes' walk to the Factory Station [Dworzec Fabryczny]. (Łódź HH 10)

Not to waste time, to be everywhere in a couple of minutes was mentioned as one of the advantages of inner-city housing by most of the Gdańsk representatives of this type as well; one of them even chose not to have a car in favour of using public transport exclusively: 'I used to have a car and now I don't. I think it would be stupid to have a car in my situation [...] Everything is within reach and it is probably one of the most important elements of [quality of] life, you just don't waste your time' (Gdańsk HH W7).

Another representative of this type said, in an even more general manner, that he could not imagine living 'anywhere else', where he has no contact to 'what is going on': '[...] I just could not imagine living in some place without direct access *[bezpośredniego kontaktu]* to all the city amenities, I am talking about cultural facilities such as theatres, the cinema, I am talking about pubs [...]' (Łódź HH 15).

Helena from Brno, together with her partner and two children, represents a core family who – contrary to the mainstream of families who wish to live in suburban locations – deliberately decided to stay in the inner city with its pre-Second World War housing stock. She again values both location and atmosphere. Coming from a large housing estate, she wanted to live in the inner city, because daily commuting took her an hour to get to work and because 'all my activities were in the city centre'. Helena gives the inner city of Brno a specific attribute: 'old-good' which for her is a single word and precisely describes her personal judgement of her residential environment. This encompasses not only the specific location and the amenities, but also the type of housing stock: 'I just wanted to live in a big flat in an old building. With higher ceilings and a little bit more space. [...] I just did not want to go back to a block *[panelák]* anymore' (Brno HH 10).

Type 1c) Functionalist Urbanite

> What does city centre mean for me? [...] that I can get everywhere within a quarter of an hour.
>
> (Brno HH 11)

The functionalist urbanite moved intentionally to the inner city, too, but in contrast to the two previous types, rational criteria played the dominant role in their decision. Residents of this type particularly value the functionality of the residential situation, such as good transport connections, the proximity to places important to them, the fabric of the residential environment with its infrastructure (schools, shops, cafes, theatre, public transport etc.) and that a car is not required for daily tasks and travelling, whether to work, leisure or shopping.

Table 9.3 Representatives of the type 'functionalist urbanite' and their characteristics

City	Number of interviews	Characteristics of the interviewee and the household
Łódź	2	· Łódź HH 5: Professional in his late 20s, lives with his wife who is also a professional, in a block of flats · Łódź HH 9: Female one-person household, in her late 20s, professional, lives in a block of flats
Gdańsk	2	· Gdańsk HH N2: Married childless couple in their 30s, both professionals, lives in a pre-1945 tenement building · Gdańsk HH N3: Patchwork family (two cohabitants and a dependent daughter), parents in their late 30s, both professionals, lives in a pre-1945 tenement building
Brno	4	· Brno HH 1: One-person household in her early 30s, living apart together (partner in Prague), lives in a pre-1945 tenement building · Brno HH 2: Woman in her late 20s, cohabiting with a partner in a flat-share with a total of five persons (two couples, one single person), lives in a pre-1945 tenement building · Brno HH 5: Woman in her early 20s, cohabiting with a partner in a flat-share with a total of five persons (two couples, one single person), lives in a pre-1945 tenement building *(prototype described in the paragraph below)* · Brno HH 11: Woman on maternity leave in her 30s, core family with two children, lives in a pre-1945 brick building

Source: Authors' work.

Hanka (Brno HH5) is a 22-year-old student. She spent her childhood in the inner city and then moved to a detached house at the urban fringe in the district Brno-Kohoutovice in her teens. Her first flat on leaving home was located in a large housing estate in the same district. But, she found it difficult to live there because of her studies:

Well, [living on the outskirts] did not bother me much because I was used to it, but the commuting [...]. At the university your timetable is so stretched, sometimes you have two or three hours off, but there is no use going back to Kohoutky [Kohoutovice] during that time. [...] Here, that is easier because everything is so close. If you want to go to the pub or meet a friend or do sports, everything is so nearby.

She and her boyfriend were deliberately searching for a flat to be shared with others in the inner city *('v centru')*. Financial restrictions meant they could not choose a flat just for themselves, thus they share it with another couple and another person (who is part of a LAT arrangement).

The two functionalist urbanites in Łódź appreciate the proximity of the flat to places they visit regularly. They stress that it is possible to reach most of them on foot. They have a pragmatic attitude towards their place of residence: they have lived here a long time and have got used to it; however, emotional bonds are not expressed which represents the most striking difference to the 'general urbanites'. Instead, proximity is an important advantage for them: 'That means I have been living here all my life, and I am used to this flat in the city centre. ... Everything is nearby. ... Close to everything. This proximity is a sort of luxury. ... Cultural and entertainment facilities are close by' (Łódź HH 9).

Similarly, the interviewees from Gdańsk and Brno particularly value the ease of commuting to work. Nina from Gdansk, for example, who has to drive a lot during the day, acknowledges that the location of her dwelling increases her mobility and at the same time is very convenient for her partner, who does not have a car of his own and who appreciates good access to public transport (Gdańsk HH N3). Sanja from Brno who lives close to the university says: '[What brought us here was] also our studies, the faculty was nearby' (Brno HH 2). One interviewee, a young mother of two children, adds another factor to this type: she stresses that as a 'starter' household, the inner city with its partly dilapidated housing stock was the only place where she had a realistic chance of getting a flat at all. She used the opportunity offered by the municipality to extend and reconstruct it at her own expense. 'Well, I would rather have chosen Komín [a district closer to the outskirts] or something like that, but I don't know [...], not some far-off housing estate but one with a good connection, I would not have minded' (Brno HH 11).

Type 1d) Urbanite by Chance

It was more or less by chance that we made that decision.

(Łódź HH 6)

For urbanites by chance, a residence in the inner city was not their primary choice. Instead, they concede that they were at least open-minded towards living in this part of the city. However, it was only with the experience of living there that they discovered the advantages of a centrally located residence. Like type 2a, the frustrated urbanite, they evaluated the situation emotionally, but with the contrary result: they became convinced urbanites, which means that these people value inner-city opportunities and value them more than the conditions in other housing environments, and this also allows them to overlook negative experiences of the inner city. The inner city is mainly described with positive criteria, similar to type 1a.

Table 9.4 Representatives of the type 'urbanite by chance' and their characteristics

City	Number of interviews	Characteristics of the interviewee and the household
Łódź	4	· Łódź HH 3: young professional in her late 20s, cohabits with her partner in a brick building · Łódź HH 6: young professional in her late 20s, cohabits with her partner in a brick building (*prototype described in the paragraph below*) · Łódź HH 11: one-person household, female, young professional in her late 20s, lives in a tenement (brick) building from the 1950s · Łódź HH 16: pensioner in her 70s, lives in a three-generation household together with her daughter and grand-son in a brick building
Gdańsk	1	· Gdańsk HH W1: One-person household (although shares the flat with a distant elderly relative), male, professional in his late 20s, lives in a pre-1945 brick building
Brno	1	· Brno HH 4: Cohabiting couple in their late 20s, professionals his brother is their subtenant, live in a pre-1945 restituted tenement building

Source: Authors' work.

Ela (Łódź HH6), who is in her late 20s, moved to the inner city one and a half years before the interview. She studied in Łódź and now works in the municipal administration. She did not explicitly look for a flat in the inner city or in a brick building, but obtained a rental flat in the inner city by chance. 'I was looking for a flat to rent … not necessarily in the city centre.' There she lives together with her partner. For now, Ela is very happy with her home situated in the inner city close to her place of work and many amenities, as well as in a brick building which she likes for its high ceilings and large rooms.

> This is a flat in a multi-family house [kamienica]. It's beautiful, because it's got high ceilings and lots of space, first of all … And the location, well, I can walk to work now and don't have to use the car anymore. And it's calm around here. It's located right in the city centre but it's in a very calm and peaceful area.

In short, she became an 'urbanite by chance'. 'Living in a multi-family house was not quite my first choice, but after I had moved in, I changed my mind.' For the future, her first choice would be to stay within the central part of Łódź, but she would prefer to live in a detached house.

Thus, these interviewees share attitudes of type 1a, b or c, although they did not choose the inner city in a preference-driven way. The motives for coming to the inner city vary. It can be the consequence of a marriage or inheritance of flats, as described in Chapter 6. Janina, a young inner-city resident from Łódź, explained that she inherited the flat from her uncle: 'I did not make a deliberate decision to live in this area, or at this place. It simply came to be like that ... It's definitely convenient that it is in the centre and close to everywhere [...]' (Łódź HH 11).

Interestingly, not only the young but also older people who moved to Łódź's inner city half a century ago tell similar stories. A pensioner who lives in a brick building with her daughter and grandson told us: 'I got to like the inner city [Śródmieście, a district of the city of Łódź] very much. After having lived at a place for 54 years, as I have, you just get so used to the atmosphere, to everything about it' (Łódź HH 16).

An interviewee from Gdańsk (Wojtek, HH W1) co-inherited the flat from his grandmother, together with his brother. He developed such a liking for inner-city living that he decided that if he and his brother should decide to separate, he would find another place to live somewhere in the same area. The interview with interviewee Brno HH 4 reveals a specific feature. The interviewed couple who are both lawyers live close to a number of different legal institutions and use the flat both as a residence and as an office:

> When we are at home, we are actually at our office, and we are in the closest possible proximity to those agencies. And if we were living someplace on the outskirts, then ah ... but we are going to that law-court twice a day sometimes, or else to those agencies, so that would just not pay off for us in terms of time.

The chance that brought them here is very post-socialist but not at all typical. In the course of the restitution, the building was restituted to the heirs of the former owner. The landlady of the interviewees lives abroad and she let them have the flat in exchange for a small sum of money. They thus became urbanites by a double chance: their profession and their contact with the landlady abroad.

Type 2a) Frustrated Inner-city Resident

> And the worst place? That is our inner city.
>
> (Łódź HH 14)

The frustrated inner city resident lives here by chance, too. The location decision was not taken by intention but because of personal or structural circumstances. The interviewees did not look explicitly for a residential location in the inner city. Instead, they came here because this was one possible solution or because there was little or no other choice. This type is frustrated with the situation, mainly for emotional reasons. According to their own housing preferences, representatives of this type would definitely prefer a different location. Characteristically, the frustrated inner city resident describes the inner city with negative attributes (noise, pollution, milieu, etc.).

Table 9.5 Representatives of the type 'frustrated inner-city resident' and their characteristics

City	Number of interviews	Characteristics of the interviewee and the household
Łódź	1	· Łódź HH 14: Married couple in their 70s, both pensioners and longer-term 'empty-nesters', lives in a brick building (in the same one as Łódź, HH 12) (*prototype described in the paragraph below*)
Gdańsk	none	
Brno	none	

Source: Authors' work.

The elderly married couple Kowalski (Łódź HH14) – both are in their early 70s – live in a brick building in the inner city. They were assigned to this after the war. They rent the flat but have to share the entrance door and hallway with two other households (typical for Łódź's inner city after the Second World War). Together with their adult children, who have since moved out, they paid for some repairs privately. They are not satisfied with the technical standard of their flat or with the residential environment of the inner city; especially, they do not like the dirt, the dilapidated housing, the noise, the high prices in the shops and the (unpleasant) social structure. They would prefer to live in a flat in a large housing estate, which they perceive as calm, clean and green:

> There is a lot of greenery, there near the blocks [przy blokach], there are those, uh, parks, walkways, greenery, and a lot of trees. … The air is totally different there … there is no noise like this … And living costs are lower, life is easier … in the housing estates ….

Among our interviewees, this was the only example of such 'frustrated' inner-city residents. But they were so different from all the others that we categorized them as a separate type. This rare occurrence in our sample is certainly due to its selectivity, as described above – only very few older long-term residents were interviewed, because of the strong focus on non-traditional households, which are mostly younger or even 'starter' households. To put it differently: we expect that this type is more numerous than shown by our sample but cannot provide any further empirical evidence.

Type 2b) Ambivalent Inner-city Resident

> I would rather live in a more peaceful place, perhaps in a back street, but on the other hand, I love to have everything close by.
>
> (Gdańsk HH W2)

Circumstances brought interviewees of type 2b to the inner city. Either they were allocated flats by administrative bodies during state socialism or they inherited a flat or could not find a flat in the preferred location and came here as second or third choice. Their evaluation of both their specific situation and the inner city as a residential environment remains ambivalent. These interviewees typically mention both positive and negative aspects that are emotionally and rationally motivated. They are not convinced that the inner city is a residential environment of choice for them at this moment; however, they see the advantages it has. The ambivalence refers to their present situation, independent of future plans and personal projections.

Table 9.6 Representatives of the type 'ambivalent inner-city resident' and their characteristics

City	Number of interviews	Characteristics of the interviewee and the household
Łódź	4	• Łódź HH 1: Student in her 20s, lives with her parents in a block of flats (*prototype described in the paragraph below*) • Łódź HH 4: Student in her 20s, lives with her parents in a block of flats (*prototype described in the paragraph below*) • Łódź HH 7: Worker in his late 40s, lives with his family (two dependent children) in a dilapidated brick building • Łódź HH 16: Professional in her 50s, lives in a three-generation household together with her mother and son in a brick building
Gdańsk	1	• Gdańsk HH W2: One-person household, female in her late 20s, professional, lives in a pre-1945 tenement building inherited from her grandmother
Brno	3	• Brno HH 8: One-person household, professional in his 40s, foreigner, lives in a pre-1945 tenement building • Brno HH 9: Divorced woman in her 50s, lives alone as an empty-nester in a large pre-1945 flat with specified tenants' rights (*dekret na byt*) • Brno HH 15: Married couple without children, young professionals in their mid 30s, lives in a pre-1945 tenement building

Source: Authors' work.

Joasia (Łódź HH1) and Ania (Łódź HH4) are students who still live with their parents in Łódź's inner city. They are both in their early 20s. Whereas Joasia lives with her parents in a block from the 1970s, in an owner-occupied flat that was formerly owned by a cooperative, Ania lives in a brick building. Her parents also bought the flat, which is located in a building owned by the municipality. Both Joasia and Ania have no emotional ties to their place of residence; they did not make the decision to live there or to move there. At the same time, they (rationally) appreciate the advantages of central living, but they also emphasize the disadvantages of Łódź's inner city, such as dirt, noise or an unpleasant social environment. Their ideas about their future homes differ, although they do not play a very important role for them at the moment, since at present they do not intend to move out of the parental flat. While Ania would prefer to live in a detached house, Joasia has not thought about this issue yet. She claims that there are many criteria which will contribute to the decision about her future home; the location and type of building are only two of them. 'It really depends on what my life will look like.'

The two ambivalent inner-city residents in Łódź who are in their 40s and 50s also look at the pros and cons of their place of residence in terms of post-socialist developments. While they appreciate central living, they complain about high costs, the bad technical state of the buildings (if they live in brick buildings), the social environment and the deterioration of neighbourly relations, in comparison with the state-socialist period. A similar experience was described by Radoslava (Brno HH 9), a long-term tenant, divorcee and empty-nester from Brno, who is faced with her flat's constant need for repairs, as well as increasing utility prices and the threat of a sharply increased rent, once they are deregulated. While she acknowledges the advantages of the inner city and is used to the spacious living provided by her flat, she considers her housing situation not as a privilege but rather as a 'brake', not allowing her to leave without a deterioration in her housing and financial situation.

The interviewee from Gdańsk inherited the (municipal) flat from her grandmother (like Wojtek; see type 1d), but even though she is currently planning to buy the flat from the municipality and appreciates the advantages of the inner city, such as good access to public transport, she is not a totally convinced urbanite. She describes her sister's fondness of new (post-1989) housing and reveals that she herself is not certain whether she would not prefer to live in a newly-built estate too, rather than in an old, unrenovated tenement building.

One of the interviewees in Brno is torn between urban life as he experiences it in Brno's inner city and his previous experiences living in complete remoteness. He is tired of personal residential changes on the one hand ('I really have had enough of moving'), but has strong financial and psychological motives (particularly a desire for calmness which was stressed time and again) on the other, which make him think carefully about leaving the inner city (Brno HH 8). Finally, for Marek, the flat he bought with his wife in the inner city was the better economic solution, even though he would have preferred a fringe-location for sport activities and she was looking for a garden. The lower costs, the functional advantages of the location and, not least, the availability of the flat in the inner city, steered the decision to buy this flat. Both made compromises: the wife now has a small garden in the backyard and, instead of jogging in the forest, he runs in a nearby park (Brno HH 15).

Type 2c) Pragmatic inner-city resident

> I guess if I were working somewhere else, I would not rent this flat here.
>
> (Gdańsk HH N5)

This type was also brought to the inner city by circumstances. The pragmatic inner-city resident now focuses on the advantages of the situation. The benefits described are pragmatic in terms of personal needs; they have less to do with personal preferences, wants or likings. Therefore, the presentation or argumentation is more rational than emotional. Compromises might be also part of the self-presentation.

Table 9.7 Representatives of the type 'pragmatic inner-city resident' and their characteristics

City	Number of interviews	Characteristics of the interviewee and the household
Łódź	1	· Łódź HH 13: three students in their early 20s who share a flat, two of them are a cohabiting couple, the other represents a one-person household, they live in a block of flats from the early 1960s
Gdańsk	2	· Gdańsk HH N4: Student flatshare, four females, all in their 20s, live in a pre-1945 tenement building · Gdańsk HH N5: Student flatshare, two females, four males, all in their 20s, live in a pre-1945 tenement building
Brno	4	· Brno HH 3: Professional in his 30s, cohabiting with a partner, owner-occupier, lives in a pre-1945 tenement building · Brno HH 6: Single in his late 20s, owner-occupier, lives in a pre-1945 tenement building (*prototype described in the paragraph below*) · Brno HH 12: Woman on maternity leave in her early 30s, core family with three children, lives in pre-1945 tenement building · Brno HH 14: Professional in her 30s, married, owner-occupier, lives in a pre-1945 tenement building

Source: Authors' work.

Ondřej (Brno HH 6) is 27 years old, a professional and single. Living in the inner city corresponds with his current life style, but is nothing permanent. His decision to come here emerged from a deliberate multi-criteria decision-making process. He could, equally, have moved to a large housing estate.

> I was intending to live somewhere calm and maybe in a housing estate, but I was thinking about that on and off. And then again, that distance from the city. Really, a brick house like this is better than a block, if you ask me, because of heating and the like. [...] I would have gone there anyway if someone had been there to convince me again. Up till now, I am not convinced.

He can be regarded as a pioneer in Brno's east end, which is characterized by strong and long-term social exclusion. But in his short occupancy of just one year, he has experienced an increase in the value of his flat of about 30 per cent. This makes him relaxed with regard to future residential location decisions, which he takes for granted.

This attitude is also found elsewhere in his neighbourhood, where another interviewee lives (Brno HH 3, already described in Chapter 8). Financial considerations, in combination with location advantages, brought Petr and his partner to the inner city, as was the case for Barbora (Brno HH 14). Anna, in contrast, came to the inner city by chance via a flat which was to be rented to her brother, who eventually did not move there. So she decided to take it for lack of an alternative: 'Otherwise we would not have decided to live there because we already knew we wanted to have babies.' While she underlines the non-appropriateness of her current residence for families several times, she also reports adaptation measures inside and outside her flat, in order to make the flat more comfortable (Brno HH 12).

The students sharing the flat in Łódź see the advantages of the flat and its location mainly in terms of their education: 'As for me, since I am studying at the polytechnic institute, it's convenient to get there, I can even walk to school ... for me, that's an advantage [...].' The main criterion for the choice of this particular flat was – apart from the rapid agreement with the owner – the housing costs, one of the most pragmatic criteria of all, and one which has nothing to do with either the location, residential environment or the type of housing and architecture: 'The main reason why I made up my mind to rent this flat? Maybe that was ... Yes, it was the low rent [...]' (Łódź HH 13). The same is true for members of the two flat shares in Gdańsk, who decided to move in because of the relatively cheap price and proximity to work, and did not pay much attention to the district or the type of housing itself. Some of them even found it hard to reply to questions regarding their attitude towards living in the district, since they treated their accommodation as only temporary. Opinions on pre-Second World War housing aesthetics varied greatly. Norbert is an example of a resident without any emotional attachment to brick architecture. Asked whether he would change something and improve the place by himself, he replied: 'I think nobody is really thinking about what to change or what is missing, everyone is seeing this as a temporary dwelling, nobody plans to stay here for long, ... don't know if this district is old, and it won't become more beautiful' (Gdańsk HH N5).

Pragmatism towards the choice of the housing location and an adaptation to the current subjective needs seem to play an important role among younger residents (without a family) who live in the inner cities of Lódz, Gdansk and Brno. This means that inner-city housing and living is preferred by them at least for a limited phase in their life in which they finish their education or start their professional career. This phenomenon has become increasingly 'common' as the interviewees told us. The next section, therefore, looks in a little more detail at this group and their impact on inner-city housing and living.

Transitory Urbanites: Major Actors of Inner-city Residential Change

As is obvious in the description of the last type (type 2c), but also of the type 1d 'urbanite by chance', many of the cases presented share two things: they describe themselves as temporary residents of the inner city. At the same time, they can be regarded as a new type of inner-city inhabitant: they deliberately decided to live in the inner city as it fits in with the current phase of their professional and private lives. But all of them consider that the future might bring a new residential location decision which may favour another location in the city or in suburbia. Lucie from Brno or Wiktor and Wiola from Gdańsk represent three of them. Lucie, for example, a highly qualified young professional, currently enjoys living in the inner city; she uses its amenities and values the short distances from there to many places in the city. But she is not convinced that she will stay here, especially if and when she enters parenthood. For this potential situation, she stresses the perceived disadvantages of the inner city:

> Hmm, as to the future, I don't know if that [the residential location] should be important, since I don't think that this entire [úplný] inner city is a good place for children. Because there is a lot of traffic here and the children don't have a place to go and play. So I don't know whether to rather rent a place outside the town, maybe. (Brno HH 1)

We call them transitory urbanites because of this notion of the actual housing arrangement being temporally limited, a transitory arrangement. Together with this transitory character of the housing arrangements, an informal, sometimes illegal status of the tenants also appears from time to time. Even owner-occupiers sometimes do not bother to register as residents, not least because they do not intend to stay for good. Lucie, for example, is still officially registered at her parents' home in the countryside, has no rental contract and pays her rent in cash to a person who is the legal tenant of this flat but lives elsewhere. The flat is owned by the municipality and still has a regulated rent. The letterbox does not display her name and 'if there is any problem, I will pretend to be his girlfriend.' Her illegality does not worry her – instead, she perceives either position as balanced. While the legal tenant is dependent upon her payments and secrecy vis-à-vis the authorities, she has to trust him not to evict her from the flat without any notice: 'I don't know anything at all about him and he doesn't about me. [...] I don't even have a lease contract. But we are more or less in the same position.' (Brno HH 1). Illegality and transitory arrangements do not necessarily go together but the former certainly reinforces the latter.

In the inner city of Łódź, transitory urbanites are also quite common. Monika, for instance, is in her late twenties, holds a university degree and works at the municipal administration. She cohabits with her partner in a consensual union and lives unofficially in a flat owned by a relative, into which she moved recently. It is her first flat after leaving her parents' home. Currently she likes living in the inner city because she regards it as appropriate for her and her partner's situation at this time of her life and the preferences related to it:

> As far as the flat in the inner city is concerned, I am close to those places I make use of [in my everyday life]. I am close to my place of work. I do not need to go to the inner city because I am already there. [...] This is very important at the moment. So far, I do not have any children and so I am not forced to think about things like the children not having a place to play. For the time being, living close to the city centre is important but it is not the most important thing in the world. Later, at some point in the future there could possibly be some changes. (Łódź HH3)

The decision and time for moving somewhere else thus depends greatly on getting married and having a child. With small children, she would prefer not to live in the place where she lives now because of the lack of greenery and playgrounds and because of noise and crime: 'In the future I would like to live in a detached family house on the edge of the city. It would be no problem to live a bit further outside of the city because you can enjoy a coffee in your back garden and enjoy walking in the woods' (Łódź HH3).

The living and housing situation of Ela who has been described in the typology above as an urbanite by chance, is similar to Monika's. She is also in her late twenties and cohabits with her partner in a pre-Second World War flat in inner-city Łódź. She works in an office of the municipality, too, and finished her university studies some years ago. Like Lucie and Monika, she represents a non-traditional household type whose housing arrangement is also clearly transitory. To live in the inner city fits into her current phase of life, but she envisions exchanging the flat for a single family house with the foundation of a family. However, in contrast to others, she would like to stay close to the inner city even when she has children (which she wants to have) in the future.

I think moving house in the future will be connected to starting a family, then I will need a bigger flat with a different layout of rooms. [...] To tell the truth, I think I will have a house of my own one day. [...] I hope to own a house close to the city centre in ten years. That is just wishful thinking because I have no chance to make these plans come true but anyway, at that time I will live in Łódź, with husband and children. (Łódź HH6)

The examples of these three interviewees show that transitory urbanites might have different housing plans for the future. Some of them might wish to leave the inner city for the suburban zone; others might (more or less voluntarily) stay because of existing preferences, pragmatic considerations, financial or other constraints. The term *transitory* applies, in this vein, to two dimensions: on the one hand, it relates to the actual *de facto* housing situation of the young inner-city dwellers, which represents a phase of limited duration in their housing biographies. On the other, it describes their general attitude to inner-city housing and living. In both cases, the term can undergo changes in relation to shifting preferences and attitudes.

Since the results presented here are based on a qualitative study, we cannot say how large the group of transitory urbanites actually is. However, these three women are certainly no exceptions, because the most striking experience from the fieldwork was the 'normality' of the young people occupying the inner city independent of the institutional framework – legal or illegal, registered or unregistered, as subtenants or flat sharers (see also Chapter 8). The interviewees themselves described their housing and living arrangements as something completely normal. The second surprise was the extent of the housing mobility of these new residents. The frequency with which they move house is very high compared to what was previously known about housing mobility in these cities, as well as in East Central Europe more generally. This certainly has to do with the uncertainty of their transitory housing arrangements. As with other interviewees in the cities investigated, Monika, Lucie and Ela represent the trend of young, non-traditional households considering the inner city as a perfect place for their current phase of life, regardless of their legal status and registration. These newcomers complement the majority population of long-term residents in the inner cities and bring completely new demographic and social characteristics with them.

Transitory urban living does not only describe an evolving form of inner-city housing. It also represents the stage of inner-city change that we find in our case-study areas at the moment: a transitory period where a variety of different residential groups populates the inner city, some of them as long-term dwellers, others as newcomers, only some of whom intend to stay for a longer period of their lives. How sustainable this period will be, and what will come after it, has to be left to future observations and research. The likely time-limited character on the level of the individual household does not necessarily imply that these arrangements are transitory on the meso-scale of the inner city. By creating new forms of taken-for-granted housing arrangements and housing careers, these young transitory urbanites might pave the way for other newcomers in the future, whether these be students, gentrifiers or other in-migrants that we cannot predict today.

When looking at the housing biographies of the transitory urbanites and discussing whether they will stay in or leave the inner city later on, we have to limit our interpretations to the logics of the cases we have analyzed and cannot reach any general conclusions. An overarching conclusion we can, however, draw is that the life cycle and attitudes to family formation and children do play an important role for the duration of transitory urbanism or the time and place when and where a household settles for a longer period in a particular case. Another factor is, of course, the purchase of a flat, because this strengthens the relationship of a household to a particular place. Natalia, a young woman from Gdańsk who now lives in a patchwork family, might serve as an example here. She – apparently – has already passed through the 'transitory' phase and has found a place to live for a longer period, unless her life circumstances change once more. She is, meanwhile, the owner of her current flat and thus represents a more 'settled' inner-city living and housing arrangement. The transitory arrangements, in her case, changed into long(er)-term arrangements as far as both the inner city as her place of residence and her current living arrangements are concerned.

Summary and Concluding Remarks

The objective of this chapter was to present and discuss the attitudes of residents towards inner-city housing and living. The qualitative data show a variety of attitudes, ranging from enthusiasm, through pragmatism and ambivalence, to frustration. A positive attitude is usually based on aesthetic or functional criteria and sometimes on both of them. Interviewees stressed the beauty and spaciousness of flats in brick buildings, the atmosphere of the inner city with its historic environment and – sometimes termed as a contrast to large housing estates – the more individual social contacts. A functional criterion frequently mentioned is proximity to places relevant to people's life, such as the place of work or leisure facilities. A negative or ambivalent attitude often has to do with a pessimistic view of the dilapidation of buildings and a stated housing preference for other locations within or outside the city.

There is no clear connection between particular household types and residential attitudes. Within the pragmatic and ambivalent types of attitude, student and one-person households actually dominate our picture. In particular, there is no evidence for the assumption that households with children have a negative or ambivalent attitude towards the inner city, whereas young, non-traditional households show almost unanimously positive attitudes. This result certainly challenges the argument that inner-city living is not appreciated by families who, instead, generally prefer suburban living. The suburban dream, nevertheless, influences the stated preferences of households. Many of our convinced urbanites merely see the inner city as a transitory residence that ideally would be followed by a suburban place of living, after they start a family. To put it differently: transitory housing, as represented by our typology of inner-city residents and their attitudes towards their housing location, represents a growing niche that merits more detailed examination.

Even though the transitory households are not likely to stay in the inner city, transitory housing as such still makes a difference to the place. It establishes new forms of use of inner-city advantages by specific types of dwellers (young students or early-stage urban professionals without children). Presently, it coexists with older forms of housing and use of the inner city by long-term inhabitants (older couples and widowed inhabitants, poorer families etc.). It is impossible to evaluate to what extent these new forms of housing will change the inner cities in the long run. But the pathways of the investigated inner cities and their neighbourhoods will be different in all cases; diversity and differentiation will increase everywhere, although the consequences and the gradient may differ. The observed in-migration of the young changes the place, no matter whether the same people stay for long or whether they function as 'gate-openers' for the next transitory – or permanent – urbanites.

Our results indicate that the prevailing picture of housing biographies, mobility and ways of life in East Central European inner cities needs to be revised. It is worth highlighting that our research adds some new facets to previous knowledge on neighbourhood change in post-socialist cities. We found young inner-city dwellers who are not foreigners and who do not live in the newly built condominiums but rather in old houses, which are often in a bad state of repair. They are fond of the central location, the cultural amenities nearby and the physical structures, that is, brick buildings, as opposed to the blocks found in the large socialist housing estates. These newcomers typically have rental agreements and are, in contrast to 'standard' housing biographies in their countries, highly mobile. From a western perspective, these newcomers in general and the transitory urbanites more specifically resemble, in many respects, the highly disputed pioneers in gentrification research. But the subjectively expressed notion of temporality, as well as the specific history and context of this hidden inner-city transformation, make us reluctant simply to apply a concept developed in a 'western' context, such as (delayed or catch-up) gentrification. Gentrification has its underlying meanings and is rooted, too, in specific contexts – thus a simple transfer of the concept might obstruct context-sensitive explanations.

Transitory urban living is possibly bound to particular circumstances, such as a housing market that does not suffer from the pressure of great demand, as it exists in the capital cities of East Central Europe, especially with respect to the inner-city housing stock. Under high(er)-pressure conditions, areas where we find transitory urbanites today could easily become gentrification areas tomorrow. Thus, transitory urban living in the inner city might be a typical characteristic of non-capital or second-order cities, at least for districts that are not a focus of interest for investors.

In a general sense, there is a close interplay between households (as actors) and the urban environment. With their housing choice, households react to given settings (housing market, priorities of professional and private life) and their alterations. They decide on the basis of their preferences and constraints. In doing so, they shape housing arrangements and biographies that impact, in turn, on the urban space, its use by other households and the shape of the urban settings mentioned above. Thus, we show how tightly the two sides of residential change – the close connections between residents on the one hand and their place of residence on the other – are interwoven.

Chapter 10

Flexible Households, Flexible Dwellings, Flexible Neighbourhoods?

Maja Grabkowska

Introduction

S1NGLETOWN was the name of an exhibition hosted at the 2008 Venice Architecture Biennale. A conceptual city designed for people living alone features its citizens: Independent Widow, Global Opportunist, 100K+ Executive, Air-bound, Seasoned Professional, Once a Mom, Solitude Seeker, Alone/Together and Recently Divorced. The makers of S1NGLETOWN, a communications agency and a design company based in Amsterdam, acknowledge that new housing design is needed for this specific group of urban dwellers, arguing that 'while housing developers in both the public and private sectors have been slow to recognize the demographic shift that has taken place across developed countries, there are signs that blind adherence to traditional housing models is on the wane' (S1NGLETOWN 2008: 2).

Yet, people living alone are just one of the many types of non-traditional households characteristic of the second demographic transition (SDT; see Chapter 3). This composite phenomenon of changing demography which it is argued characterizes post-modern societies, manifests itself in such trends as declining birth rates, postponement of marriage and rising rates of cohabitation, as well as increasingly varied types of household arrangements as an alternative to the traditional core-family model (van de Kaa 1987, 1994, Lesthaeghe 1995). Since the resulting effects on the size and structure of contemporary urban households have substantial socio-spatial consequences, the argument made by the S1NGLETOWN architects seems to apply as much to one-parent households, patchwork families or flat sharers as to one-person non-traditional households. Hence, an interesting research question arises: in what ways do the housing needs and preferences of non-traditional households differ from those attributed to traditional nuclear families and how do these needs and preferences translate into residential change in the inner city? Referring to the concept of residential change as introduced in Chapter 2, this chapter sheds light on residents' behaviour and how it relates to the physical characteristics of the place.

Several recent studies examine the interrelations between demographic change and reurbanization tendencies, the authors suggesting that non-traditional households have a distinct inclination towards inner-city living (Ogden and Hall 2000, Haase et al. 2005, Bromley et al. 2007, Buzar et al. 2007). It is therefore worth investigating the factors underpinning the attraction of this particular group to the inner city and exploring how they take advantage of the benefits of inner-city living on a daily basis. A possible explanation is that there is a link between the individualism, fluctuation and instability of the non-traditional households and their quest for *residential flexibility*, understood as the degree of facility with which a household is able to change its use of, or movement through, the built environment, in response to altered social, economic or political circumstances (Buzar and Grabkowska 2006: 161). This chapter aims to examine this association, as well as providing empirical evidence that it remains valid in the case of non-traditional households living in old built-up neighbourhoods of Gdańsk, one of the case-study cities of this volume as introduced in Chapter 6.

Theorizing Residential Flexibility

One of the most significant social spaces is home. Its multiple functions, ranging from protection against the elements, through to provision of places for rest, entertainment, study, work and other activities, to serving as a space for self-expression and identity formation, can be classified as biological, mental, cultural and economic (Wallis 1977: 8). These functions and their hierarchy keep altering throughout the lifespan of the household. As Jałowiecki (1980: 26) points out, the substance of 'home' equates with the capacity for ceaseless adaptations of a dwelling, according to the number, age and changing requirements of household members. This adaptability has gained even more importance in the current 'age of flexibility' associated with demographic change, a quickly evolving post-industrial work environment and a remodelling of lifestyle patterns.

In view of the fact that the 'new' types of households appear to be particularly susceptible to instability and operate in a relative state of flux, it is clear that their residential preferences do not necessarily conform to those of the traditional core family. For instance, it is expected that a young childless cohabiting couple who are freelance workers are more likely to appreciate living in a converted loft within walking distance of the city centre rather than enjoy life in an attractive, though far-off, newly-built suburban housing estate. Thus, the old inner-city built-up areas seemingly have the potential to offer a flexible environment for those urban dwellers who value and seek after accessibility, versatility, functionalism, as well as the ambiance and appeal of historic architecture.

Although the state of preservation of old building structures often entails sizeable investment, the trade-off is their extraordinary convertibility. Their large size, well-designed yet adjustable division of living space and endless possibilities for rearrangements by means of erecting and demolishing partition walls, adding mezzanines, annexing adjacent flats, attics or basements, may be regarded as prerequisites for adaptability. This accords with the findings of Altaş and Özsoy (1998) which show that users' positive evaluations of satisfaction with the dwelling size can be determined by its potential to create new uses (ibid.: 319).

Spacious interiors allow for and facilitate work-at-home or flat-share arrangements – a matter of importance for young urban professionals, single parents with pre-school children, students and the like.

Furthermore, also important are the aesthetic and sentimental values of old architecture. Results of a survey examining public perception and evaluation of new-built and pre-1945 housing structures in Poland reveal a remarkable level of liking among the younger generations for the latter (Kaltenberg-Kwiatkowska 2004). In response to the question about the appropriate way to treat historic buildings in very bad condition a quarter of respondents opted for their renovation regardless of the cost.[1] The supportive attitude however appears to correlate with the age of the respondent since the share rises to 36 per cent in the cohort under 25 years old. Even if this kind of differentiation can be partly explained by the pragmatism of older age groups as opposed to the younger age groups' idealism and romantic assessment of old buildings, it is evident that the stereotype of older neighbourhoods as ramshackle and only suitable for slum clearance is gradually disappearing.

Certainly, the inner-city areas of European cities dominated by pre-1945 housing, typically turned into neglected areas of social and urban decay during the second half of the 20th century. Nevertheless, mostly because of its central location, the inner city boasts a number of strong points such as easy access to public transport or a range of urban amenities. It seems that after an extended period of relative disregard and neglect, the long-underrated advantages are now once more being recognized as signs of a revival of old residential centrally located districts that can be seen in many cities across Europe (see Chapter 5). Given that the ongoing processes of socio-demographic inner-city transformation are not only quantitative, but also more importantly qualitative, they must be seen as more than just a simple repopulation.

Residential change in the inner city is intricately connected with the residential flexibility of inner-city dwellings, that is the adaptability of flats in old buildings (connected to the way they are built and the associated possibilities of use for housing, work and everyday life) together with the advantages resulting from a central location. Accordingly, households inhabiting inner-city dwellings can benefit from the elastic capacity for as-the-need-arises housing adjustments as well as the infrastructural richness of the immediate surroundings. The notion of residential flexibility, even if not named directly, appears as early as in the 1940s. Svend Riemer, author of *Sociological Theory of Home Adjustment* (1943), recognized the composition of the family as 'a fundamental determinant of housing needs' (ibid.: 276) and develops his line of reasoning that

---

1   The remaining options included: 'renew only when the costs are not higher than construction of a new building of similar character' (37 per cent of all answers), 'renew only such historical monuments like churches, town halls, palaces, but not the housing stock, unless one has money for it' (28 per cent), 'destroy old buildings and build modern houses instead' (7 per cent) and 'I have no opinion' (3 per cent).

neither the housing needs of the family nor the technical means of the architect can be formulated in terms of fixed standards or quantities. Housing needs are relative. [...] [Meanwhile, the] satisfactory home adjustment can be achieved in two ways: 1) via the tangible, objective part of the physical shelter, and 2) via the more subjective part of individual attitudes and family behavior. (ibid.: 277)

Elaborating on this idea, Turowski (1979) proceeds to argue for the 'creation of "flexible" dwellings, allowing their inhabitants for adjustment to the diversifying and changing needs over time' (ibid.: 9). Other followers of Riemer's theory, Morris and Winter (1975), present three strategies for coping when the housing needs of a family can no longer be supplied. These include: residential mobility (moving to a more suitable dwelling), residential adaptation (introduction of functional and other changes in the structure of the current dwelling) and family adaptation (conforming to the existing housing situation). The choice of strategy is expected to be based on *a process of weighing alternatives* (ibid.: 84) and although moving is expected to be the typical 'first choice', it is underlined that 'individual families and classes of families may order the behaviour preferences differently' (ibid.: 85).

Since the 1970s there has been a proliferation of literature on residential mobility and housing adjustments of traditional families (Speare 1970, McAuley and Nutty 1982, Clark et al. 1984, Deane 1990). Despite the unquestionable value of these studies, the adopted linear life-cycle approach with successive stages preceding, coincident with and following marriage totally neglects non-traditional families. An attempt to differentiate between the residential behaviour of various household types was made by Frey and Kobrin (1982), who demonstrated that married couples with children tended to be more attracted to the suburbs than central cities of American large metropolitan areas than 'other families and primary individual households and showing that the latter represent the greatest potential source of city household increase' (ibid.: 270).

Even though non-traditional households have gradually entered the scope of research on residential rationales and conditions, few of the studies explicitly cover the issue of the motive forces behind non-traditional households' attraction to the inner city. A partial explanation of this phenomenon is provided by Bromley et al. (2007) who investigated the attitudes and opinions towards city-centre living among inhabitants of centrally located neighbourhoods in Birmingham, Bristol, Cardiff and Swansea, differentiating between newcomers and long-time residents to trace the dynamics of change. According to the results, a large proportion of the newly arrived are young adult one-person households, which is only partially associated with the growth of student populations. Quite unsurprisingly the mundane location advantages of city-centre dwelling, such as proximity to facilities and services or convenience of living in general, appear to be most valued among young adults and older residents alike. The differences are to be found in the domains of work and social life, in perceptions of stylishness of the city centre as well as in the commitment to living in the current accommodation and are generally attributed to a generational lifestyle transformation. Nevertheless, the description of the 'new populations' is limited and no further analysis of the household change implications is provided. Neither are the housing preferences of newcomers taken into account, apart from a short discussion of property tenure issues and the recognition of the importance of renting.

The standpoint adopted here follows from the outcomes of a pilot research project undertaken in two districts of inner-city Gdańsk, Wrzeszcz Dolny and Nowy Port (see also Buzar and Grabkowska 2006). Aimed at uncovering the residential behaviour of households inhabiting the two case-study neighbourhoods, forty two structured interviews were conducted with themes ranging from household configuration, housing biographies and changes to the layout of the flat to daily living arrangements and general perception of the built and social environment. The results allowed for a differentiation of housing strategies according to the choice between residential mobility (capacity for removal from one home to another) and spatial flexibility (ability to adapt the home to changing needs). Four possible permutations (mobility/immobility versus flexibility/inflexibility) meant four typical household situations could be summarized in a square of oppositions (Figure 10.1).

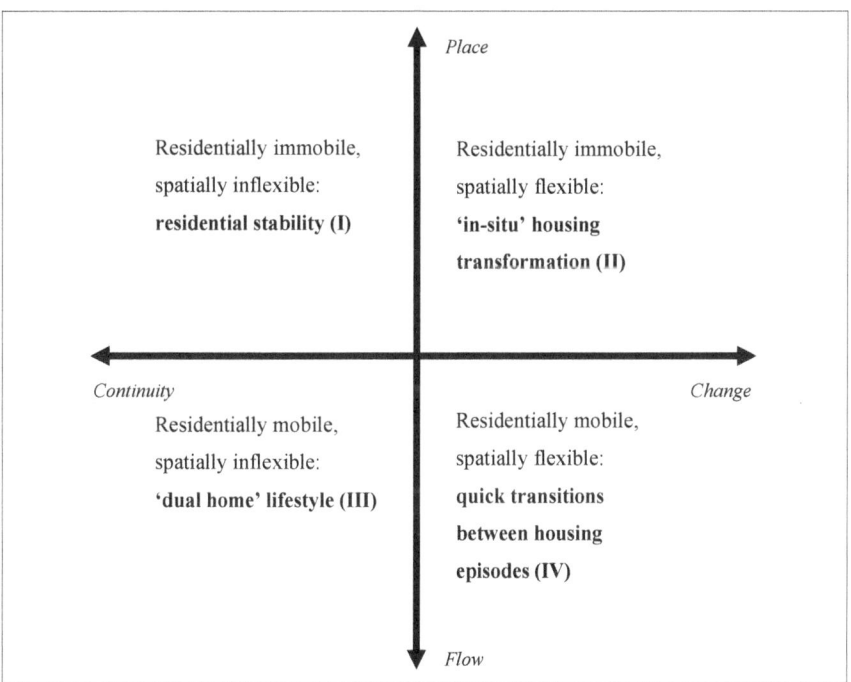

Figure 10.1 Residential mobility and spatial flexibility patterns: A square of oppositions

Source: Buzar and Grabkowska (2006: 165).

The follow-up research discussed in this chapter focused on the two cases involving spatial flexibility (types II and IV) with the recognition of *'in-situ'* housing transformation as characteristic of the post-socialist context and rapid transitions between different housing as commonly executed by young nest-leavers.

The former argument is in line with the reasoning of Mandič (2001) who shows the prevalence of 'in-place dynamism' in her study of residential behaviour of households in Slovenia attributing it to the specifics of the housing market during post-socialist transition. Issues like housing shortage and resulting low mobility rates, the dominant preference for ownership and the persistent 'move-to-stay' approach to residential relocation can be used to account for the similarity between the Slovenian and Polish housing patterns which are not the same as the mobility-centred Anglo-Saxon perspective.

However, some patterns of household behaviour typical of North America and Western Europe seem to be found in post-socialist cities and have been adopted especially by the younger generations of urban dwellers. Young Dutch adults' household and residential decisions made on leaving their parental home are investigated by Mulder and Manting (1994). The term *flexibility* is used by the authors to describe the strategy which resolves itself into the postponement of settling down (that is entering marriage and homeownership) so allowing the individual to 'keep all options open' and at the same time giving temporary priority to his or her educational and occupational career over the household and housing career (ibid.: 157). Traces of ongoing studentification in Polish inner cities suggest that such is the typical path followed by flat-sharing students who represent residentially mobile and spatially flexible households of the fourth type (see also Chapters 8 and 9).

The following section presents the methods and outcomes of a study undertaken in Wrzeszcz Dolny and Nowy Port, the two case-study districts of Gdańsk previously investigated in the pilot study, with the aim of supplementing the results of the preliminary research and supplying missing information on the ways in which non-traditional households articulate both types of flexibility, residentially mobile and immobile, in their inner-city dwellings.

Inner-city Dwelling in Gdańsk:
Residential Flexibility for Flexible Residents?

The post-socialist city of Gdańsk, a seaport and part of the largest urban agglomeration in northern Poland, appears as an interesting setting for a study of residential flexibility for a number of reasons. First of all, it is a city with a long and complex past resulting in the multifaceted social and cultural landscape of today intertwined with the palimpsest of built structures from different periods of time. Secondly, the multiplicity of socio-economic processes which have been taking place over the past two decades following the systemic change renders the city an excellent urban laboratory. For instance, as illustrated in the previous chapters, according to the statistical evidence the second demographic transition is well under way, even if not all of its characteristics are detectable through census data. Thirdly and not least, despite the fact that the inner city of Gdańsk suffered relatively little war damage (except for the historic city core) and therefore fulfilled a key role in the post-Second World War period, it has since then experienced great change. The inner-city neighbourhoods with their high proportion of municipal flats have gradually fallen behind the newer districts because of lack of investment in the old buildings. Nonetheless, the inner city currently seems to be experiencing a certain renaissance, being increasingly recognized as well-located and simply convenient if not (yet) fashionable.

In order to capture these initial reurbanization processes, which is almost impossible to do from the statistical records, a purely qualitative research method was employed, involving detailed semi-structured interviews with selected households from Wrzeszcz Dolny and Nowy Port, the two case-study inner-city districts as introduced in Chapter 6.

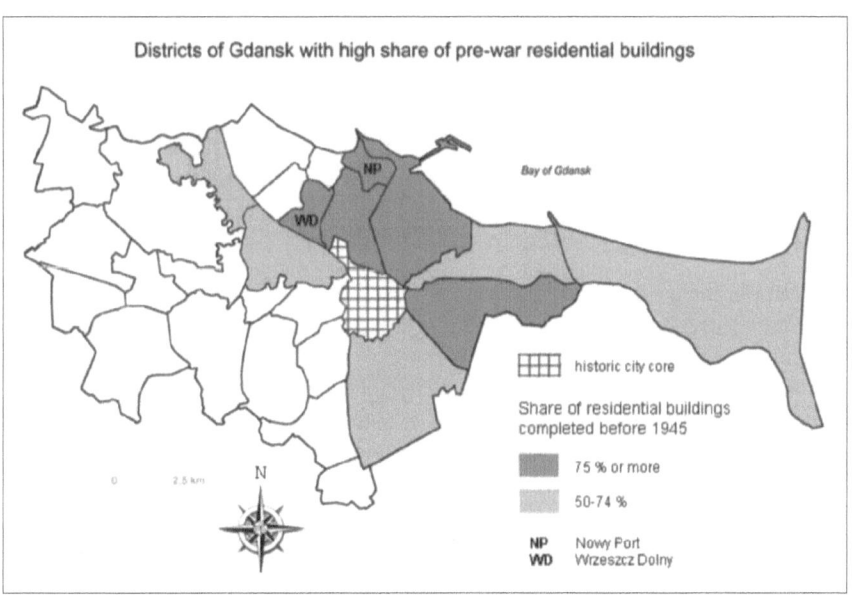

Figure 10.2 Location of Nowy Port and Wrzeszcz Dolny within the inner city of Gdańsk

Source: Author's work.

Two particular criteria were used for the selection of the interviewees: the households under investigation had to be *inhabiting flats in pre-Second World War tenement buildings* (that is housing structures representative of the inner city) *after having moved into them since 1989* (that is their location choice was not a result of the regulations of the state-socialist era). Traditional and non-traditional households alike were to be approached on equal terms to provide grounds for comparison, although in spite of this the latter were overrepresented in the sample interviewed. The following chapter is thus based mainly on their accounts in order to capture the assumed relationship between demographic and residential flexibility of non-traditional households as the agents of inner-city residential change.

Interviewees were contacted through snowball sampling and interviewed usually at their homes with the use of a digital recorder. The choice of interviewees was quite selective but reflected the aims of the study, which focus on identifying patterns of households' residential behaviour rather than constructing a fully representative image of residential change in the inner city of Gdańsk. In total, members of nine households from Wrzeszcz Dolny and six from Nowy Port were approached and interviewed on a number of issues connected with residential flexibility.[2] In addition, eight expert interviews were conducted with city planners, representatives of housing associations, local leaders of non-governmental organizations and estate agents in order to complement the inhabitants' accounts with the viewpoints of urban practitioners dealing with housing issues in Gdańsk on a professional basis.[3] The subsequent analysis aimed to uncover residential flexibility interrelations through their deconstruction with regard to households, dwellings and neighbourhoods.

Flexible Households: The 'Ever-searching People'

The sample households are dominated by various types of non-traditional households, predominantly young, highly educated, relatively well paid and pursuing high status professions – in other words, they are both socially and economically advantaged. The housing biographies of the interviewees show high levels of residential mobility, although its extent varies between different types of households. In the following discussion the question of the flexibility of inner-city inhabitants is examined, that is the households' needs and preferences together with their financial ability to move in their quest for flexible living.

---

2   In the following text the first letters of interviewees' nicknames indicate the name of the case-study district (N for Nowy Port and W for Wrzeszcz Dolny), while the endings specify gender (all feminine names end in 'a'). As in the other chapters, HH stands for household and E for expert.

3   It should be noted that in a number of cases the experts were former or present residents of the districts under investigation.

Results of the investigation into the interviewees' housing biographies confirm that residential mobility is often, but not necessarily, connected with transition from one household arrangement to another. Using the example of Natalia, aged 31, whose dwelling at the time of the interview was her seventh address (see Chapter 8), the stages of such transition included leaving the family home when she married, her subsequent divorce and remarriage. The story gets even more complicated in the case of Walery, aged 42, who left his family home to live in multiple flat shares during his studies outside of Gdańsk; he then rented a flat in Wrzeszcz with his newly-wed wife, bought a flat together with her as a married couple and finally stayed there as a single after divorce. Currently he is cohabiting but is undecided about the future of his relationship with his girlfriend. Nevertheless, it should be noted that the duration of the current flat occupancy amounts to two years in Natalia's case and is more than eight in Walery's.

As the household configurations are closely linked to housing possibilities it needs to be acknowledged that housing strategies are hugely dependent on the financial resources that are available as well as the material help provided by relatives. Especially where younger households are concerned, *inheritance of flats* appears to be a common way of obtaining a place of residence of one's own, alongside *purchase* and, to a lesser extent, *rent* (see Grabkowska 2007 and Chapter 8).

Young adults more usually inherit flats from their grandparents, rather than parents, on the principle of generational alternation.[4] A number of makeshift arrangements are also frequently applied, especially the practice of 'flat juggling' within a family. A single flat owned by a member of the extended family is then treated as available to all and used according to the current needs of individuals within the wider family group. For instance, several interviewees experienced living with an older relative both in order to provide him or her with company and to help with daily activities as well as to gain more personal space and to prepare to inherit the flat following the relative's death. Such was the rationale of Wojtek, aged 28 and unmarried, who moved into the flat of his late grandmother after his older brother moved out in order to provide company for their great-aunt. Wojtek is happy with such an arrangement since after sharing a room in his parents' home with his brother for almost 20 years he can now enjoy having a room of his own. His housing needs are thus satisfied for the time being, especially since his aunt's presence is only a temporary situation.[5]

Nearly all of the interviewed flat owners recognized low price as an important criterion for choosing an inner-city dwelling. Although soaring housing prices throughout the years 2004–2007 affected all types of flats, regardless of their

---

4   By the time the housing resources of grandparents are 'released', the housing needs of the parents are usually supplied and so grandchildren become the main beneficiaries of the bequeathed flats.

5   According to Wojtek much greater inconvenience is caused by his father's usage of part of the kitchen as office space for his advertising enterprise – yet another example of 'flat-juggling' arrangement.

location, the pre-Second World War housing stock of Nowy Port and Wrzeszcz Dolny retained their competitive edge. Some interviewees admit that in order to lower the cost of purchase even further they looked for a flat in poor condition with the intention of restoring it in stages at a later date (see below). However, this does not relate to Wanda, a divorcee and mother of two teenagers, who wanted a flat in good condition in order to move straight in so as to save money and time, both of which are rather limited resources in a one-parent household.

Price-connected pull factors are also important for tenants, represented in the case study by flat-sharing students. The location of a number of higher education institutions in Wrzeszcz, including a technical university and a medical academy, accounts for the considerable 'studentification' of its immediate neighbourhood (Gdańsk HH W9) whereas the quite well-connected Nowy Port is regarded as a cheaper second-best option (Gdańsk HH N4, N5). The large size of the pre-Second World War flats in both case-study districts permits coexistence of many occupants which keeps the rent low, while the relatively central location and accessibility minimizes or simply eliminates commuting costs. Overcrowding and lack of personal space, as well as the basic quality of interior design are among the necessary compromises. Nonetheless, the temporariness and transitional nature of flat sharing seem to lessen the students' current expectations, as opposed to those exhibited towards the idealized future 'real' adult home (see also Kenyon 1999). As concluded by Norbert, one of the flat sharers, 'I don't think any of us thinks of what (s)he could change here or what is missing. Everyone treats it as a transitional place and does not intend to stay here for long' (Gdańsk HH N5).

Quite understandably, most of the interviewed students have unspecified plans for the future and admit to *ad-hoc* decision-making. As both the flat-sharing and makeshift arrangements are transitory by definition, it is not very likely that they will last for extended periods of time.[6] Wojtek's outlook for the near future is, for example, very vague:

> I don't know how much longer I will stay here, my brother [currently living abroad] has recently told me that he plans to come back [to Poland and to the flat] in August, so maybe I will stay here until August and then, … I don't know, … I also plan to go [and live] abroad sometime. … Perhaps I will join my friend who lives and works in London. (Gdańsk HH W1)

These examples show how the choice of residential strategies may be influenced by household configuration. Single people for instance seem to be fairly autonomous and unfettered in their decision-making, while a single mother

---

6 Some of the flat sharers however declare that they have become emotionally involved with the district, as for instance Winona, a student of architecture, who cherishes the idea that she might continue to live in Wrzeszcz Dolny after graduation. Together with her flat mates she dreams of jointly owning an old tenement house with their common flat in the attic, design office on the ground floor and flats for rent in between (Gdańsk HH W9).

is forced to find a middle way between her own housing needs and the needs of her offspring. Consequently, the presence of children in a household appears to be an impediment to residential mobility, especially in the case of younger schoolchildren who have their own social networks in the original place of residence.

This particular reason is perhaps one of the key factors determining traditional households' attachment to suburban environments once they have decided to settle down there and have children. Such attitudes are apparent from remarks made by a representative of one of the local property companies, who recently launched a development in Nowy Port and who admitted to having thought about the (eventually rejected) possibility of moving there from an estate in southern Gdańsk. Although the relocation would have made the daily commuting of the interviewee's husband much easier, the argument which tipped the scale against it was the fact that her son had already started attending school in the district and she thought that moving him away from his friends would be wrong (Gdańsk E2).

The high residential mobility revealed in the housing biographies of most of the interviewed non-traditional households does not though necessarily mean a low attachment to their current place of residence. On the contrary, the interviewed owners had made well-thought-out and careful choices of their flats (see Chapter 9) and simply do 'not think about leaving' (Gdańsk HH W7). Furthermore, Wiola, who has lived with her husband in their current flat for eight years, maintains that 'something very, very strange would have to happen in order to make us change our mind and move out from here' (Gdańsk HH W3). All the same, she underlines that

> we are ever-searching people and we still don't know what we will be doing in life. ... If we, for example, set up an ostrich farm or engage in dog breeding then we will have to change the place of our dwelling. At present we are 30 years old and we don't have a clear vision whatsoever. (Gdańsk HH W3)

In summary, the residential mobility patterns vary between the interviewees and show a close connection to household type and form of tenure. The most flexible housing strategies are employed by the renting and flat-sharing students while owners with children seem to be the most restricted in their housing choices. The possible reasons behind the high level of residential satisfaction are presented in the next section which provides a view of the adaptability of inner-city dwellings.

Flexible Dwellings: 'According to Our Own Ideas'

Some of the interviewed households' reasons for moving to pre-Second World War flats in inner-city Gdańsk are introduced in Chapter 9. However, these reasons appear to be significantly enhanced in the interviewees' accounts of their daily geographies by the concrete examples they give of the ways in which the flexible setting of home facilitates the everyday practice of living.

The notion that 'lifestyle matters' is found throughout the interviews as respondents reveal their desire for self-expression, individualism, creativity, work fulfilment and a thriving social life. The attitudes and corresponding activities of the 'ever-searching' city-dwellers are all reflected in the design and the ways they use their flats. For instance, Wiktor and Wiola, a childless married couple, appreciate the 79 m² capacity of their flat which allows them to display their large and much prized collection of books and historic artefacts. The open-plan kitchen and adjacent antique-furnished sitting room serve both for family and social life, with regular get-togethers. In the case of Walery, even the balcony plays a significant role being appropriated at party time as extra floor space, an extension of home. Since Walery is engaged in numerous civic movements, his spacious flat is also frequently used as an activist operational venue.

The flats under examination very often contain working spaces, either in the form of separate study rooms or delimited 'professional corners'. Considering the occupational profile of the interviewees it is unsurprising that many of them work from home. The fact that their flats double as workplaces is valued because of the convenience and comfort of working from home, even if it requires a lot of motivation and good organization skills:

> The advantages [of working from home] are such that I can freely organize my time according to my own needs. Some things may be done later or I can work during the night. (Gdańsk HH W3)

For several interviewees the equipment of the domestic workplace is limited to a portable computer and a table or desk hence 'in the era of the laptop the working space is kind of more liquid', as stated by Wawrzyniec (Gdańsk HH W6). However certain types of work activities require special equipment and settings. For example, moving into her pre-Second World War tenement flat finally enabled Natalia to pursue her childhood dream of becoming a furniture restorer. Her previous dwellings lacked the necessary space and adaptability which are the main advantages of her current 85 m² living space augmented by a 60 m² basement. Recently the basement was renovated and tailored to meet the standards of a workroom while a writing desk in the living room functions as an office for her website-based furniture restoration company. She also employs her living room as a showcase – it is filled with restored old furniture and curios, most of it found and bought cut-price on internet online auctions. One side of the room is occupied by a 100-year-old Gdańsk neo-baroque concert piano, featuring an ivory keyboard and stylish candelabra. Natalia remarked that she has more and more work to do as 'a lot of people move, furnish, redecorate and so the number of commissions is increasing' (Gdańsk HH N1) which provides a hint that *'in-situ'* adaptation and change is a common social practice.

In fact, a distinguishing mark of most of the flats visited during the interviews is that their designs merge new look with old architecture. The walls may be painted in bright colours, IKEA furniture may be introduced, walk-in wardrobes installed but at the same time old ceramic stoves are displayed and paint is removed from the wooden window and door frames to expose their natural beauty. Some interviewees emphasize their personal contribution to the appearance of their dwellings, for example, Walery boasts about designing and carrying out (with help from his father and brother) all the interior decoration of his kitchen (featuring wooden handmade cupboards, fragments of wall with uncovered brickwork or textured plaster and a beamed ceiling adorned with wicker twigs).

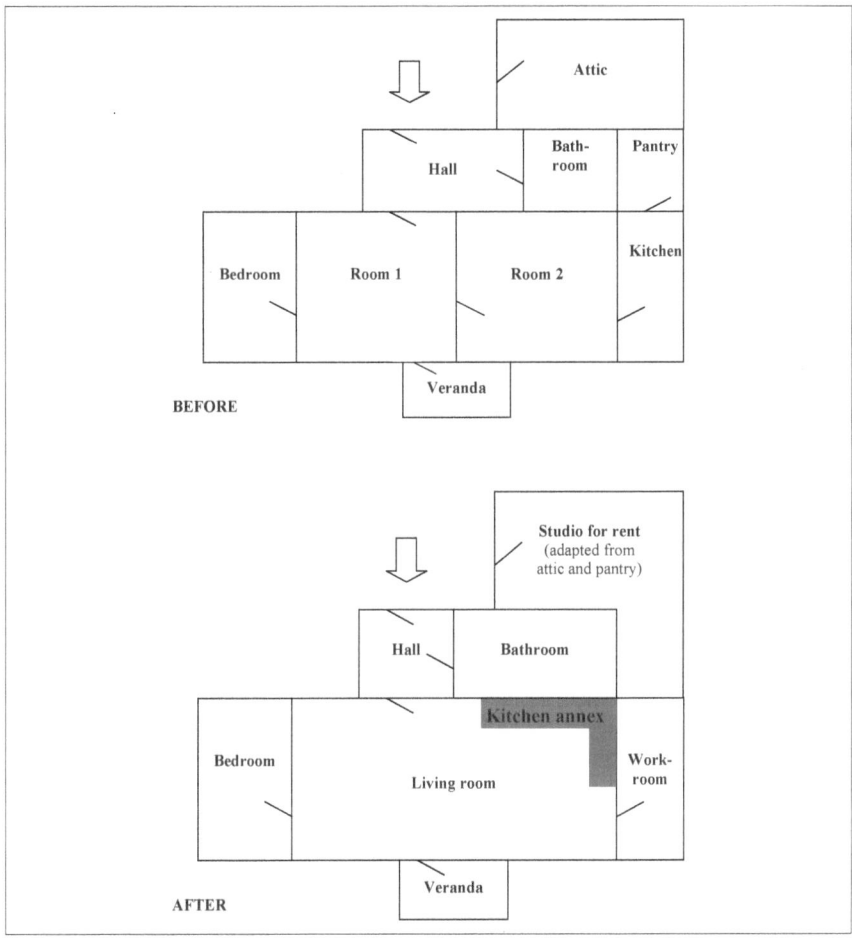

Figure 10.3 Changes of the layout in Witold's flat
Source: Author's work.

Individualism and creativity seem to be shared approaches towards the appearance of flats, but the most spectacular manifestation of this is evident in relation to changes in the construction and layout of flats. Many flats have undergone massive transformations aimed at adjusting them to the needs and tastes of their inhabitants. One of the methods of transformation can be described as 'once and for all' and is exemplified by Nina:

> After moving in we demolished everything [inside of the flat] and then rebuilt it from scratch according to our own ideas. […] The only thing that now keeps changing is the interior design and decoration […], apart from that nothing will change now. (Gdańsk HH N3)

Witold speaks of two phases of major repairs, 'the first was general repairs, as this flat was in ruins … and the second dealt with finishing off' (Gdańsk HH W5). The extent of work done was immense as it included a complete makeover of the original layout – turning two small rooms into one large room, the construction of a kitchen annex in the enlarged living room, the conversion of the former kitchen into a workroom and extension of the bathroom at the cost of the hall, all of it followed by adaptation of the adjoining attic space into a separate studio for rent (Figure 10.3). Asked about the reasons for such dramatic changes Witold explains that 'the two small rooms were in fact like dog kennels and it was very impractical for me not to have direct access to the kitchen [previously it was located at the far end of the flat]' (Gdańsk HH W5: 82). He also admits that the design of the living room dictated by his 'open house' policy turns out to be very guest-oriented.

The integration of the kitchen into the rest of the flat proved to be of special importance to the interviewees. It allows for the person preparing the meals to be with the rest of the household and visiting guests and very often is the centre of family and social life. In the flat of Nora and Norman it is the big kitchen table which attracts their visitors rather than their fancy living room. A number of interviewees admit that the function of rooms tends to be flexible, among them Walery:

> What are the [room] functions? Er, well this room, … I treat this room as a room where I have my computer, where I sleep, work, etcetera, and the other room is, … I don't know, … more social [that is intended for socializing]. Anyway, at this moment it's my partner's realm. (Gdańsk HH W6)

The room's functions also tend to fluctuate along with the household's ebbs and flows. Natalia's patchwork family has reached the stage where her stepson is about to leave home for university in another city:

> His room will be cleared and we will move in there with our bedroom, whereas the current bedroom will become an office and an occasional guestroom. In

point of fact it is only a matter of some rearrangement of the furniture. (Gdańsk HH N1)

The interviewees' flats vary in size from 45 up to 140 m², most of them consisting of three rooms, and so they are regarded as relatively big. Despite the conventional opinion that one-person households need less living space, the research shows that taking into account their specific lifestyle requirements they may demand more space than a nuclear family.[7] The size of flat also makes a difference to renting flat sharers, who seem to gain more from economies of scale than from benefits connected to adaptability of dwellings. Namely, since they are not owners, they usually cannot introduce many physical changes to the flats and thus make full use of their adaptability. However, the bigger size of flats allows more persons to live together in a flat share and thus 'save' money on the fixed costs.

There are obvious restrictions as to how far a rented flat can be transformed. However, the members of the three flat shares who were interviewed, appear to have managed to carry out some fine-tuning. For example, Winona and her friends painted the walls decorating them with colourful pictures and brought some furniture from home. As the flat was not in particularly good condition at the time of moving in they also undertook a few minor repairs and upgrades, including replacement of a broken boiler, wallpapering the kitchen and tiling the bathroom, all at their own expense. By contrast the flat inhabited by Nela and her flatmates was thoroughly renovated by the owners before they rented it out and any necessary repairs are dealt with directly by them. The rooms were well-furnished and the slight additions by the tenants were only aimed at domestication of the interior design.

Although the layout of the shared flat is fixed, different arrangements among the household members are put into practice. In Winona's case the two bedrooms are divided between two subgroups of friends with a shared kitchen and bathroom. A similar division has been applied in the mixed flat share – of the four rooms, two are individual and the other two are shared. In the less crowded four-person flat share of Nela it has been possible to arrange the space into two bedrooms, each for two girls, and a communal living room with a TV set and a big table serving as a desk for the laptops. It is, nevertheless, generally agreed among all the flat sharers that the greatest advantage offered by their dwellings, apart from 'economies of size', is their location in the inner city.

Before moving on to discuss the location-oriented flexibility of inner-city flats it is important to emphasize the relative 'interior-flexibility' advantage of pre-Second World War dwellings over those situated in newer buildings. An insight to the issue of the small size of flats in prefabricated housing estates, a consequence

---

7   According to the SINGLETOWN experts single households are not satisfied with 25 m² which is the normal amount of space you would have for a person if you are sharing – a family of four might normally have 120 m². Single people tend to have a minimum of 75 m² but they want 100 or 200 (SINGLETOWN 2008: 7).

of rigorous norms and regulations of the socialist period (see also Figure 3.1 in Chapter 3), is provided by Wojtek, who mentions that he had to share a room with his brother almost until he graduated. The very same norms and regulations are responsible for the fact that in many blocks of flats built after the Second World War the kitchens are relatively small and sometimes windowless which substantially reduces their comfort of use. Similar limitations are either nonexistent or much more easily overcome in the pre-war flats. For example, the lack of bathrooms not included in the original layout was dealt with, often illegally, through their construction by the self-reliant inhabitants.[8] Even in comparison with the newest housing developments the old flats are much more adaptable. Such at least is the opinion of the estate agent who, when interviewed, criticized the current primary market offer in Gdańsk as ready-made flats created according to the architect's vision and which did not take into account the needs and preferences of potential buyers (Gdańsk E8). It is, therefore, something of a misconception to think that the rigidity of the socialist cost-effective prefabricated architecture has been eliminated with the arrival of capitalism, as private property companies in Gdansk at least seem to be acting in a similarly rigid way.

Flexible Neighbourhoods: 'You just don't waste your time'

The major benefit of an inner-city location, mentioned by the majority of the interviewees, is the proximity to work. For example, a five-minute walking distance to his workplace was the primary reason for Noel to join a flat share in Nowy Port. For several other interviewees from both Wrzeszcz Dolny and Nowy Port alike, the notion of proximity to work does not necessarily relate just to physical distance. Nina's job, for example, involves a lot of driving around the Tricity area during the day, from one big shopping centre to another, where she organizes promotional events.[9] For her it is a blessing not having to be stuck in the traffic jams on the roads connecting the upper-terrace part of the city with the lower terrace, which would be inevitable were she to live in the suburbs.[10] Likewise, Wiola would not wish to live in any of the suburban estates as she maintains that commuting from there to the inner city is "a horror" (Gdańsk HH W3). Nikodem, deliberating on a possible alternative location of his dwelling, concludes that

---

8   Several interviewees speak of such operations having been undertaken by the previous owners or tenants, while a couple of them describe their own actions aimed at enlarging the existing bathrooms.

9   The Tricity agglomeration consists of the neighbouring cities of Gdańsk, Sopot and Gdynia (see Chapter 6).

10   The terms *upper-* and *lower-terrace* refer to the specific physiography of the city of Gdańsk, which is divided along the edge of a moraine plateau on a NW-SE axis. Because of this natural barrier the earliest spatial development of the city was restricted to the lower terrace while the urban expansion on the upper terrace has intensified only in recent decades (see Chapter 8).

there are things which I find totally unacceptable, like, for example, having to spend one and a half hours in the car every morning in order to get to work. For this reason the southern districts [of Gdańsk] are in my case completely out of the question. (Gdańsk HH N6)

The inner-city location is thus viewed as advantageous compared to the upper-terrace peripheries and since many of the households do not own a car the accessibility to the transportation systems rather than just nearness of home to work is important.[11] To Walery his home has an excellent location in terms of communication (Gdańsk HH W4) and this view is shared by most of those interviewed in Wrzeszcz. Even Witold who mainly moves around by car, admits to using public transport during bad weather or roadwork-induced heavy traffic and considers the presence of the light railway (called SKM), two tram lines and multiple bus lines to be 'a very big plus' (Gdańsk HH W5).[12]

Access to the SKM is not possible directly from Nowy Port though, which generally adds to its relative separation from the functional centre of Gdańsk.[13] However, most of the inhabitants of Nowy Port do not complain about its location within the Tricity communication system and some even speak of 'an underestimation' of the district's location as 'after all, either by tram or car access is quite efficient, which isn't the case for the southern districts' (Gdańsk HH N6).

Closely linked to the issue of inner-city accessibility is another flexibility factor – the more general notion of 'closeness' (for example, Gdańsk HH W1, W8, N2). This can be understood as proximity to public transport, shopping and services and is described by Wanda 'as one of the key elements of [her household's] quality of life'. Similarly, having 'all necessary [urban facilities] at hand' means that 'you just don't waste your time' (Gdańsk HH W7). The ease of moving about on foot is an additional advantage. According to Wojtek

It is always possible to walk out of the door and get everything. [...] Here we have a 24-hour food shop, over there is the butcher's. ... Somewhere there you can find the cobbler ... and there is the hairdresser. [...] I like the fact that this network is so dense ... because back in Morena [a large high-rise housing estate] you had to walk quite a way even to get to the grocer's. Here I can do this without taking my slippers off. (Gdańsk HH W1)

---

11   Some respondents explain their not having a car by pro-ecological attitude and lifestyle prerequisites ('SKM or a bicycle – these are my most frequent means of communication'; Gdańsk HH W4), while others openly speak of economic factors ('I think it would be just stupid to [live here] and [pay to] maintain a car'; Gdańsk HH W7).

12   It is worth underlining that the role of SKM as an alternative not only facilitates communication within Gdańsk but also provides a convenient link to Sopot and Gdynia.

13   The branch line between Nowy Port and the Gdańsk Main Station was closed in 2002 due to unprofitability. Interestingly enough, it was the first branch of the SKM (opened in 1951), which only later was extended along the Tricity main communication axis.

The inner city therefore, according to the interviewees, appears to offer a better residential environment than the high-rise estates. Wrzeszcz Dolny however ranks higher, being described by Walery as 'the commercial base for the whole city of Gdańsk' (Gdańsk HH W4), especially as in recent years the range of trades and services on offer has been supplemented by two large shopping centres.[14] Apart from being a shopping area, Wrzeszcz (as a whole) is also seen as a fairly attractive entertainment and eating-out area, although many interviewees said an increase in the number of pubs, cafés and restaurants was needed.

Nowy Port on the other hand, is less vibrant, but its inhabitants claim that it provides all of their basic needs and in recompense for the district's more peripheral location they benefit from peace and quietness and a sense of community:

> It is specific for a small neighbourhood that everyone knows each other here. If someone lives here for a while and goes to the same shops, then people start addressing you by your first name, the shopkeepers know your shopping preferences, you can have a chat with everyone. ... The atmosphere is like that of a small town and old-world traditional at the same time, I like it very much. (Gdańsk HH N1)

> Such are the advantages in the eyes of the owners. The students who were interviewed as previously stated value proximity to the university campus which has to be balanced by the higher price for flats in the case of Wrzeszcz. Students of architecture from Wrzeszcz Dolny justify their choice of location because of the need to carry bulky models and one of them admits that her determination to find accommodation close to the university at the beginning of her studies was dictated by unfamiliarity with the layout of the city (Gdańsk HH W6).[15]

The initial unfamiliarity with the city experienced by newcomers limits their daily movements and seems to influence their choice of location. For example, Nela, a first-year student, does not yet know her way around Gdańsk nor has she developed social networks, and therefore spends most of the day at home studying or communicating with her family and friends at home through the internet. Her longer resident flatmates, however, tend to spend a lot of time away from the flat, travelling around Tricity to attend various social activities. Hence, access to public transport is of key importance to them. The role of SKM as a link between Gdańsk and other parts of the agglomeration is also recognized as an advantage.

---

14   One of them, Galeria Bałtycka, contains 45,000 m² of commercial surface and has been advertised as the biggest shopping centre in northern Poland.

15   Her determination was so strong that it led to her renting, for almost a year, a small under-heated room to which the only entrance was via the en suite bedroom of the owner. This particular case also proves how the students' first-time residential choices may not be too picky.

To flat sharers from both case-study districts alike Wrzeszcz appears as a major city hub ('because this is where we spend most of the time at the uni or hanging out with friends', Gdańsk HH N5; 'Wrzeszcz is an important focal point in Gdańsk, it lies within the central axis', Gdańsk HH W9). Its old built-environment is also perceived as more attractive in comparison with the adjacent large prefabricated housing neighbourhoods, underlined, for example, by Wilbur:

> It's a cool district. I think it's much better to live here than in those [relatively] new block-housing estates, like Morena or Niedźwiednik (Gdańsk HH W9).

Some members of the flat-sharing households in Nowy Port do complain a little about the district's accessibility, especially those who remember the time when there was a SKM connection ('when I moved in here in 2000 this place was well connected [...] the SKM ran quite frequently and [...] it was the alternative means of communication [to trams]', Gdańsk HH N5). Nonetheless, opinions are again mixed, as Nela's flatmates, for example, accept a full hour as a tolerable time for 'short-duration commuting within the Tricity area' (Gdańsk HH N4).

With the exception of the mixed flat-share tenants (Gdańsk HH N5) the interviewees are generally enthusiastic about living in inner-city neighbourhoods. The most frequent complaints about both districts apply to the neglected physical condition of the pre-Second World War buildings and the passive mindset of the local inhabitants. However, the respondents who have been resident longer agree that the utterly negative stereotypes of Wrzeszcz Dolny and Nowy Port no longer hold true as both the material and social environment have been undergoing changes:

> It used to happen that [...] people thought of Wrzeszcz Dolny as a tricky district on account of the type of people living here ... but it's just a cliché which perhaps had its raison d'être some ten years ago, ... now it is all changing for the better. (Gdańsk HH W4)

> There have been a lot of changes. The turnaround is huge and when you walk in the streets, you only need to look at the new windows and cars. At the moment it is a district of huge contrasts. (Gdańsk HH N1)

The generally positive attitude towards both neighbourhoods was additionally confirmed in the answers to the 'good friend question'. Almost all interviewees claimed they could recommend buying a flat in 'their' district to a close friend and indeed, several had already done so.

In the case of Wrzeszcz Dolny its advantages are recognized by all of the interviewed experts. City planners, because of its central and lower-terrace location, see it as meeting the requirements of different types of household and becoming increasingly 'inhabitant friendly' (Gdańsk E1). The same interviewees did not think so highly of Nowy Port; however, the local investor rated it very highly regarding it as 'ideal for [housing] investment' and predicted that in future its location would be more favourable than that of Wrzeszcz Dolny, which 'will become traffic-jammed, excessively dense and its accessibility will decrease' (Gdańsk E2).

Apparently though, judging from the opinions of the interviewees, the case-study districts at present seem to win the competition with both upper-terrace peripheral suburbs and lower-terrace estates of prefabricated blocks of flats when it comes to a comparison of their locational and infrastructural assets. This component of the inner-city residential flexibility is also most commonly appreciated by all types of households examined in the case study.

Conclusions

The aim of this chapter was to investigate inner-city attractiveness in the context of residential flexibility, described as the aggregate of qualities concerning the adaptability of inner-city housing and availability of related locational advantages. It was assumed that non-traditional households would be especially attracted by the inner city as it seems to offer a flexible residential environment for those urban dwellers who value and seek after accessibility, versatility and functionalism, but also the particular ambiance and appeal of old-world architecture. This assumption was tested by research in two case-study neighbourhoods situated within the inner areas of Gdańsk. The adopted ethnographic approach included semi-structured interviews with households who moved into pre-Second World War flats in the chosen districts after 1989.

On account of the very specific post-socialist conditions it was expected that the spatial flexibility of the households under investigation would often be accompanied by residential immobility, resulting in the employment of an '*in-situ* housing transformation' strategy, while the young nest-leavers, including flat sharers in particular, would typically follow 'quick transitions between households', a strategy combining spatial flexibility with residential mobility. The outcomes of the study generally confirmed those tendencies and at the same time allowed for a number of additional observations.

To begin with, the interviews showed that despite the high levels of residential mobility recorded in the respondents' housing biographies, their current, inner-city, place of residence was in most cases regarded as satisfactory and was unlikely to change in the near future. Thus, the residential-flexibility potential of the physical structures not only attracts households to the inner city, but also encourages them to stay there, that is they support a certain residential behaviour. In other words, while one of the reasons for the migration of 'traditional' households to the inner-city neighbourhoods lies in their desire to supplement individual flexibility with spatial flexibility, moving there simultaneously decreases their need for residential mobility. Limited residential mobility however does not apply to the transitional arrangements of non-owners represented in the sample of interviews by renting students and the 'makeshift' occupant of an inherited family flat. The interplay of residential behaviour and characteristics of the place, as conceptualized in Chapter 2, depends on the tenure of residents, on their age and household structures.

Accordingly, articulations of flexibility connected to the adaptability of the dwellings vary greatly between owning and transitional households. While the former appear to undertake considerable investment in the structure of their dwellings in order to adapt them to their specific and perpetually changing needs and preferences, the latter, having less room for manoeuvre in this field, are relatively more bound to change their place of residence. Being constrained in their customizing capabilities the flat sharers and other renting young nest-leavers appreciate the locational advantages of the inner city even more. This component of spatial flexibility is more pronounced in Wrzeszcz Dolny, considered by owners and renters alike as more attractive and offering a wider spectrum of urban facilities than Nowy Port. Nonetheless, against a background of state-socialist housing estates and the new suburban developments, housing opportunities in both case-study neighbourhoods are rated very positively.

Another important finding of the study is that, apart from the markedly specific residential flexibility patterns of the young non-owners, the expected distinctiveness of housing needs and preferences resulting from a non-traditional household composition appears to be less significant than the particular lifestyle patterns. It therefore needs to be underlined that the residential flexibility features discussed render inner-city living advantageous to more than the restricted category of non-traditional households. Most of the conveniences are of universal scope, as for instance, proximity to shopping facilities and availability of public transport. Nevertheless, the representatives of the newcomer non-traditional households with their specific lifestyle patterns can be regarded as the harbingers of residential change. This change, if sustained and not developed into gentrification-led displacement of the long-established inhabitants, might translate into a genuine renaissance of Gdańsk inner city. Follow-up research needs to be conducted in several years' time to test this hypothesis.

Chapter 11

Brick or Block – Housing Preferences and the Urban Fabric

Katrin Grossmann, Annegret Haase

Introduction

Inner cities are not just another housing location. They are associated with the cultural heritage of cities, including their old housing stock. Starting from the Western European debate, inner cities, including the historic core, are nowadays regarded as the culturally most valuable part of the urban fabric, giving identity to the place. This was not always the case; for long periods of the 20th century, modern, newly built housing was the design of choice. But from around the late 1970s onwards, inner cities slowly regained the reputation of being desirable housing locations.

Often, old built-up areas – which are referred to as brick in this chapter – are contrasted with both monotonous suburban areas and post-Second World War large housing estates – which are referred to as blocks. In the western debate, the old built-up neighbourhoods are the 'good' or preferred locations that need to be preserved. Prominent pioneering voices like Jane Jacobs (1961) for the US or Alexander Mitscherlich (1965) for Europe spoke out for both urban development and research that emphasizes the potentials of old built-up structures, of 'grown' inner cities, for urban development. The criteria of historical preservation have shifted a great deal in the post-war decades. As a consequence, today entire housing areas are to be found on preservation lists. This emphasis on old built-up structures is also reflected in the discourse on the characteristics of the European city and in the debates about gentrification and reurbanization, which all assume that, apart from their location, inner cities are highly attractive for their brick qualities: high-quality old built-up areas with valuable building stock, large flats and individual ground plans and interiors.

Our approach to the perception of physical structures is certainly influenced by the German debate that shows a special affirmative bias for old built-up structures. Here, the demolition of inner cities, as well as the demolition of pre-Second World War housing stock through urban planning in the post-war period, together with the neglect of existing pre-war housing stock in the socialist era, led to a widely-held high regard for historical preservation efforts concerned with anything built before the First World War. In contrast, modern high-rises, housing estates from the 1960s, 70s and 80s have a rather poor reputation. Their image is that of undesirable neighbourhoods, characterized by social problems and ugly architecture (see Grossmann 2005, Richter 2006).

The poor image of block housing, of post-Second World War housing estates, is a phenomenon present throughout Western Europe. They are usually seen as places built with good intentions but bad outcomes. The literature on post-socialist cities generally agrees with this evaluation. Many scholars involved in the international debate assume that the large housing estates of post-socialist cities would see degradation similar to that experienced in the housing estates of western cities (Szelenyi 1996, Sailer-Fliege 1999, van Kempen et al. 2005: 360). Out-migration from the large housing estates to other parts of the city and the suburban hinterland is assumed to be highly socially selective. It was expected that highly educated people would leave the estates, or as Szelenyi put it in 1996:

> Those who can afford it are beginning to escape from them, leaving the poor and ethnic minorities to concentrate in them (a process of residualization familiar in the similar estates on the periphery of many Western European cities). As a result, the whole belt of "new housing estates" is likely to become the slums of the early twenty-first century. (Szelenyi 1996: 315)

Inner cities, by contrast, were expected in most cases to gain in attractiveness and, sooner or later, become the most preferred residential locations, together with newly evolving suburban areas (Karsten 2003, Seo 2002, Buzar et al. 2007, Haase et al. 2010; for Poland Parysek 2005, for the Czech Republic Sýkora 2005). Declining housing estates were expected to become increasingly de-mixed and areas of concentrations of poorer households, while the better-off moved to the upgraded inner cities and growing suburbs.

This chapter analyzes the role played by brick and block in the perception of neighbourhoods, as well as possible or implemented relocation decisions, or, to put it in the context of our model of residential change in Chapter 2: which residents prefer which places of residence, both in terms of built structures and location. Brick and block stand for different types of building stock. While brick buildings represent both the 'bourgeois' and working class architecture of the late period of industrialization up to the 1930s, blocks represent the mass housing built in the state-socialist period from the 1960s to the 1980s. These words are also synonyms for certain neighbourhoods. Whereas historical inner-city neighbourhoods are usually associated only with brick construction, large housing estates in the outer urban areas are associated almost entirely with prefab construction. However, it is necessary to stress that brick or block does not stand exclusively for a specific location in the city. In many cities, the physical fabric of neighbourhoods consists of a variety of building structures. In inner cities, brick and block housing can often be found in close proximity. Especially in cities affected by war damage and that were redeveloped in the second half of the 20th century, block infill development or replacement is quite common. In Brno, large housing estates were often built adjacent to a historical village; thus they also consist of mixed structures (see the information on Brno below).

Although this chapter is written from the perspective of the inner city, of necessity it goes beyond the boundaries of the inner city and uses empirical evidence from other neighbourhoods. The following questions are discussed:

- What role does the type of building stock play in housing preferences?
- How do preferences for a certain building material interfere with other preferences, with needs and constraints?
- What is more decisive, location or building material? How do types of building stock/physical features interact with location? Does it make a difference for preferences whether bricks and blocks appear in large clusters or in mixed physical environments?
- Can observed preferences be related to certain household types?
- Do we find evidence for the assumption of a selective migration to the inner city?

The chapter is structured as follows: it starts with a short introduction to housing preferences in post-socialist cities. Secondly, the physical characteristics of the two cities considered in this chapter, Łódź and Brno, are introduced. Thirdly, we present a survey of preference patterns found in the interviews. Finally the questions stated above are discussed and conclusions drawn.

Understanding Housing Preferences in Post-socialist Cities

Housing preferences are discussed as a combination of attributes of desired housing environments on the one hand and characteristics, goals and values of residents on the other. The attributes of the desired places usually comprise a bundle of factors such as location, facilities, transportation and so forth. The characteristics of residents are their social status, their economic resources and their attitudes and values. It is common to distinguish between stated and revealed preferences. Whereas stated preferences refer to the set of desired attributes of the residential environment, revealed preferences are those visible in the actual process of decision-making and the housing choice, where constraints and contexts come in. This branch of research is related to mobility research, housing market conditions, social segregation and lifestyles (Timmermans et al. 1994, Mulder 1996, overview in Coolen 2008).

It is not the purpose, and indeed, it would be far beyond the scope of this chapter, to provide a definitive and exhaustive summary of this debate. This section, by contrast, focuses on the question of the role of the physical structure of buildings in both revealed and stated preferences. Since research on housing preferences usually uses the perspective of individuals or groups, this question is somewhat unusual. In addition, the literature on housing preferences and mobility in East Central Europe tends to focus on suburbanization and aggregate spatial patterns. Therefore, we briefly lay out the framework conditions that are necessary to understand housing preferences in post-socialist cities, with particular respect to the type of building stock, that is brick or block. Four issues are discussed:

- brick and block as housing market segments in East Central European cities during state socialism,
- choice as a matter of constraint during state socialism,
- the impact of post-socialist transition on housing markets and
- housing careers in transition.

Brick and Block as Housing Market Segments During State Socialism.

As mentioned in Chapter 5, large parts of inner cities in East Central Europe, especially in Poland, still suffer from physical decay caused by a lack of investment throughout the socialist period. Most investment went into the building of large housing estates on the peripheries of cities. In many cities, prefabricated blocks were also erected within the inner city, either on brownfield sites or to replace demolished buildings or run down old built-up stock. Therefore, the proportion of prefabricated blocks is much higher in East Central European cities than in Western European cities, where only 10 per cent of the housing stock is in large housing estates; in Eastern Europe, large housing estates provide 40 per cent of the housing (van Kempen et al. 2005). Thus, it is the dominant segment, especially in the large cities (Rietdorf et al. 2001).

Choice During State Socialism as a Matter of Constraints, Allocation and Chance.

During state socialism, housing location decisions were not so much a matter of choice but of allocation by state authorities. Local housing markets were tightly regulated and the legacies of these regulation policies are still apparent today (see Chapter 12). In general, in order to obtain accommodation, people had to apply for a flat and then wait to hear from the local housing administration which assigned people to a flat wherever one was available. The applicant usually did not have a choice of location, as is the case in a housing market where one acts as a customer. Another way of getting a flat was to inherit the so-called right to a flat from relatives. Today, it is still common practice to move to a flat of a grandparent or other relative. The location decision is still, therefore, often made by chance rather than by choice.

Moreover, due to the severe shortage of housing in most East Central European countries throughout the 20th century, residential location decisions were made quite pragmatically during the period of state socialism. A flat in a large housing estate was regarded both as normal and, indeed, a good choice because of its comparatively high standard and modern facilities (hot water supply, a bathroom with a toilet in the flat, a lift, district heating). As a result, flats in large housing estates were occupied by nearly all types of households and social groups: one-person households and families, younger and older people, higher- and lower-income recipients. Thus, the level of residential segregation in large housing estates was relatively low, whereas in the old built-up areas in the inner cities, in spite of decline and physical dilapidation, socio-spatial differentiation was higher throughout the socialist period (as a relict of structures that were already there before the Second World War; see Hamilton 1979b: 227).

The Impact of Post-socialist Transition on Housing Market Segments.

In the post-socialist urban world, the discussion about the role and importance of brick and block needs to take account of the specifics of the post-socialist urban condition (Chapter 4). The situation after the political turnaround in 1989 has been very different from that in Western European cities. The inner-city areas with old built-up housing stock have continued to be neglected since 1989 and refurbishment efforts have been very selective. Only in eastern Germany was there legislation that supported the refurbishment of old houses through tax incentives as well as plenty of (speculative) investment in the old housing stock, which led to a massive refurbishment and substantial upgrading of old built-up inner-city areas. As explained in Chapter 5, in Poland and the Czech Republic, this has not been the case.

We observed in our field work that, in recent years, many of the prefabricated blocks in the inner cities or in estates have seen privatization, refurbishment and upgrading of the housing environment financed by the owner-occupiers themselves or also (partly) by funds of the European Union. Depending on local policies, refurbishment ranges from mere maintenance to experimental architectural remaking of the blocks. Whereas in Western Europe large housing estates often struggle with stigmatization, in Eastern Europe, living in large housing estates is considered to be a good and normal living arrangement for most people, including those of higher social status (Maier 2003, Demszky von der Hagen 2006, Barvíková 2009, Szafrańska 2009b). In the east, old built-up areas, on the contrary, have been long neglected and therefore housing conditions in many older brick buildings are poor, especially in less prestigious areas.

After the revolutions of 1989, the post-socialist cities started off with weaker or very different residential segregation patterns, and, as a result, the symbolic constraints of housing preferences for a particular type of housing stock were not as influential for location. Even now, the prevailing housing shortage across submarkets continues to foster these inherited patterns. In addition, the pre-conditions for choice and the constraints have changed. While the liberalization of the housing markets and the creation of new segments (suburban housing, new multi-storey condominiums) have led to an increasing choice, constraints have increased, too (due to rising housing costs, the impoverishment of particular groups and more transitory arrangements that often have income limits). Therefore, we must look critically at the explanatory value of housing preferences by categories of building stock as a decisive factor for relocation choices and for revealed preferences.

Consequences of Changing Housing Arrangements for Housing Careers and Preferences.

Post-socialism has brought about considerable changes in household and family formation as well as a rearrangement of private life scripts and life phases (Chapters 3, 8 and 9). *Firstly*, urban societies in Poland and the Czech Republic are undergoing processes of individualization and singularization, similar to what has been happening in Western European societies. There is, *secondly*, a general postponement of demographic decisions such as marriage and childbearing (see also Coleman 2005) as well as a trend towards smaller households and a pluralization of living and housing arrangements ('new' household types). This creates, *thirdly*, an increasing need for specific housing to meet the needs of different types of households (bigger flats without room hierarchy for flat shares, small flats for singles etc.). And it leads, *finally*, to a change in housing occupancy duration and housing careers. As a result, there are some groups that are highly mobile and move very often. They may live in various housing arrangements in a relatively short time. It seems that the housing choice of urban households is increasingly dependent on subjective and individual decisions and that such people are more likely to move than to adapt to a place. To put it differently: housing careers are becoming more diverse and the transitory phases of housing arrangements are both increasing in number and becoming more prolonged (Chapter 9). This changing housing demography (see Chapter 3) is shaped by the specific housing market conditions of post-socialism as well as by the socio-economic resources and constraints of those households who are its players.

The Case Studies

The case studies of Łódź and Brno provided a more detailed analysis of the physical dimensions of housing preferences. These two cities show interesting similarities and differences. Both have large brick-dominated areas and also a large proportion of housing stock in large housing estates. The main difference is that in Łódź the physical features of housing in the inner city are very varied, whereas in Brno, physical structures are more homogenous. In Łódź, the inner city saw some infill developments using prefabricated technology during the 1970s and 1980s. Consequently, brick buildings and prefab blocks are now located next to each other in some places. The large prefab housing estates are, however, situated in the outer city districts. There are, therefore, areas dominated by brick housing (in the inner city) or prefab housing (in the outer parts of the city). In this way, the cities contrast in an interesting way and we can discuss a variety of questions.

Brick and Block in Łódź[1]

Łódź's housing stock is characterized by two large components: the old built-up, brick inner city and large estates of blocks that were built during the 1970s and 1980s around it. Apart from these predominating structures, there are also structures built in the period between the First and Second World War and during the 1950s and 1960s; these are mainly small estates located next to the brick housing areas.

The inner city of Łódź represents one of the largest old built-up inner-city areas in Poland and was only partly damaged during the Second World War. About 65 per cent of the inner-city housing stock is old and built of brick, and it displays a wide range of qualities (from simple workers' housing to generous bourgeois flats). Additionally, there is an area of prefab buildings from the 1970s and 1980s in the inner city, called 'Manhattan'. These offer mostly smaller and medium-sized flats of varying quality, but they are of higher standard than the simple pre-Second World War workers' housing. In the pre-war buildings, the physical conditions are in many cases very poor. The buildings have not been renovated for decades. As a consequence, some of the old built-up housing stock is no longer habitable. While most of the uninhabitable buildings are vacant and barred, some of them are still inhabited. Most of them await demolition. Tenure structures in the inner city are mixed. The buildings and flats are partly municipally or privately owned and are partly owner-occupied and rented out. The share of housing owned by cooperatives is, in contrast to the overall city level, very low. Cooperatives have most of their housing stock in the large housing estates (blocks). Renovated housing stock can be found only occasionally; most of the backyard houses and courtyards typical of the inner city of Łódź are not refurbished.

A large proportion of the flats in Łódź is situated in the large housing estates. There are several of these in the city; the two biggest are Retkinia in the western part of the city centre and Widzew in the eastern part. Both estates were planned for 70,000–80,000 inhabitants. Most of the buildings and flats are owned by cooperatives, although there is a high ownership rate, which is increasing, because cooperatives are trying to sell the flats to the sitting tenants. In addition to the cooperative housing, the municipality still owns some flats in the large housing estates. In contrast to the old built-up stock, the majority of the buildings in the estates were renovated during the 1990s and 2000s. Moreover, there are many examples of infill developments which often represent high-quality housing for better-off households, some of which are in fenced, guarded or even gated complexes.

---

1   This section overlaps partly with the introduction of the inner city of Łódź in Chapter 6. The authors feel this partial repetition will help readers to orient quickly without having to read the earlier chapter.

Figure 11.1 Inner-city brick building in Łódź
Source: Photo, A. Haase.

Figure 11.2 Inner-city block in Łódź
Source: Photo, A. Haase 2007.

Figure 11.3 Large housing estate block in Łódź
Source: Photo, A. Haase 2007.

Brick and Block in Brno

Brno has a long-standing historic inner city with a 'first ring' of secessionist housing stock. This first ring is divided into a more prestigious western or north-western part, with villas and spacious architecture, and an eastern and south-eastern part that is dominated by industry and working-class neighbourhoods, parts of which have been home to Brno's Roma community for decades. Today, in many of the old built-up areas in the north-west, renovation is continuing. There is very little investment in the south-east section of the inner city, and the building stock here is in very poor condition (Vaishar and Zapletalova 2003; see also Chapter 6).

Large housing estates cluster around the compact, old built-up city in every direction. About 43 per cent of occupied flats in Brno were constructed using prefabricated technology (Table 11.1). They were often constructed around former villages, which gave their names to the newly built estates. The villages usually remained and form a further segment of brick housing.

The estates differ in size, age, location, environmental quality and ownership structures. The eight larger estates that form administrative districts of the city[2] consist of at least 4,000 and up to 10,000 flats (MMB 2008: 55). Block housing can be found throughout the city as infill or additional development, but not in the inner city.

The perception of local residents – the east-west (north-west/south-east) divide distinguishes good from bad addresses in the inner city – also dominates the symbolic differentiation of the city as a whole. This has nothing to do with brick or block construction but with scenery and environmental quality. The north of Brno forms 'the foot of the Moravian Karsts', a protected landscape with caves and gorges. The proximity of northern neighbourhoods to recreational areas has long made them the better addresses in Brno. The southern and eastern districts are associated with the dirt and smell of industrial sites of the past and present. Today, they are cut off from the surrounding landscape by the motorway to Prague and suffer from the through traffic. These south-eastern districts are perceived as the less desirable addresses. Proximity to green spaces and recreational areas are the key to neighbourhood images, especially when it comes to large housing estates and suburbs. These differences are as present and alive today as they were during state socialism (Steinführer 2004b: 254–61).

The ownership structures in Brno are diverse. Following privatization, many of the flats throughout the city are now owned privately, many of them owner-occupied. These owners today form small owner associations to manage refurbishment, both on the small scale of blocks or just single buildings. The older, large housing cooperatives still exist and form another major segment, especially in the large housing estates. The municipality holds the third segment, both in the inner city and in the large estates. These flats are rented out through the district housing offices and they are much less expensive than private rentals. Therefore, long waiting lists exist for them. Housing estates also differ according to the percentage of municipal, cooperative and private housing stock they contain (see also Chapter 6).

---

2   The housing estate Lesná is an exception here. Being the oldest large estate, it is part of the district Brno-Sever.

Table 11.1 Characteristics of flats in Brno

	flats	%
Total	151,724	100
... by construction method		
brick and stone construction *(cihly, kámen)*	82,924	55
prefabricated (panel)	65,508	43
combinations and others	3,292	2
... by ownership		
communal ownership	165,882	24
listed cooperatives	22,069	13
other ownership	103,966	63

Source: Magistrát města Brna (MMB) 2008, referring to census data for 2001.

Figure 11.4 Inner-city brick building in Brno
Source: Photo, K. Grossmann 2008.

Figure 11.5 Large housing estate block in Brno
Source: Photo, K. Grossmann 2008.

Figure 11.6 Large housing estate and village structures in Brno
Source: Photo, K. Grossmann 2008.

Materials and Methods

The findings presented in this chapter represent an explorative, qualitative study in the two cities, using the evidence from interviews with residents. In order not to misinterpret phenomena found in the inner city, the research findings need to be supported by material from other parts of the city. Therefore, in the Brno case study, 15 interviews with inner-city residents are supported by 13 interviews with residents of other urban areas, mainly in large housing estates and some in suburbs. The interviews with Łódź inner-city residents are particularly relevant, because the 16 households interviewed live in both brick and block buildings. For the purpose of this chapter, the housing preferences in the physical dimension were cross-referenced with location preferences. From this starting point, we organized the cases into a typology of stated preferences, employing the same procedures as described in Chapter 9. Basically, we identified dimensions relevant for the housings preferences of interviewees, contrasted the statements of interviewees along these dimensions, clustered the results and, from here, identified patterns of stated housing preferences.

Brick or Block in the Light of Housing Preferences

As Figure 11.7 shows, the stated preferences were organized in two dimensions, namely, the preference for a certain location and the preference for a certain physical structure or type of building. Some interviewees showed clear preferences for certain physical structures, such as a preference for access to green recreational spaces or a preference for brick housing. Other interviewees did not show clear preferences for either brick or block housing. Instead, they expressed somewhat pragmatic views concerning their current flat or home. But even though they did not express explicit preferences concerning physical structures, they had clear preferences for certain locations such as the inner city, the urban fringe or a suburban location. From these two dimensions, we identified a number of preference patterns that are described in the following section, starting with the preference for a central location and brick structures.

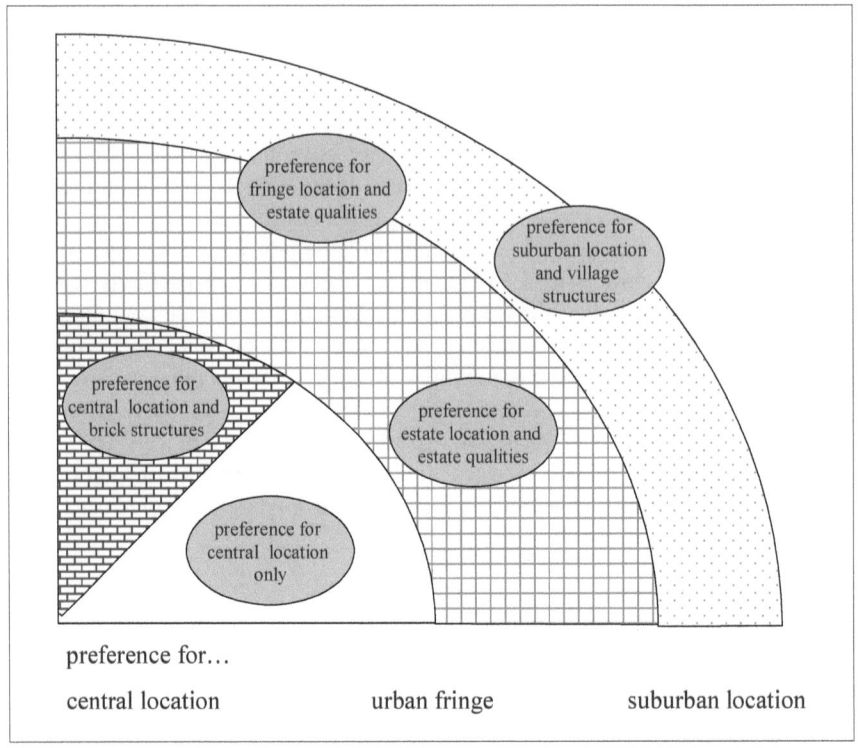

Figure 11.7 Patterns of stated housing preferences of interviewees in Brno and Łódź

Source: Authors' research.

A Preference for Central Location and Brick Structures

There is a group who prefer brick and who deliberately chose to live in old built-up stock in a central location. They like brick buildings because of their generous room sizes, high ceilings and spaciousness. An example of this is Basia, a woman in her 50s, who lives with her family in the inner city of Łódź:

> Such a tenement house, there is something [...] don't know, something special about it for me. Something that makes you feel freer. Even when I visit my parents. They live in a block, I always feel stifled there at first. I have to open a window or something, [...] I just got used to living in a spacious flat. (Łódź HH2)

The appreciation refers to the building, but also to the ensemble. Maria, originally from Austria and now living apart from her partner, who works in Prague, says:

'In Brno, things have been nicely refurbished and they try to renew the old places without losing the old character. And I like that very much. It is a bit of that spirit and – talking about spirit: it is cosier' (Brno HH16).

Often, the preference for bricks is related to a rejection of block housing. These people reject living in blocks because of a sense of anonymity, low ceilings, lack of character in mass housing, etc. Many of them have lived in a block and explain their rejection as a result of this experience. Interestingly, in particular young people who are occupying their first flat and live in a non-renovated brick building, reject the block, despite the better physical housing conditions of the blocks (heating, bathroom etc.). Ela from Łódź, for instance, a woman in her late 20s, cohabits with her partner in a brick building and says about her flat:

> It's a flat in a tenement house. The flat is nice because, first of all, there are high ceilings and large rooms. But I used to live in blocks only, and therefore I just could not imagine living anywhere else, because I was so used to it, you know. But now I think I would say that has changed completely: I cannot imagine moving back to a block again. Even when they are more or less the same size as in a block, the flats in an old tenement building are definitely more spacious, that is to say they are because of the high ceilings, these are really gorgeous. (Łódź HH6)

Most of those who prefer bricks are quite critical of the technical disadvantages of brick buildings, but the disadvantages did not stop them from deciding to live in a brick building or stating that they prefer to live in brick buildings – like Janina, a young woman from Łódź who lives on her own.

> To be honest, I like old tenement houses more, even though their maintenance is more expensive. […] I like these houses because I used to live in one during my childhood. That's for sure, there are larger and higher rooms, that is important too, technically speaking, the rooms are higher. […] I dislike blocks, maybe because I am used to living in a tenement house as I said before, and there are higher rooms, the windows are of a different size, which is also important and better, if you ask me […] apart from that, tenement houses are better insulated against noise than blocks, […] I am not keen on knowing everything the neighbour is doing. (Łódź HH11)

A preference for brick housing in a central location was also the case for an interviewee who currently lives in a large housing estate. He is in his 50s and represents an empty-nester household. Up to the mid 1990s, he lived in the inner city, then the house he lived in was reprivatized and he was forced to move to a flat in a large housing estate. He is still unhappy with the situation and criticizes the block structure for being large and anonymous with conflicts between different kinds of social strata in the immediate neighbourhood, in the building and even on the same floor. Today, he is a frustrated estate resident who would have preferred to stay in the brick house in the centre (Brno HH26).

As the quoted interviewees show, fans of brick represent a mixed group in terms of age and household type (including singles, couples and families). Whereas some of them belong to the group we identified as 'transitory urbanites' (see Chapter 9), that is younger, non-family households who see their current flat as a transitory residence because it fits their current wants and needs, others have decided to stay in the inner city for a long(er) time.

A Preference for Central Location Only

For inner-city residents, the central location was, in many cases, very important in the list of advantages of the particular flat; often, it was more important than the type of building stock. Instead of aesthetic aspects, here the functional aspects of everyday life dominate the perception of neighbourhoods and serve as evaluation criteria for the differentiation among housing areas.

Adam, a young inner-city resident from Łódź living in a high-rise block, for example, stated that the advantages and disadvantages of a flat in this type of block are 'not so different from the pros and cons of living in a block. So we are living as in a beehive, trampling on each other's toes, you have to be silent when the neighbours want you to be' (Łódź HH5).

Thus, the rejection of the block type does not, in all cases, refer to the type of building itself but just to the type of settlement. Blocks in the inner city for this group of interviewees are perceived as much more acceptable than blocks in a large housing estate. For these dwellers, the location plays a more important role than the type of building stock. This opinion was shared by interviewees from different household types and age groups, but mainly by those who were identified as functional urbanites in the typology of Chapter 9. A good example for this is Tomek, a young inner-city dweller from Łódź who lives in a high-rise block:

> Tomek: No, I don't like these large estates, these blocks.
> Interviewer: Why don't you like them?
> Tomek: Don't know, they are too monotonous and uniform, there is too much concrete, too little diversity. Here in the centre, there is something […]. Apart from that, those estates are far away from the city centre. Once I felt that sometime I would like to live downtown. If I live in a city I want to live downtown, but if I cannot live there it should be somewhere at a distance from the city, where I don't have the feeling I am living in a city. (Łódź HH15)

Interviewees might therefore reject living in large housing estates but they do accept block living in the inner city – even if they have a 'general' preference for brick buildings. Tomek, for example, stated that he generally prefers to live in a brick building; however, he rationally and functionally decided for a flat in a block, since he wanted to live in the inner city: 'I was looking for a flat. [...] I would have preferred a flat in an old building. [...] But no flat like this was available then. In here, the nice views [he lives on a top floor] make up for that a little [...]' (Łódź HH15).

A Preference for Estate Location and Estate Qualities

In this preference pattern, the dominant theme in the evaluation of residential environments was the functional fit of the characteristics of the place and the needs of daily life. Jana, a woman in her forties who lives in a large housing estate in Brno, said:

> The advantage [of Bystryc district] is that everything you need in your everyday life is close by. All sorts of shops, you don't have to go downtown. We do not live close to a street, so the children can play in front of the house. Also, the city is close enough to commute to work, there are bus and tram lines going there, that is quite okay. [...] For me, Bystryc is the number one place. (Brno HH20)

Just like the interviewees who have a preference for the centre as a location, these interviewees do not state preferences along physical characteristics; rather they are indifferent towards the categories of brick and block. The close availability of basic functions, such as transport, schools and services, shopping facilities as well as recreational facilities, make the large housing estates places of choice, especially for families with young children, middle-aged professionals and older people in our sample. Strong social networks, like relatives or friends living in the same estate, or good neighbourly relations, increase the identification with an estate.

But there are also people living in the inner city in old building stock who explicitly reject living in old brick buildings. They are unhappy with their housing situation. Similar to those who reject living in a block because they lived in one with their parents, these people were either assigned their flat and would like to change, or they are dissatisfied but realize that they cannot afford better housing, for example, in a flat in a large housing estate or a flat in a block in the centre.

The rejection of inner-city brick living relates to both the housing situation and the residential environment. It is difficult to distinguish what the main reason for such a rejection might be. Often, the affected interviewees were dissatisfied because they could not change their situation. Mr and Mrs Kowalski, for example, an older couple (in their early 70s), lived in a brick building in the inner city of Łódź and were allocated their flat during state socialism. (They share the entrance door and the floor with two other households.) This couple would prefer to move to a large housing estate, but this is impossible for them:

Mrs Kowalska:	No, as for now I don't like living here. I would prefer to move to [the large housing estates of] Widzew.
Interviewer:	Why?
Mrs Kowalska:	Because I like it much more there. There is so much greenery, the air is cleaner […] it is less noisy there […] things like that. Living in those estates is less expensive and easier. […] And the woods there, you can just have walk in a small wood. […] It's just wonderful there. And Widzew is the best housing estate.
Interviewer:	And what is the worst place for living in the city?
Mrs Kowalska:	All the inner city. (Łódź HH14)

Another interviewee in Łódź, Stanisław, representing the head of a low-income family living in a (dilapidated) brick building in the inner city, introduces another aspect that is important mainly for poor inner-city dwellers living in bad housing conditions. He claims that he could not afford to pay the housing costs for a flat in a block, and he also states that this would apply to many with a precarious income situation:

Interviewer:	Does that mean that housing conditions in the inner city are not good?
Stanisław:	Well, they are not, especially in the old buildings.
Interviewer:	And in the prefab areas?
Stanisław:	In the blocks, well, that's quite a different thing. But who can afford to live in a block? (Łódź HH7)

A Preference for Fringe Location and Estate Qualities

Finally, we had cases of estate residents who expressed a preference for a housing environment that is characterized by both natural and urban assets. Their ideal place to live, therefore, is a housing estate at the edge of the city, the urban fringe. This perspective is neither a suburban perspective nor a truly urban one. It is instead characterized by a simultaneous liking of the urban assets of large housing estates, their infrastructure and good connections, as well as their natural assets, both within and around the estates at the urban fringe. Jiří, a young academic who lives much of the time apart from his girlfriend (she lives in Ostrava), explains his location decision for Brno-Bystrc like this: 'We decided for a district close to nature, so that we can take the bike and go to the woods and the lakeside in five minutes, and that was what made our decision' (Brno HH22).

Jiří grew up in another large housing estate in Brno, so he says for him it is 'natural' to live in a block, a suburb would not be an alternative. Asked whether some prestigious parts of the centre would be an alternative residential location, he again disagreed: 'And if you live in a flat with a window onto the street, […] that's horrible for me because I am used to sleeping when all is silent. There you still hear the noise of cars the entire night, and of big cars and buses' (Brno HH22).

An interviewee from Brno-Kohoutovice, a large housing estate in the west that is surrounded by green areas, shows the same rejection of the inner city for its poor environmental qualities. He moved out of the inner city because his young son started to have health problems, the diagnosis blamed the bad air quality of the inner city. He chose a fringe location, in order to overcome these problems. Today, he is a convinced resident of a large housing estate at the urban fringe (Brno HH19). An older inhabitant of Lesná, the oldest large housing estate in Brno, spontaneously answers the question, which neighbourhood would he recommend a good friend to move to: 'Lesná. Lesná is nice and green with large trees and it is well connected to the centre by public transport' (Brno HH24).

A Preference for a Suburban Location and Village Structures

Another group with yet another preference pattern neither wants to live in the inner city nor in housing estates; these people prefer suburban living. They like family houses, newly built or refurbished old, as well as the proximity to greenery. Pavel from Brno, who currently lives in a large housing estate, preferred a suburban location irrespective of the age and type of construction of the family house: 'Probably we will reconstruct an old house or we will build a new one' (Brno HH21). The main consideration for all is proximity to green space, be it forest or a private garden: Vaclav, a professional in his 30s who had just refurbished an old house in a suburban village close to Brno for his family (one child), stated:

> We would like to have a garden. That is the first criterion, and let's put it this way, it would be a safer place for the children. […] I think this village is a safer place for them to grow up in and move about freely. People are more focussed, I mean they care more about their neighbours and about people living in the same street, for example, so for me that's much better because I think there are nice people here. (Brno HH27)

He clearly rejected a central location because of the poor environmental conditions:

> I don't like city centres because they are crowded, I like to have more space, and because of the noise, and, don't know, air pollution and so on, so, if I have the choice it will be the outskirts of the city. That's for one, and of course I would like to have woods nearby, trees and nature. […] Actually, living downtown does not mean anything to me. (Brno HH27)

Discussion

The following section first discusses the main findings before we answer the questions raised in the introduction of the chapter.

Attitudes or preferences are only one dimension that comes into play when decisions have to be made. Location decisions made 'by chance' or as a result of the many interacting factors that have characterized the housing market during the post-socialist transition are still very important. Generally speaking, we can say that in the majority of cases, pragmatism – driven by constraints – dominates, rather than preferences. For example, for those who are first and foremost inner-city oriented, the type of building stock is in many cases negotiable – all the more so since parts of the old building stock in both cities are in a poor state of repair and do not meet the housing quality standards wanted by the majority of inhabitants. In many cases, just living centrally is important, no matter if it is a brick building or a block. Other priorities and pragmatism 'outweigh' the issue of the type of building stock in such cases, so that its importance should not be overestimated.

An inner-city perspective is not identical with a brick perspective. Since many second-order inner cities in Poland and the Czech Republic have a mixed housing stock made up of both brick buildings and prefabricated blocks, an inner-city perspective always has as much to do with the location as with the brick building stock. The inner cities of Łódź and Brno and the interviews we carried out there provide good examples of this. We found not only a particular appreciation for brick housing within a smaller group of interviewees, but also a critical stance towards brick housing because of its poor technical state and high repair costs, at least in the case of Łódź. Brick housing is, for some interviewees, a pull factor, but it is by no means dominant within the sample as a whole. Since the sample was very selective (see Chapter 2), further research would be needed to determine the proportion of people for whom brick buildings are themselves the pull factor.

The same applies to blocks and large housing estates. When talking about blocks, interviewees did not automatically relate to housing estates. And, what is more, when talking about housing estates, they did not automatically mean or emphasize their location at the urban fringe, but rather their good connections and shopping facilities and services. This was particularly true in the case of Łódź, where blocks are also found in desirable inner-city locations. Contrary to assumptions made in urban research literature, block housing is still widely accepted; especially block housing within the inner city. Even though there was no explicit preference for block architecture as such, there was an appreciation of the large amount of green space within and around the estates, the calm and healthy housing environment, the safe places for children to play outside and the limited amount of traffic noise. 'New' household types appreciated these features as much as more traditional types.

There is no clear relation between particular household types and age groups and a preference for a specific type of building stock or location: brick and block buildings, inner cities and estates, are inhabited by a variety of residential groups in terms of household types, living arrangements, age groups, educational backgrounds and lifestyles. The assumption that young people generally have a preference for, and actually move to, the inner city, cannot be supported with the data available for either Łódź or Brno. The fact that brick housing is preferred by a restricted group of young people is related to the poor physical state and generally poor housing conditions in many brick buildings. The fact that block housing is widely accepted undoubtedly relates to both the physical quality of the flats and the residential environment – in the case of inner-city blocks, it is the central location, short distances to travel and the amenities available.

Transitory housing as described in Chapter 9 is, in our view, an example of the pragmatism shown in housing choice. Transitory urbanites choose to live in the inner city since it fits in with their current life situation. They are also clear that there will be changes in the future (although of course these changes may not necessarily come about), which will be dependent upon their particular preferences, wants and needs. The interviews do not allow us to say whether the majority of these people will stay in or leave the inner city. But we are able to conclude that pragmatism relates mainly to specific life situations and only secondarily to general attitudes of the interviewees towards particular housing qualities, environments and types of building stock and equipment, and these are not necessarily related to their present home.

In both Brno and Łódź, large housing estates are unlikely to undergo significant social change resulting in them becoming the home of predominantly poorer households, while the better off move to the inner city or suburban areas. Large housing estates are well-established housing areas in Poland and the Czech Republic. The inner city of Łódź, by contrast, is in danger of losing more inhabitants to the large housing estates and the suburban zone because of the very poor housing conditions of the brick buildings. Infill developments in large housing estates are, for many inhabitants with medium or higher incomes, more attractive places to move to. This leads to a socio-spatial differentiation and selective upgrading of these residential areas, with population decline being the only effect for the inner city. The repopulation of the inner city by young people reported on in Chapters 7–9 as well as here will not provide a general or 'ultimate' solution to the massive dilapidation of, and the resulting exodus from, Łódź's inner city.

Conclusion

Returning to the research questions at the start of the chapter, the following conclusions can be drawn:

- Brick and block building stock, as a part of housing preferences, do play a role but must not be overestimated as a pull factor in the choice of an inner-city residential location.
- Housing location is first and foremost a pragmatic choice, and trade-offs

between opportunities, constraints, needs and wants ultimately force decisions – choices are made with reference to current life situations and re-adapted much more flexibly than formerly.

- For most of the interviewees, inner-city location is more important than the physical fabric. In some cases, single inner-city blocks are more acceptable than large clusters of blocks in large housing estates – a fact that is in line with location being the top priority.
- Preferences for brick and block housing cannot be related to particular household types, age groups, living arrangements or lifestyles; the population of both inner-city brick and block buildings is mixed – a statement that also applies to the large housing estates.
- We found some evidence for a selective in-migration to the inner city by younger households (students and early-stage professionals) without children and better-off middle-aged households. To make an exhaustive statement on the character of in-migration to the inner city in Łódź or Brno, however, we need more in-depth studies.

Attitudes towards brick and block housing represent just one aspect of preferences and decision-making; they are not the sole criterion. Residential change is driven less by preferences or attitudes and more by a weighing up of advantages and disadvantages, subjective priorities and constraints, as well as general ownership and price conditions of the housing market and patterns of socio-spatial segregation (the 'intermediate structures' of our model; see Chapter 2). These factors always have to be discussed along with other preferences and constraints, wants and needs. The two case studies are specific, but they also show some overarching problems such as the wide-spread dilapidation of inner-city brick housing in large Polish cities and the continuing social mix of the residential population. Blocks are unlikely, therefore, to become the 'slums' of the future; indeed, some inner-city brick areas are far more likely to endure as concentrations of low-income and minority households.

Chapter 12
Tenure Change and Sociability: Transformation of Neighbourly Relations

Jana Mair, Jana Pospíšilová

Introduction

During the post-socialist transition in the Czech Republic and in other East Central European countries, residential tenure underwent major changes that have influenced neighbourly relations. However, these changes have not been uniform, nor have they been characterized by a single process. Individual residents have been affected by the changes differently, as have various types of residents (for example, members of the Czech majority, in contrast to the Roma population).

This chapter[1] describes neighbourly relations in the inner cities of Brno and Ostrava, with particular focus on sociability and tenure change, shedding light on the role of social networks in residential change as well as those influences on residential patterns which stem from changes in the local and societal context (Chapter 2, Figure 2.1). The residents are the actors who shape neighbourly relations, but the resulting neighbourliness becomes a characteristic of the place, the residence. Neighbourly relations are defined as relations among residents of a locality – in our case, of a building. We make this distinction because most of our respondents considered their neighbours to be only those people sharing the same building (as opposed to the same street or quarter).

We examine the particular forms neighbourly relations take and how they have been influenced by the post-socialist transition, especially by the transformation of housing policy and property arrangements. However, we are not only interested in the present state of affairs and those of recent times, but also in the residual effects of state socialism and even of the inter-war period. The ethnicity of residents was not the focus of the study. Even so, this topic also emerges as an important factor distinguishing and defining particular areas of the inner city, particularly as a Czech majority versus Roma minority issue.

Sociability, in general terms, is the tendency to associate with others and form social groups. In the context of the social sciences, however, the notion of sociability

---

1   The text describes some of the results of *Cultural Identity and Cultural Regionalism during the Formation Process of the Ethnic Landscape of Europe*, a research project of the Institute of Ethnology of the Czech Academy of Sciences *(Kulturní identita a kulturní regionalismus v procesu formování etnického obrazu Evropy)*, no. AV0Z90580513.

is rather ambiguous (Ninomiya 2001). Simmel and Hughes (1949: 254), for example, see sociability *(Geselligkeit)* as social relations regardless of their content. They focus on the form of socializing and speak of the 'impulse to sociability in man' (ibid.) or the tendency to associate for its own sake. Gurvitch (1949: 17) understands sociability as various 'micro-sociological elements' or 'ways of being bound in the whole and by the whole' (ibid.). Many forms of sociability exist (for example, organized or spontaneous sociability) and 'every group is a microcosm of forms of sociability' (ibid.). The term is also used by historians in reference to 'different forms of social relationships, in particular interpersonal bonds that are initiated either consciously or unconsciously in a given context' (Ninomiya 2001: 208). In this sense, formal and informal sociability are differentiated, and various social activities, both in rural and in urban environments (for example, neighbourhood, social life in a street or city quarter), are studied (ibid.). This study focuses on neighbourly relations and how they vary, respondents' attitudes towards and interactions with their neighbours, and the degree to which these interactions coincide with their expectations.

The research presented in this chapter is largely ethnographical and emic[2] in perspective, since we aim to consider neighbourly relations from the inside – that is from the point of view of the community under investigation. The emphasis is on the residents of the inner cities of Brno and Ostrava themselves and on the meanings they ascribe to their neighbourhoods, neighbourly relations and residential communities. Moreover, we do not see residents merely as passive recipients of change, but also as actors who shape their environments. To put it differently: it is the agency of residents, using their space of action that creates change (see Chapter 2). We also adopt a historical perspective, considering how changes in housing policy and other factors have influenced the way people perceive their neighbourly relations.

The research aims to answer the following questions. Have neighbourly relations been influenced by the post-socialist transition, especially by changes in housing policy? How has the transition changed the sociability and attitudes of neighbours towards their neighbourhoods? Is there any relationship between tenure change and relations among neighbours? Are there differences between Roma and non-Roma households?

A neighbourhood can be considered spatially, as a specific residential locality, or socially, as a group of people living in this locality. What their interrelationships look like and whether they constitute a community depends on various factors, including the homogeneity or variability of household types, the socio-economic status of individual residents (that is determined by age, education, occupation and so on), whether they share a common identity or unique customs and their level of

---

2   According to Malina (2009: 2478), the term *emic* was introduced to anthropology by the US-American linguist Kenneth Lee Pike and derived from the linguistic concept *fonemic*. Emic method involves studying a culture from an inside perspective, that is on that culture's own terms. The opposite approach is called *etic* (derived from *fonetic*) (ibid.).

social interaction (for example, friendship and kinship inside the locality) (Musil 1967: 223, 2002: 237–95, Carmon 2001: 10, 490–92, Mueller 2005: 326–7).

With this in mind, changes in neighbourly relations are examined on two levels. *Firstly,* how changes in residential environments have affected communities and what such changes have meant for residents and their attitudes toward their flats, buildings and localities. In particular, we address this topic in connection with tenure change, considering alterations of residential behaviour as well as shifts in residents' attitudes towards their neighbourhoods after tenure changes.

Secondly, we address sociability, which encompasses various social relations, contacts and interactions among the studied population. However, to examine such changes in neighbourly relations, it is necessary to discuss the institutional landscape in housing allocation and policy, not only during the post-socialist transition but also during state socialism. This is the aim of the first part of this chapter, where we summarize earlier research. In the second part, which focuses on post-socialist housing and neighbourhoods, we present the results of our own research:

- two case studies of the inhabitants of buildings in the inner city of Brno, which are part of ongoing ethnological research in the city and inspired by ethnological literature such as Karel Fojtík's *Dům na předměstí* (1963) and the anthology *Ein Haus in Europa* (1997), and
- results of a questionnaire survey and follow-up interviews undertaken in the inner cities of Brno and Ostrava, which are part of a PhD thesis focusing on everyday life, sociability and social relations in these locations during the post-socialist transition.

Although two separate research efforts, they have proceeded in parallel. The authors were in constant touch, discussing particular questions, selecting informants and analysing the results. While the testimony of the inhabitants of the two buildings traces that community's mutual relations in detail, the questionnaire survey, based on a quota selection and followed by interviews, paints a broader picture of neighbourly relations in the inner city as a whole. Thus, the research data and conclusions are complementary.

Material and Methods

Our research uses a mixed-methods approach, with a strong focus on qualitative methods. During the fieldwork, various methods were used, including a questionnaire survey and semi-structured interviews.

For the questionnaire survey, a sample population of 92 respondents in the inner cities of Brno and Ostrava was chosen by quota selection. Quotas were determined according to age, gender and education, on the basis of the 2001 census. The questionnaire included both open and closed questions concerning housing, social networks and value orientation.

The interviews were conducted at several levels. *First,* to complement the questionnaire survey, 20 semi-structured interviews were carried out with inhabitants of the inner cities (10 in Brno, 10 in Ostrava). Informants were chosen from previously surveyed respondents. However, the aim was not, as in the survey, to reproduce the real distribution of inhabitants of the inner city, but rather to encompass various categories of inhabitants and, thereby, their lifestyle diversity. Topics included everyday life in the inner city as well as neighbourly and kin relations. *Second,* semi-structured interviews were conducted with the inhabitants of the buildings that were the subject of the case studies carried out in Brno: one in a middle-class area on the edge of the inner city, close to a villa quarter (Stojanova Street), the other in the socially marginal locality of Cejl Street in the eastern parts of Brno. Thus, the sample reflected different property arrangements, various ethnic and social groups and diverse attitudes toward flats and localities. Semi-structured interviews were also carried out with a sample population covering both traditional and non-traditional households in the inner city (see Chapter 3). *Finally,* expert interviews were conducted with officials of the housing authority of the urban district in Brno-Sever.

Other methods applied included the autobiographical method, which was used during interviews with inhabitants of the two buildings. Narratives of people about their lives are used to understand the issue under investigation (Brednich 1982).

Participatory observation, a fundamental method of ethnographical research, took place over a long period at the building on Stojanova Street.

Housing Policy During State Socialism and its Impact on Neighbourly Relations

Czech state-socialist housing policy was based on the Housing Management Act of 1964 (*Zákon o hospodaření s byty*), which defined the rules and procedures for assigning and using flats. Four main categories of housing tenure existed: state, company, cooperative and private (for example, family) housing,[3] all of which were under the purview of the state. State and company flats were assigned to applicants who were registered on a housing list, which in turn served as the basis for a waiting list compiled by the local authorities (*národní výbory*, 'national committees', equivalent to today's municipal or county governments). These

---

3   State housing included both newly built, state-financed flats and formerly privately owned or rented flats, which were gradually nationalized from the 1950s onwards. Company (in some literature, 'enterprise') housing was designated for employees of (state-owned) companies. Cooperative housing was owned and built by cooperatives. Private housing was usually built in smaller towns and villages. Older privately owned housing stock (in villas or detached homes) persisted from the period before the Second World War.

waiting lists were supposed to take into account the housing needs of applicants (for example, their current housing situation, health, marital status and so on).[4] However, applicants' political or professional status also played an important role in determining the priority of their access to housing (Zákon o hospodaření s byty 1964, Raban 1986: 40, Musil 1992a, Sýkora 1996b and 2003, Kostelecký 2000, Lux 2000: 9, Donner 2006: 35–6). Moreover, due to the generally low availability of flats (supply not keeping pace with demand), which resulted in long waiting periods for state and cooperative housing, decisions on housing assignments frequently served as a means of political control. Thus, while the law designated state dwellings 'primarily for families with lower incomes and a larger number of children' (Zákon o hospodaření s byty 1964: 249), housing authorities did not always comply with these guidelines. Party officials usually received preferential treatment, and opportunities for corruption, especially bribery, multiplied (Sýkora 1996b, Telgarsky and Struyk 1991: 156,[5] Donner 2006).

The law also regulated rents, which were based on the living space and furnishings of a flat, rather than on its location or the household income of its occupants. Rents were fixed and, until 1991, they remained at the same low rates, which did not even cover the operating costs of buildings (Telgarsky and Struyk 1991: 157, Kostelecký 2000: 178, Lux 2000: 9, Donner 2006: 37). Making matters worse, preference was given to building new housing estates (*sídliště*) over repairing or maintaining existing buildings. As a result, most of the older housing stock in Czech inner cities deteriorated, ending up in very poor condition or even in a state of collapse (see Telgarsky and Struyk 1991: 158, Kuča 2000, Musil 2002, Steinführer 2003, Sýkora 2003 and Chapter 5).

As in other spheres of the state-socialist economy, social networks played an important role in housing. The right to use a state or company flat was confirmed by the so-called right of tenancy (*dekret práva na užívání*), which was *de facto* unlimited and, in certain cases, seems to have been a form of private ownership (for example, the *dekret* was inheritable and treated as a commodity with a black market value) (Šmídová 1999: 172, Kostelecký 2000: 178). Due to the housing shortage, which impeded the acquisition of housing in the prescribed way, a flat, once obtained, became an important family asset, or 'family estate' (Šmídová 1999: 175–6). For example, people could be assigned flats with their grandparents as co-users, thus guaranteeing retention after the grandparents' death. A similar

---

4 Housing policy thus influenced demographic development as well as the lives of people. Priority was given to married couples over singles or divorcees. Therefore, marriage was necessary to become entitled to a dwelling and, not least for this reason, the average age at the time of first marriages was very low (from the 1960s to the 1980s, around 24.5 years for men and 21.5 years for women) (Kučera and Fialová 1996: 12–14). Also, the birth of a child improved one's claim for a suitable flat (the average age of women at the birth of their first child was 22.5 years during this period) (ibid.: 12).

5 These authors refer to Pisova, Eva. 1990. *Problems of Housing in Czechoslovakia in the Period of Transition to the Market Economy*, paper presented at the World Bank Policy and Research Seminar on Housing Reforms in Socialist Economies, June, Washington, DC.

practice was common in cooperative housing, with membership shares being passed on (Donner 2006 and Chapter 8).

These conditions also affected relations among residents. Neither neighbours nor neighbourly relations are ever entirely subject to choice. However, as Beate Völker and Henk Flap point out, 'if there is a housing market, one can select a particular neighbourhood or sometimes get situated next to a particular neighbour, but in a state- or party-controlled system, this is not feasible' (1997: 243–4). Although the literature on neighbourhoods during socialism is not extensive, existing studies do shed light on this issue.

The study of Völker and Flap (1997), carried out in eastern Germany (and relating to the GDR) in 1991–1994, is one of the more recent investigations of the influence of communist ideology and socialist housing policy on neighbourly relations. In the GDR, the allocation of housing allowed the local authorities to influence the private lives of citizens: for example, by mixing people in neighbourhoods and establishing building and neighbourhood committees to organize collective activities and support the 'development of a socialist lifestyle, thus bringing them under control' (ibid.).

Nevertheless, Völker and Flap show that, in Leipzig and Dresden, even though targeted housing policy succeeded in creating socially mixed neighbourhoods, it did not succeed in creating friendly neighbourly relations: 'Meeting did not lead to mating: next-door neighbours were socially distinct and hardly socialized with each other.' (ibid.: 241). Ties and contacts may sometimes have been made among neighbours with similar social status or lifestyles, but socially dissimilar people failed to forge close ties, even though they lived near each other. On the contrary, physical proximity and mutual knowledge of their private lives contributed to mutual scepticism (ibid.: 248). However, as Völker and Flap indicate, their research was concerned with the final stage of socialism, and neighbourly relations might have been markedly different in earlier periods, when the social distance between neighbours might not have been as great (ibid.: 259–60).

Moreover, other studies – carried out, for example, by Harth in Halle or Kames in a building in Leipzig – come up with different findings, showing very close ties and relationships among neighbours (Harth 1994: 147–70, Kames 2009: 84–108). Generally, the problem is that most of these studies were carried out later, so people were remembering the past *ex post* with a certain temporal distance.

Similarly, in Czechoslovakia housing policy also became an important vehicle of political control and state paternalism (Šmídová 1999: 180), which influenced the demographic composition of individual quarters and built-up areas and therefore their neighbourly relations. As Musil shows (1967, 2002), based on research carried out in Czech cities by the Institute of Development and Architecture *(Výzkumný ústav výstavby a architektury)*,[6] striking differences existed between particular

---

6   This research institute was established in 1963 by the Department of Urban and Housing Sociology *(Oddělení sociologie města a bydlení)*, which in later years carried out research on city environments (for example, in 1963–1965 in Kladno, České Budějovice,

spatial types of housing. For example, large housing estates became 'more open local communit[ies] with a higher degree of sociability [*družnost*] and with readiness for social communication than in old settled city quarters' (Musil 2002: 295), despite the fact that people usually knew each other less than in other parts of a city. To a certain extent, this was a result of clustering residents in the same life phase, mostly families with small children. Moreover, housing cooperatives and local authorities strengthened mutual relations through community activities (for example, constructing and maintaining playgrounds, cleaning buildings and their surroundings, hosting children's events or balls). These facilitated contacts among residents and led to close-knit communities. Though relations among residents remained primarily utilitarian (Musil 1985, 2002: 296–7), not all residents were amenable to such efforts: 'The degree of social life [...] was so high sometimes that some families felt it as a threat to their privacy' (Musil 2002: 295). In addition, many housing estates were inhabited by the employees of a single company, resulting in further 'overlap of the public (professional) and private sphere and further consolidation of social control' (Šmídová 1999: 180–81).

By contrast, mutual assistance and interaction were not well developed in inner-city quarters. However, according to Musil, informative and ceremonial functions (for example, participation in family events) were common. This means, on the one hand, that people usually knew each other, greeted each other and expected social rules and etiquette to be enforced. On the other hand, they maintained a high degree of social distance in order to preserve privacy (Musil 2002: 294–5). Musil describes this form of relationship as highly formal.

In general, an increasing proportion of older and less educated people, households with lower incomes, one-person households, childless couples and one-parent families lived in the inner cities (Musil 2002, 1992b: 455). Roma[7] also

---

Plzeň and Prague, as well as a survey in Ostrava in 1966; Musil 1967, 1985, 2002, 2004). The institute applied a socio-ecological approach that focused on the socio-psychological consequences of various housing and urban environments, lifestyles in old parts of cities or in new large housing estates, housing problems, relations between the urban environment and human behaviour and so on. However, these studies – conducted outside the official stream of sociology, which was influenced by official ideology – were strictly empirical, 'necessarily lack[ing] broader socio-political connections and consequences' (Musil: 2004).

7   During the Second World War, almost all Czech and Moravian Roma were exterminated, and only several scores of Roma family survived (Nečas 1997, Haišman 1999). During state socialism, Roma migrated from Slovakia. They went mostly to the industrial centres and to the Sudeten borderlands. Nevertheless, the migration was not without problems. Newcomer Roma had not been accustomed to living in a city environment, and in their new destinations, social and cultural conflicts arose. State authorities decided that, in order to solve 'the Gypsy issue', Roma would be re-educated and assimilated. They created a program to eliminate their cultural and social 'backwardness' and to concentrate on Roma cultural, health and educational issues. From the mid 1960s, the state also pursued the politics of organized dispersion in order to remove streets, quarters and settlements where Roma were concentrated. Simultaneously, these officials had minimal interest in

congregated there: 'Flats in old deteriorating housing [...] were often allotted to Roma, sometimes as lower category flats[8] or as more spacious ones corresponding with the size of their families. In combination with natural migration, this concentrated Roma in certain parts of a city' (Baršová 2002: 12).

In Brno, such neighbourhoods included Bratislavská and Francouzská Street and their surroundings (that are also parts of the inner city). Several Roma families had already lived there in the 1920s and 1930s. From 1947, Roma from the Spišská Nová Ves area in Slovakia were settled mainly in these neighbourhoods. In addition, other socially disadvantaged people were relocated to the nationalized, formerly blue-collar apartment blocks in this area (Cejl Street). In the 1970s, Roma from dissolved Slovak Roma settlements moved there (Souralová 2008: 23). Overcrowded flats and bad living conditions are still typical of such neighbourhoods. For example, an eight-member family that we met during our fieldwork was living on Cejl Street in a one-bedroom flat in a typical courtyard-based apartment block, where they were without sanitation, a bathroom and hot water – conditions which have not changed since the early 1900s.

Ostrava became a focus of the immigration of Roma (mainly from Slovakia) at the beginning of the 1950s, in response to demand for unqualified labour to support the developing industry. This migration was both spontaneous and organized. At the start of the 1970s, a new concentration of Roma in Vítkovice sprang up, together with the emergence and spread of Roma settlements in older parts of the city, where they have remained localized (Pavelčíková 1999). Two of them – Přívoz and Moravská Ostrava – are located within the inner city (see also Chapter 6).

Tenure Change During Post-socialism and its Impact on Neighbourly Relations

Regardless of the far-reaching changes of housing ownership as described in Chapter 6, protection of the occupation rights of tenants – both in municipal housing and in privately rented flats – has remained strong. Rent control has continued, and despite gradual increases, regulated rents have often remained far below market prices. As Lux (2000: 22–35) shows, this situation has given rise to tensions between households living in rent-regulated tenant housing and households renting flats at market rates with almost no tenant protection. Regulated rents are not determined

---

developing, preserving or using Roma culture. However, intentions differed from reality. Even though the forced spatial dispersions of Roma were partially successful, specific types of Roma areas and concentrations emerged, mainly as a consequence of their spontaneous migration. Among them were areas scattered among old, deteriorating, built-up inner-city areas (Haišman 1999, Pavelčíková 1999, Víšek 1999, Baršová 2002).

8   These are flats of lower quality, for example, without central heating and/or without facilities (bathroom, toilet).

by the income of residents; therefore, many well-off and long-term households have been living in low-rent flats, which have thus remained unavailable to socially needy tenants. In reality, tenants often live elsewhere while renting out their rent-regulated flats illegally and pocketing the difference (Kostelecký 2000: 179–80, Donner 2006: 40–54; see also Chapter 8).

As a result of rent regulation, relations between landlords and tenants have worsened. Naturally, tenants have not wanted to surrender their quasi-property rights (for example, the possibility of exchanging rented dwellings or passing them on to their relatives). Nor have they been interested in vacating flats with regulated rents, buying them or signing new rental contracts at market prices (Lux 2000: 23, Donner 2006: 79). Since regulation does not end until a dwelling is abandoned, some landlords have tried to evict tenants by whatever means possible, even unethical ones.[9] Moreover, the gradually increasing regulated rents have become too high for certain socially disadvantaged people (for example, pensioners), who, due to their lower mobility, often become trapped. Such social inequalities can be observed not only in the building but in the locality, where affordable stores and services are often unavailable (Musil 1992b: 456; for the specific housing situation and stock privatization in Brno and Ostrava see Chapter 6).

Two Worlds in one City: Empirical Research at Two Addresses in Brno

What these changes might mean for residents, their attitudes toward their building, locality and their relationship to each other is shown in the two following examples from the inner city of Brno.

A House on Stojanova Street

This rental house, situated near the western edge of the inner city, close to a park and villa quarter, is an example of what can be described as housing with a 'better address'. Numerous small stores, restaurants, good public transport, schools and kindergartens and other services are located in the neighbourhood. The house, built in the early 1930s by a Jewish investor, was nationalized after the Second World War and, in 2000, reprivatized. Twenty eight two- and three-bedroom flats, an atelier and two garages are in the house. Two flats with three tenants each remain in municipal ownership. The other residents have established an owner's

---

9   In order to abrogate a regulated rent contract, a release must be approved by a court for a valid reason (for example, personal housing need of the owner, flat renovation, expansion of adjacent non-residential space). In any case, the tenant has the right to an alternative dwelling. Moreover, because of the highly inefficient judicial system and the lack of alternative housing, a trial can last several years, even in clear-cut cases (for example, tenants breaking a contract), and the owner must bear all the related expenses (Sýkora 1996b: 287–8, Lux 2000: 2003, Donner 2006: 47).

association (*sdružení vlastníků jednotek*), which is convened by a chairman and an elected committee at least once a year. At first, the committee's work was not easy: the house was among the first privatized in Brno, and the municipal government officials had neither knowledge of nor experience with this form of ownership.

At the time of privatization, the house was in bad condition, both externally and internally. First, the residents cleaned out rubbish from the basement and reconstructed the house, including the installation of new windows and building a new façade. They did the first cleaning of the communal space on their own, thus getting to know each other better and cementing their relationships. Most see the direct financing of the reconstruction, together with their control of the process, as the greatest advantage of privatization. Such control would not have been possible if done through an outside agency. They also voted to purchase new energy-saving products (for example, automatic lights, new insulation). The municipal government contributed 40 per cent of the cost of the new façade, which is the general rule in Brno.

Nowadays, 57 inhabitants, including children, live in the house. Their number fluctuates slightly, according to the number of inhabitants in the rented flats, usually students or people living alone. In 2009, four flats were rented. The living and housing arrangements vary: senior citizens, families with children, unmarried couples, both with and without children, and student households. Dogs and cats, along with small children, often serve as pretexts for communication. Animals are generally perceived positively, with residents complaining only about dog excrement on the streets around the house.

People moved there for various reasons, but nobody chose the inner city as their ideal locality. Either chance got them there or family ties (see also the typology in Chapter 9). For instance, many 'inherited' their flats from their parents or grandparents. Two women have lived there since birth. People's attitudes toward the locality and the building became more positive after they became owners of their flats. Some claim they would not want to live elsewhere and that they are completely satisfied there. Factors increasing satisfaction include access to the city centre, schools, stores and parks, lack of noise and so on. The residents are proud of the pleasant appearance of the building, and they all help to keep the communal space clean. They take care of the front garden, which they planted with flowers after the reconstruction of the façade. In tandem with their sense of being co-owners and residents of a nice building, their feeling of solidarity has apparently increased.

Neighbourly relations come in many varieties, from friendly through indifferent to hostile, as the following statements show: 'Our children play together.' 'From time to time we visit each other.' 'My neighbour visits me to cut her cat's claws.' 'Fortunately, Mr X has now died.' Relations toward student households, which do not seem to fully 'belong' to the residential community, are ambivalent. Such households change yearly; they are therefore unstable and, for certain other residents, even appear 'strange'. An 'us versus them' mentality also pervaded with respect to a Roma family that lived there for a time.

Two Houses on Cejl Street

The second case study involved two three-storey houses, centred on a courtyard, on Cejl Street and in the municipal ownership of the urban district of Brno-Sever. Mainly inhabited by Roma, they were chosen as a contrast to the first case study.

Cejl Street is located in the eastern part of the inner city of Brno. By the beginning of the 1900s, seven big textile factories had been built in the area, and at that time multi-storey tenement buildings were built to provide accommodation for textile workers coming from the countryside. The neighbourhood has retained its blue-collar character to the present day. From the mid-1990s onwards, an area with a high concentration of socially excluded people (mostly Roma) – called 'The Bronx' by other people in Brno – has developed. However, nowadays, alongside gambling establishments and pawnshops, there are several newly built and reconstructed houses. Companies and government offices are also situated nearby, and even a few tentative signs of gentrification are apparent (for example, a squash hall). The municipal government is currently preparing an EU-funded project to improve the environment in problematic areas in this part of the city.

All residents of the two buildings, with the exception of two or three households, are Roma. They mostly live in flats with one bedroom and a separate kitchen (so-called 1+1 flats), centred on courtyards, with their entrances accessible by common inner galleries. The only exceptions are two bigger flats created by merging several small flats into one. One of them is a flat reconstructed at the residents' own expense, with a shower and chemical toilet. Other flats have neither bathrooms, nor toilets, nor hot water. Communal toilets are situated on the courtyard gallery. The owner of the house, the municipality, has made no repairs, apart from dealing with acute problems. One inhabitant is paid by the municipality to clean the house. A housing department official told us that the municipal authorities assign flats at a 'better' locality, Černá Pole or Lesná, to 'more decent tenants'.

The concentration and segregation of Roma in old and often unsuitable or inappropriate municipal housing (for example, small or poorly equipped flats, dilapidated buildings and buildings far from amenities) is increasing, and they are often unable to change their place of residence or find other housing. They depend on the local authorities and social workers for help. They move out if the municipal government assigns them a better flat. Moreover, they cannot rent flats in another locality because of the widespread racial prejudice of owners. Roma often occupy dwellings illegally and live without rental contracts. The owners of the buildings and flats tend to abuse the inability of Roma to find their way around the official bureaucracy and their tendency to accept any offer that seems beneficial at the time, even if it is completely disadvantageous to them in the long term or sometimes even borders on the illegal (Šimíková and Navrátil 2002: 14).

Neighbourly Relations in the Inner Cities of Brno and Ostrava

As discussed earlier in this chapter, other studies found that neighbourly relations in the inner city differed from those in other parts of the city (for example, large housing estates) in its low level of community intimacy and mutual assistance. The inner city was also marked by a certain degree of formality and social distance in neighbourly relations. Furthermore, our research revealed that neighbourly relations in the inner cities have retained certain characteristics from the state-socialist period.

The questionnaire survey and complementary interviews confirm a similar degree of formality today. Knowledge of, and interactions with, neighbours are often limited. Even though many respondents claim to know their neighbours, their familiarity is usually superficial. The results indicate that the fewer the dwelling units (according to the 2001 census, in the inner cities of Brno and Ostrava, almost 60 per cent of the buildings have less than 11 units), the higher the probability that neighbours know each other. People in larger buildings (for example, corner ones) need more time to get to know each other, while the situation is completely different in big prefab buildings with hundreds of inhabitants, even those in the inner city. A respondent from such a building in Ostrava said he considers his neighbours to be 'only people from the same floor, not others in the building'. However, even short-term residents in all building types can usually identify next-door neighbours on the same floor, even though their knowledge of them is typically not very deep or detailed. Their familiarity is usually based solely on mutual observation and interaction in the common building space (that is recognition as neighbours, knowledge of names and household composition and so on).

Similarly, the questionnaire survey reveals that interpersonal relationships are central to people's idea of neighbourhood and that most people consider their neighbourly relations good. However, they view such relationships as defined by peaceful coexistence, cooperation, mutual politeness and occasional hallway greetings – and nothing more.

Close ties, reserved for family and friends, are uncommon with neighbours, as the following statements indicate:

> Neighbourhood? I don't need a neighbourhood. People should guard their privacy. [...] It is polite to exchange greetings, of course; we shouldn't close our doors on people. But I don't need to go out with everyone in the building for coffee. I have my own privacy, and I want to maintain it. I prioritize my family over my neighbourhood. (woman, 29 years old, pre-high school education, Brno)

> I'm not such a community type. I don't need to visit my neighbours every day, help each other cook and all that. The ideal neighbourhood for me should take a more relaxed form. People should know each other, and know things about each other, to a degree. They should be able to talk. If they agree to meet somewhere

in the evening, they should have something to say. They should help each other with minor things. But I don't mean gregarious building life. I don't mean that. (woman, 46 years old, university education, Brno)

I'm not a very neighbourly minded person. I don't go in for it much. I prefer peace and quiet. […] We are on a first-name basis. We talk to each other. But we are not friends. […] We are acquaintances, not friends. And it suits me this way. I don't want closer contacts. I have my friends, they have theirs, and it suits us this way. Of course, we help each other out when we have to. Maybe I'm expecting a delivery, and I know I won't be at home, but [neighbour] will be. So, I ring the bell at her place, and she helps me out. Or maybe she's baking, but she forgot to buy something, so she comes over to borrow it. […] But we are not close friends. (woman, 43 years old, high school education, Brno)

Only a fifth of respondents cited mutual assistance, including helping older residents, assisting with small repairs, favours, guarding flats, holding spare keys, watering plants during holidays, looking after a pet or a child, or taking a child to or from school. But such neighbourly behaviour was appreciated.

The same number (one fifth) described close relations, including mutual visits, though usually with only one or two other households and typically with those having similar interests, lifestyles, age structures or social positions (for example, families with small children, long-term neighbours, older widowed women, people active in the same church): 'It doesn't matter where you live, but it's important to have someone there who you can rely on' (woman, 36 years old, university education, Ostrava).

This does not, however, preclude close relations between dissimilar people. For example, in the survey sample there are several cases of friendships between young respondents and older people in the same building, marked by reciprocal relations: each visiting the other from time to time, with older residents telling their young neighbours about the old times, the history of the house, its current inhabitants, who lived there in the past and so on. In return, the young neighbours keep them company and help them out occasionally with their day-to-day affairs. These relations thus perform a socially beneficial function, as the following example illustrates:

I have established a friendship with a neighbour – a lady who is over 80, maybe 90 years old – who lives one balcony higher. We talked a little bit when I was on the balcony and after a while I invited her over for coffee or tea. And she came. It has been three weeks since she was at my place. We talked for two hours about the house's past: the people who used to live here, what the flats looked like back then. It was very nice, so we agreed that I would visit her next. She lives alone, and I expect she will need something from time to time. There's no lift in the building, and she lives on the highest floor. She's very nice. (man, 39 years old, university education, Brno)

Neighbourly relationships are also often utilitarian in purpose. For example, building upkeep and administration frequently serve as the basis for initiating relations. Building committee chairpersons, or even cleaners and caretakers (for example, furnace repairmen), necessarily interact with all, or at least many, of a building's inhabitants. In the course of performing their functions, they may form closer bonds with certain people or even introduce people to each other and help them to get along. Nevertheless, such relationships are usually purely utilitarian, based on the requirements of cooperation, and generally do not cross that frontier:

> It is necessary to establish relations with other tenants, because I am on the committee. I have to take care of the building, so I have to be in touch with them. And they have to be in touch with me, because if they want something they have to contact me. (man, 28 years old, university education, Ostrava)

Children can provide another strong context for establishing relations. Households with similarly aged children naturally have similar interests, reasons and opportunities for closer communication and contact. For example, families with small children may go for walks or to playgrounds together. Despite the fact that these relations are quite close and intimate, when the children grow up, the relations tend to weaken or end completely:

> Our children are about the same age, so we know each other. When we were on maternity leave, we would go to the park together and the children would play together. But after we both went back to work, we stopped hanging out. We don't have much time anymore. (woman, 43 years old, high school education, Brno)

Even though most respondents do not cross a certain threshold in their neighbourly relations and, in spite of physical proximity, seek to maintain their privacy and a certain distance, this does not mean that no one in the inner city would want, or does not attempt, to forge closer ties. Their attempts to build a closer neighbourhood community might not be reciprocated, however. This was the case for a respondent and her family. One way they tried to foster closer relations with their neighbours was by trying to build a playground in the inner courtyard of their building. They obtained some old equipment from a kindergarten and hoped to repair it with their neighbours and set up the playground together. Unfortunately, the equipment gradually disappeared or was destroyed. They are now moving to a new housing complex in a suburban location, where they hope it will be easier to fulfil their dream of a close neighbourhood community and where they expect to find a more homogeneous social environment among predominantly young families. Thus, the quality of neighbourly relations is an important factor in determining people's satisfaction with a given building or locality; indeed, it can be a reason for changing residence. In the following example, the respondent is clearly dissatisfied:

> I think it is related to Czech society, which is not generally very open. [...] It is about better communication among the people. Simply not staying locked in one's flat, but also being willing to communicate more. My ideal would be to create a community centre in a tenant building, where people could meet in a common space. (woman, 32 years, high school education, Brno)

On the other hand, only a minimum number of respondents consider their neighbourly relations pronouncedly negative. Likewise, when respondents mention conflicts or quarrels, these usually only involve one other household, not the inhabitants of the whole building. Many negative experiences simply arise from mutual proximity, which makes neighbours witnesses to each other's private lives: for example, household disagreements, repairs and other noise can affect other people in the building. The following is an example of such *un*neighbourly relations:

> Those two women are very unhappy. They are so acrimonious. I understand how this can happen. They have probably had a bad experience with something or other, but it influences my life quite significantly. But I can't ring them and say, "Look, it bothers me how you yell at your son and your dog. It destroys my sense of comfort." There are times when you can ring and say: "I don't like the fact that you leave rubbish here. It smells, so don't leave it on the balcony." I can say that. It's OK. But about dogs and sons, I can't. One cannot choose one's neighbours. It's not possible. I have to accept them. (man, 39 years old, university education, Brno)

New Aspects of Neighbourly Relations

In addition, having been significantly influenced by the transformation of property relationships (for example, privatization), neighbourly relations in the inner city have acquired certain new features during the transition period. The impact has been both positive and, in certain aspects, negative.

Positive results are often associated with the transfer of ownership to residents through owner associations *(společenství vlastníků)*. In such cases, common concern for, and upkeep of, communal spaces in buildings creates an atmosphere of cooperation and trust (Steinführer et al. 2009: 143). This is true of the building on Stojanova Street presented above and of the prefab building that was the subject of Bazac-Billaud's ethnological research (1996) on a large housing estate in Prague. His study confirms that privatization can lead to solidarity and responsibility among residents and intensified neighbourly relations. Private ownership usually results in cooperation among residents, including building reconstruction, which was usually neglected during state socialism. During our research, major repairs were being carried out in many recently privatized buildings in Brno and Ostrava (for example, windows were being replaced, buildings insulated). In the process, close interpersonal contacts among residents are often established, with new or younger neighbours being put into touch with older residents:

> These were municipal flats before, [...] then we bought them and started to take care of them. How our home looks depends on us. It is better now, by far. Things have improved since I moved here. We've owned the building for four years now, and we've been working on the building all that time. (woman, 36 years old, university education, Ostrava)

> I am actually one of the owners, so we interact more often. It's different now than it was. [...] We have to stay in contact now. (man, 28 years old, university education, Ostrava)

On the other hand, property changes can also cause tensions among residents. This can occur, for instance, in buildings with diverse rental and property arrangements: for instance, where half the tenants pay regulated rents and the other half pay market rents, their different socio-economic positions are highlighted. This kind of coexistence can feel threatening to the original tenants. They might worry about their ability to pay the increasing rent, accompanied by pressure to move out, though they may be unable to do so. Thus, property changes can underscore group discrepancies:

> Those who are moving into flats with market rents are mainly younger. Among those I know, some are probably IT people. One is English. There are two young girls. They're mostly young people. And those who have remained from the old times are older people. Some do not work anymore, obviously, [...] and they are probably making ends meet from their pensions and however else they can. [...] The neighbourhood is rejuvenating, and it's basically young people who make enough money to rent a flat in this building, in this locality. (man, 39 years old, university education, Brno)

Additionally, as mentioned above, disagreements and conflicts can surface between building owners and their tenants with regulated rent, further exacerbating ill will in the building.

The transformation has often meant greater resident churn,[10] which despite the traditionally long period of residence in Brno and Ostrava[11] can affect neighbourly

---

10   Resident churn can be defined as the number of residents moving into or out of a building compared to the total number of inhabitants. The term *churn rate* is used most often in business, but it can also be used in other areas, such as demographic and housing topics (for example, Robson et al. 2009).

11   According to previous research conducted in Brno, average periods of residence are rather high (around 25 years in one inner-city area and 16 in another one with a high share

relations. Regardless of how long they have lived there, people living in a locality characterized by high residential turnover have fewer opportunities for establishing relationships, and they realize that any relationships established will not last long (Sampson 1988: 766–8). Changes in housing policy and property ownership have led to more intense resident churn which has undermined the stability of pre-established communities. Even long-term residents may lose their previous contacts and their knowledge of other residents.

Individual buildings differ in their resident churn rate. For instance, unsolved problems of ownership structure or the establishment or expansion of commercial space, which is financially more profitable for the owner, do not necessarily accord with the housing function of the other units in the building and can, therefore, stimulate a higher churn rate. According to an employee of a private company that manages apartment buildings, it is ideal to have stores on the ground floor of an inner-city building, offices on the first floor and flats on higher floors: 'This model minimizes possible market swings' (man, 60 years old, university education, Brno). This conflict of interests can lead to a high turnover of tenants and thus prevent the establishment of attachments for a particular building and relationships among residents. The following is an example of commercial space clashing with tenants' interests:

> Our neighbours change a lot. There are actually four or five flats here, but people left because […] in this building, in the basement, there is a discotheque – or, how to describe it, an amusement facility? – and it is quite noisy. I think they certainly exceed the legal limit. Everything is up in the air here – with the building, with everything – because the owner is not actually the owner anymore, he did not manage to get the necessary approvals, and he did not actually pay the mortgage. Therefore, the house belongs to the bank and we cannot even purchase it. But the reason those people left, the ones next door – two or three of them moved, I don't even recall anymore how many left – was the noise from the basement. But the club is […] a regular supply of money for the owner, so he closes his eyes to it. (woman, 34 years old, high school education, Brno)

Relations Between Student Households and other Residents

The number of student households is growing due to the rising number of university students and the chronic shortage of dormitory space (see also Chapters 8 and 9), but they have not yet had a clear and substantial impact on neighbourly relations in the inner cities of Brno and Ostrava. Previous studies have shown

---

of Roma people; Steinführer et al. 2009: 143). In our questionnaire survey (2007/2008), the average period of residence was 21 years in Brno and 15 years in Ostrava. In both cities, however, 28 per cent of respondents have lived in their current residence less than five years.

they can negatively affect neighbourly relations. For instance, a study carried out in Sunderland in the UK by Elizabeth L. Kenyon indicates that students living in privately rented flats are perceived by other residents as a special community that negatively influences neighbourhoods. Other residents are suspicious of student households, whose occupants are away for long periods during holidays and breaks, leaving their unoccupied flats vulnerable to burglaries, and whose behaviour is thought to downgrade the physical environment. This last aspect results in owners having difficulties in maintaining properties with student tenants, due to students' reputed 'untidiness' (Kenyon 1997).

Our work uncovers neither serious complaints (only minor ones, about noise) nor evidence of students disturbing neighbourly relations. Respondents do, however, confirm a lack of relations with student households. On several occasions, respondents were not able to identify the members of student households, nor did they have a clear idea of how many students lived in any particular dwelling. This is due mainly to the character, structure and size of student households, with big flats often inhabited by a large number of students. Moreover, student households are very changeable in both size and composition. Not everyone has signed a contract to live in them, prolonged visits of friends are not uncommon and so on. Nor is passing flats on to the next generation of students unusual. Student households are also diverse. Some are characterized by close relations among members, sometimes partnerships, while in other cases inhabitants of a single flat may not know each other well or even by name. The following are two statements of respondents who have had experience with student households:

> Last year a huge number of people lived there [...]. There were some students [...], maybe 15 of them, [...] and they were so loud [...]. We did not have any contact with them. [...] They were so distant and probably not inclined to become closer. They had iron stairs, and when they walked up and down it was very unpleasant. Or they used to – and it was crazy – slam the door, the entrance door. But really, every time, all of them, when they were closing the door, they slammed it. (woman, 35 years, high school education, Brno)

> It is a building with nine housing units. Half of them have been rented to students, who are always changing. The other half are fairly permanent residents with whom we exchange greetings, talk and sometimes we meet as well. (man, 36 years old, university education, Ostrava)

Roma Neighbourhoods

What holds for the majority does not necessarily hold for ethnic minorities such as the Roma population. While the issue of the majority versus the Roma minority was not originally part of our research scope, it became increasingly important, because areas inhabited mostly by Roma contrast sharply with other areas of the inner city and consist of socially excluded localities. The inner cities of both Brno and Ostrava have areas with high concentrations of Roma, who face problems connected with social marginalization. Therefore, we find it important to emphasize the specificity of these parts of the inner cities.

Political and economic changes during post-socialism have resulted in generally higher differentiation within the population: for instance, the social, economic or spatial exclusion of certain groups, such as minorities, the unemployed or the disabled (Pospíšilová 2009). The Roma community especially suffers from a high risk of social exclusion, due to their ethnic and cultural dissimilarity, as well as the negative perception and discrimination they face from the majority (Šimíková and Navrátil 2002: 11). Moreover, ambiguous state or municipal policy and authorities' negative attitudes or inexperience with Roma do not help to solve the problem.

Roma are mostly concentrated in certain localities. These homogenous environments further deepen their social and ethnic segregation, removing them from majority values and social norms, thus impeding integration (Mareš 2000: 292, Vašečka 2002: 245). Left unresolved, the situation may foster 'culture of poverty' elements (unemployment, criminality, dependency and so on) (Pospíšilová 2009). On the other hand, the cities themselves, through their housing and social policy, could play an important role in improving the situation. Authorities these days usually realize the need to reduce the spatial concentration of ethnically homogenous populations. However, no single system is in place that they can use, nor is there a global solution (Frištenská 2002: 51). The non-governmental sector can play an important role in these areas, and after 1989 many Roma and pro-Roma organizations were established. Most have focused on improving the education of Roma as well as Roma work qualifications and housing. Still, it is often hard to evaluate the effectiveness of their activities (Šiklová 1999).

In Brno, the locality with the highest concentration of Roma is situated in the eastern part of the inner city, around Cejl Street, where one of the studied houses is located. This area contrasts sharply with other areas of the inner city. The activities of the Museum of Roma Culture *(Muzeum Romské kultury),* which is situated nearby, is paradoxical, because the Roma themselves rarely visit it. On the other hand, NGOs, such as the Association of Roma in Moravia *(Společenství Romů na Moravě),* the Church and a community housing project are bringing social and housing improvements to the area. However, the future of the house on Cejl Street remains uncertain. On the one hand, Cejl Street is partly being gentrified; on the other hand, changes in political orientation in the wake of the municipal elections might alter the urban landscape and result in the suspension of currently planned EU-supported projects.

In Ostrava, the areas with the highest concentration of Roma are also located in older parts of the city: for example, in Přívoz and one part of Moravská Ostrava (that is in the inner city) as well as in other quarters such as Mariánské Hory, Hrušov, Slezská Ostrava, Kunčičky, Poruba or Radvanice. People in these areas often occupy flats and buildings which are in very poor condition. Many people are unemployed, have debts, are behind in rent payments or depend on social support. Some criminality is also apparent: for instance, involvement in the grey economy, gambling, drugs, usury and so on (John 1998, Hajská and Poduška 2004, Pavelčíková 2008).

Conclusions

In our chapter, we described neighbourly relations in the inner cities of Brno and Ostrava, focusing on sociability and tenure change and examining how neighbourly relations have been influenced by the post-socialist transition. Referring back to the concept of residential change presented in Chapter 2, the post-socialist condition has been shown to be very influential in shaping neighbourly relations. It is mainly tenure change that has resulted in higher residential turnover rates, leading to the superficial and rather formal character of neighbourly relations.

Neighbourly relations in these inner cities combine features that characterized them in the previous period with new features that spring mainly from changes in housing policy. Our research shows that neighbourly relations and sociability among neighbours in the inner cities of Brno and Ostrava continue to exhibit a certain degree of formality that was typical prior to the transition. Although interpersonal relationships are central to most people's idea of neighbourhood, peaceful coexistence and cooperation – particularly among people who share property, space and therefore interests – are sufficient for respondents to consider neighbourly relations as good.

Knowledge of and contacts with neighbours are often superficial, and close ties with neighbours are uncommon. Mutual assistance, visits and closer bonds characterize a few cases. However, they are usually between two or three households, not within neighbourhoods as a whole, even though people value occasional mutual help and rely on one another. Moreover, neighbourly relations are often based on utilitarianism (for example, building management) and do not necessarily cross this frontier, as many people prefer to retain their privacy.

In addition, neighbourly relations exhibit certain new features related to changes in housing policy and which have had a variety of impacts. The effect of transferring flats to private ownership and establishing owner associations has been generally positive, promoting cooperation among neighbours. It has further strengthened housing continuity: families have often lived in their flats for two or three generations and ownership enhances continuity. On the emotional level, memories of childhood, youth and living jointly with parents or grandparents strengthen ties to the flat and locality as well. An important role is also played by the building's administrator and committee, which facilitates contact between residents. An owner association works on democratic principles: every co-owner has voting and decision rights about substantial investments.

However, several other changes have affected neighbourly relations rather negatively. For instance, the combination of regulated and unregulated rents creates tension among neighbours, and higher residential churn rates, caused by the property transformation and strengthened by unresolved problems of ownership structure and unsuitable non-residential space (for example, discotheques, pubs and noisy restaurants), have weakened neighbourly ties.

The consequences of the increasing number of student households remain ambiguous. Even though we did not come across any serious complaints, relations between inhabitants of student households and other residents are usually non-existent. We assume the situation will evolve along with the rising number of student households and certainly deserves further research.

PART III
Summary, Conclusion and Outlook

Chapter 13

Conclusion: Findings and Reflections

Annegret Haase, Annett Steinführer, Sigrun Kabisch,
Katrin Grossmann, Ray Hall

This chapter reflects more generally upon the small-scale and fine-grained findings of the research presented in the book from three perspectives. *First*, by referring back to the conceptual model developed in Chapter 2, we summarize the results of the investigations on inner-city residential change presented in the case studies. *Secondly*, we consider how the research has contributed to a better understanding of the current and likely future development of East Central European cities, together with their related challenges for further research. *Finally*, lessons learnt for future research on residential and wider urban changes in East Central Europe are discussed.

Residential Change in East Central European Inner Cities

Residential change is a continuous process of interrelated changes of residents and the places they inhabit, all happening in specific settings and circumstances. Such change is a function of intermediary structures such as the housing market, existing patterns of socio-spatial differentiation, institutional frameworks and local governance arrangements, all of which contribute to new and ever changing residential patterns. Residential change occurs at different temporal and spatial scales, most prominently at the scale of neighbourhoods but also at the scale of individual dwellings. Its outcomes are residential patterns which may exist for quite some time or immediately become subject to further change. We found three broad residential patterns in East Central European inner cities:

- *persisting residential patterns* – that is patterns which, in spite of changed institutional frameworks, still exhibit characteristics considered to belong to a previous era (for example, the small-scale clustering of old people in large rental dwellings in municipal ownership, cross- and three-generation housing arrangements or small-scale clusters of the urban poor in all cities, which in the Czech case strongly overlaps with the ethnic dimension),
- *rearranged residential patterns* – that is patterns which existed in some form during state socialism or even before the Second World War but whose character was reinforced or changed during the post-socialist transition (for example, the socio-ethnic exclusion of Roma in the Czech cities or multi-

household housing arrangements of unrelated others forced to share one common dwelling), and

- *post-socialist residential patterns* – that is patterns which arose as a result of the post-socialist transition (for example, the new small-scale concentrations of students in private rental dwellings or the increasing diversity of housing and living arrangements, including non-family households).

Residential change occurs in two dimensions: in the socio-demographic dimension (referring to the residents) as well as in the physical and the tenure dimensions (residence). Both are influenced by intermediary structures and actors. Residents and residence, together with the intermediary structures and actors, are discussed here; thus we are able to summarize our main findings by referring back to our model of residential change (see also Figure 13.1).

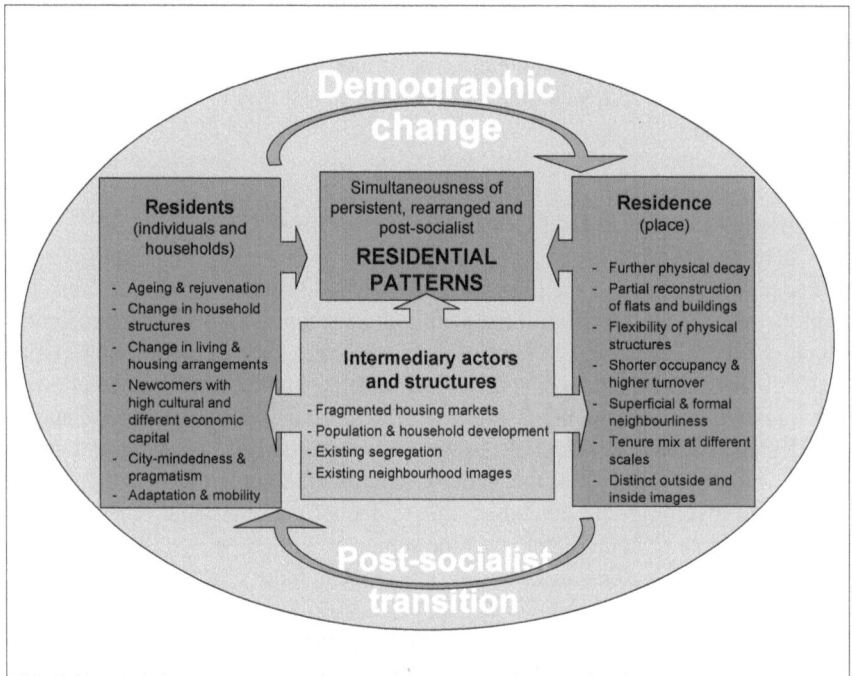

**Figure 13.1 Residential change in East Central European inner cities
– main findings of our research**

Source: Authors' work.

Residents

Population decline and ageing are no longer the only demographic changes taking place in the inner cities of Polish and Czech cities. We found evidence of repopulation, which, since it is driven mainly by younger and middle-aged incomers, can also be seen as rejuvenation. Rejuvenation and ageing are therefore occurring simultaneously and in close proximity to each other. Although the influx of younger households is, so far, the dominant characteristic of the inner city, these households have started to change its socio-demographic make-up. The changes might be even more significant than is revealed by the analysis of the census statistics, since many of the, mainly young, newcomers are registered elsewhere. The interviews portrayed the small-scale dynamics of changing inner-city populations, along with the transformation of household structures and related living and housing arrangements.

However, residential change occurs not only as a result of residential mobility but also as a result of transformations in place through adaptation and renovation of the housing stock. Residents make use of flexible physical structures and adapt them according to their current needs and personal preferences. Therefore, a merely mobility-centred perspective would result in a limited understanding of residential change in East Central European inner cities today.

As a consequence of recent in-migration, the socio-demographic make-up of the inner-city population in the four cities is becoming increasingly diverse. New household types and their corresponding housing arrangements (for example, flat sharing and cohabitation) have become more wide-spread while, at the same time, shorter periods of housing occupancy are more common. High levels of housing mobility are now seen as normal for a certain segment of the residential population that is in transition between different phases of their personal lives. Many of the newcomers are characterized by cultural but not necessarily economic capital and, as a result, in-migration has, so far, not led to large-scale gentrification in any of the four inner cities. Some essential reconstruction has led to partial upgrading and spot regeneration, but these are not overarching inner-city trends.

Inner-city choices are not based solely on preferences but rather on a variety of both emotional and rational criteria. Although we found evidence for a group of city-minded residents who, in many cases, preferred brick-built housing, residential location decisions in favour of the inner city arise out of weighing different resources and preferences and trying to find, if not a best, at least an acceptable solution. The principal argument for households to move to the inner city or to stay there seems to be the functional advantages of the central location. Other motives are less important. Among the types we identified, 'transitory urbanites' are particularly significant, because they best mirror the contradictory development in Polish and Czech inner cities today: although the inner city at present suits their everyday needs and housing preferences, they are actually contributing to its profound transformation; for them, the inner city is, in reality, simply a convenient place to live.

Residence

Simultaneous change and inertia not only describe the social structure of the inner city, but also characterize its function as a place of residence. All four inner cities have, to varying extents, decaying and upgraded built environments; however, each has a different potential for regeneration. The current activities of the new owner-occupiers of the former municipal housing stock are bringing about the reconstruction and refurbishment of single flats and sometimes entire buildings. As yet, however, this is not resulting in large-scale inner-city regeneration. Renovation and decay can be experienced at the scale of the individual building, where diverging interests and various types of tenure come together. New investment relates mainly to single buildings, blocks or streets and contributes further to the fragmentation of urban housing markets. To date, neither investment nor in-migration has led to a larger-scale change of the residential environment, but there are signs of incipient change in shopping, social and cultural amenities.

As already emphasized, residents are responding to the flexibility of the apartments by modernizing, adapting and so transforming their physical fabric. Flats are changed in response to individual needs, and floor plans are adapted to particular housing arrangements. This strategy has its roots in the long-term housing shortage. But it is more than just a pattern persisting from state socialism since, in many cases, adaptations are now being made in owner-occupied dwellings and thus in 'real' and not just 'quasi'-property. However, there are a number of constraints to flexibility that relate, firstly, to the specific conditions of (inner-city) housing markets in East Central European cities and, secondly, to the limited financial resources of many residents in terms of possible housing choices. Although we found a tenure mix at different scales (building, neighbourhood, inner city as a whole), the openness of local housing markets for the flexible use of dwellings and transitory arrangements is still limited and gives rise to persistent residential patterns, including adult children still living with their parents, older people living alone in large flats or several households using one common flat for cost reasons.

The specific segment of highly mobile urban dwellers who, in their short housing careers, have already moved several times in order to adapt their housing arrangements to their current life circumstances, brings about shorter housing occupancies and increased residential turnover. Also, physical changes now occur more rapidly than during state socialism. The increased mobility also leads to more formal and superficial neighbourly relations among inner-city residents. Tenure impacts on these relations, too, because from a structural point of view, new owner-occupiers and remaining long-term tenants have distinct duties and roles which, in many cases, hinder neighbourliness and sociability. The differences in tenure can also cause conflicts (for example, about reconstruction measures). Not least, the illegality of a number of housing arrangements, particularly with regard to the black rental market, discourages close and open-minded interactions with neighbours.

In the post-socialist period, the inner city as a place of residence has become increasingly differentiated with respect to place images. While it is still frequently evaluated as a comparatively 'bad' housing area from outside, many of those who have moved there recently emphasize its location advantages and they have a much more positive view of inner-city housing and living. We also found evidence for positive images of central living through long-term familiarization ('got used to it and like it'). In contrast, those who dislike it generally live in very poor housing and have not chosen to live there, or face strong restrictions to leaving. Residents view the inner city from different perspectives and distinguish between more and less attractive places; their own residential environment is often evaluated positively, even if it is situated in a less attractive area.

Intermediary Structures and Actors

Residential change in East Central European cities today is occurring in highly fragmented housing markets. The public rental market has shrunk as a consequence of large-scale privatizations, while the private rental sector is negligible. Thus, owner-occupation is widely regarded as the best housing option. But when deciding to buy property, suburban locations are strong competitors to inner-city housing, with their promise of fulfilling the wide-spread dream of 'one's own home', a housing ideal that has gained significance during post-socialism in Poland and the Czech Republic. Most of the new housing has been built in the suburbs while, in the inner cities, single condominium projects with high-standard flats and spacious floor plans targeted at more affluent households have been realized. Even with the end of state socialism, the shortage of housing has continued, with few vacancies in existing housing to cope with the new demand. The pressure on making housing available has increased with the growth in household numbers, in spite of declining or stagnating urban populations in the four cities under investigation. For many new households, therefore, there is no real choice about where to live; rather a decision is made about which is the best choice among a very poor range of options. Housing in the inner city is one of them.

Residential change in the inner city is thus not occurring in isolation but rather is happening in the context of existing and persisting residential patterns of socio-spatial segregation. In-migrants, such as transitory urbanites, choose affordable rental housing or search for niches in the housing market that are available for short- or mid-term arrangements. In today's East Central European inner cities, residents with completely different attitudes towards their housing environments typically live together in one building, such as long-term and recent residents, tenants and new owner-occupiers or those who have deliberately chosen to move there together with those who were assigned their flats.

Small-scale residential change mirrors overall societal trends of population and household development, such as the declining size of households, falling levels of fertility and the postponement of marriage and childbearing. The prolonged phase of post-adolescence results in the increase of transitory housing, leading both to a new group of urban dwellers and housing market actors. The trend toward smaller households increases the demand for smaller flats and leads under current conditions to new forms of housing arrangements (such as flat sharing with other households).

Finally, it has to be emphasized that residential change in the inner city occurs partly *because of* its advantages and partly *despite* its disadvantages. While city-minded urban dwellers deliberately choose central living, others have moved to the inner city because they sought cheap housing, or inherited or used the flat of relatives or friends. There is a group of households who ignores existing patterns of residential segregation and its mental representations in terms of neighbourhood images. These people move, for example, to highly stigmatized places, called the 'Bronx' in Brno, and use the term themselves and, by so doing, they change the existing social make-up. Others initially might not have had a positive image of the inner city but – at least in some cases – after living there, changed their views. The symbolic upgrading of inner-city housing areas at the time of our investigations is variable; some areas might see a change in their image, others may not.

General Reflections

Inner-city Residential and Demographic Change

Residential change is an appropriate approach for conceptualizing and understanding the small scale and long-term changes in post-socialist cities, neighbourhoods and actors. In this volume, the specific aim was to study the link between inner-city residential change and more general societal trends, particularly demographic change. Below, we draw three general conclusions with respect to our approach.

(1) Residential change of the inner city in post-socialist East Central Europe is more than just 'spectacular' transformations of the residential population and their places of residence. Post-socialist inner cities should not be described simply in terms of a polarized geography of socially excluded places and places of exclusiveness. While we have challenged the argument that decline and ageing in Polish and Czech inner cities are the two overarching socio-demographic processes, we also argue for a range of processes happening at the same time. Decline and ageing should be seen as interspersed with other socio-spatial processes of residential change. There is, therefore, not just one new story to be told, for example, that of gentrification or up-market change, but rather a number of new stories about residential change, many of them still at an incipient stage and hard to detect with official data. Although there is a need for research on 'spectacular transformations', such as gentrification and the influx of western foreigners or deprived immigrants into East Central European second-order cities, we argue that the social groups in between the extremes of residential segregation also need attention, since the majority of residents in these cities continues to live in

neighbourhoods that are more heterogeneous and difficult to categorize. Focusing only on the poles of residential segregation carries the risk of misrepresenting present-day realities in post-socialist cities.

(2) The post-socialist inner city today is a mosaic of new and old residents, new and old housing arrangements, as well as new and old housing aspirations.   The inner cities of the East Central European cities under investigation are inhabited by a diversity of social, educational and age groups. The term 'old-new diversity' refers to the socio-demographic and socio-economic characteristics of the residents as well as to their attitudes towards living and housing in the inner city. Various types of residents – from the frustrated to the convinced urbanites – can be found in post-socialist inner cities today. They have arrived in the inner city by choice, chance or by property allocation, or because it was the second- or third-best option. The diversity is created by a juxtaposition of persistent residential arrangements and structures originating from before the post-socialist transition, on the one hand, and new in-migrants with a variety of social backgrounds, household and housing arrangements, as well as housing attitudes, including sentiment-driven in-migration to the inner city and brick-built housing, on the other. This results, at least for the time being, in simultaneous processes of up- and downgrading in close vicinity as well as to small-scale social and demographic fragmentation. The inner cities in East Central Europe will certainly continue to differentiate so that small-scale patterns of social exclusion, gentrification, studentification, revitalization and social mix are also imaginable in the future.

(3) Younger non-traditional households are among the most important actors in current inner-city change in East Central Europe.   Urban demographic change in many European cities is most apparent with respect to household change. Certainly, in East Central Europe, increasing household numbers and diversity are two of the major drivers of inner-city residential change. The influx of so-called new or non-traditional households to the inner city, demonstrated by the research presented in this volume, indeed alters the socio-economic and demographic structure of the population of the inner city and contributes to rejuvenation and social diversity on the meso-scale. These household types are not simply a copy of those already known in Western Europe. We have found housing arrangements specifically adapted to the local housing markets (for example flat shares with a specific composition).

Yet, in spite of the new significance of non-traditional households in East Central European inner cities, 'family flight', that is the movement of middle- and upper-class families from the cities to the urban fringe and to suburbia, as well as ageing, remain major challenges for inner-city neighbourhoods. The non-traditional households of today may well become the suburbanites of tomorrow.

Conceptual Issues of Urban Development and Urban Research in Europe

Although our research focused on four cities in two specific countries, it required the wider perspective of urban research in post-socialist East Central Europe, and the comparison of inner-city change between the four case-study cities enabled us to look at analogies and differences at different scales. The three conclusions that follow summarize our findings.

(1) Before applying a concept to a new context, it is absolutely necessary to critically consider its origin and question its explanatory power for another context.   Understanding urban processes also means finding appropriate terms and explanatory approaches. Urban research during the last few decades has developed a number of such overarching concepts, among them gentrification, segregation and polarization. These have been tested in and applied to different contexts. In our research, we sometimes encountered difficulties in applying them, since they did not describe adequately the variety of processes we found. As a result, we have had to think about either new terms and concepts or new contents for the 'established' concepts, in order to bring them into line with the empirical realities in contexts that are qualitatively different from those for which the concepts had been developed (usually in the US-American or British urban context).

We decided to go back one step and systematize the ingredients of residential change in a model that could then describe the results without pre-supposing a certain type of process.

(2) Understanding and interpreting urban processes is highly dependent on the theoretical concept used.   Examining processes of socio-demographic change in a particular city through the lens of well-established concepts, such as those mentioned above, means that it is likely that evidence for the processes described by the selected concepts will be found. When, however, the processes are examined from an 'outcome is open' perspective, it becomes apparent that such concepts are not applicable. For example, in all of the case-study cities, evidence for processes that encompass the Western understanding of gentrification and displacement was found, particularly in very small areas, for single streets or complexes of buildings, or even a single building. At the same time, there has been an influx of very different residential groups to the same area, producing a more mixed residential structure. This is also true for tendencies towards segregation and fragmentation. When we look only at processes of concentration (of different groups of the population), we are able to identify processes of differentiation and (resulting) patterns of segregation. When, however, we look at the scale of these processes, we often find a strong spatial fragmentation that might result in new mixes of population at the scale of a district or even of a neighbourhood. Therefore, it is possible that contradictory processes and contrasting patterns coexist simultaneously. We found evidence for this in all four of our case-study cities. How durable this 'coexistence'

will be depends on many factors, such as housing market dynamics and ownership issues, as well as the socioeconomic situation of the population as a whole.

(3) The post-socialist transition remains important but it is no longer a complete explanatory factor with respect to East Central European urban change. We argue for the use of wider explanatory approaches that examine East Central European cities more comprehensively (for example, their increasing involvement in, and dependency on, European and global processes) while not ignoring the importance of the post-socialist condition. Current and future research faces the challenge of neither over- nor underestimating the impact of the socialist past and the post-socialist transition on East Central European cities. The impacts of global and European processes, as well as of overarching processes, such as demographic change, interact with the long-term consequences of the systemic changes after 1989. In most cases, the exact causalities of the complex tangle of post-socialist transition, globalization and demographic shifts are hardly detectable. Often, post-socialist transition accelerated processes that had already started before 1989, such as population losses and ageing in the cities. But, for an understanding of change and persistencies in East Central European cities, it is very important to consider the different factors and to emphasize their interplay with the subsequent developments of post-socialist transition.

Methodologies for Research on Residential Change

Using a mix of methods has helped to give a more nuanced picture of the changes taking place. Using quantitative information alone, the dynamics of change might have been underestimated, while they might have been exaggerated using only the qualitative results, with their selective sample. With the data gathered, valid conclusions cannot go further than the ones presented above. And we hold that there is a methodological aspect to this. There are four major methodological conclusions to be drawn.

(1) To detect and understand change, research at different scales from the house or flat to the city and beyond is needed. Mixed methods are most appropriate for this. Change can occur in a whole region or in individual flats. Change can be rapid or gradual. This means it is hard to detect the diverse patterns of change by a research design using a single method. Statistical analyses are needed to provide basic knowledge about processes and patterns, together with their change and persistence over time. In post-socialist settings, residential change is best reflected in small-scale data that make visible the small-scale fragmentations characteristic of the transition. But to detect processes that are hidden from, or only partly reflected by statistics for a variety of reasons, qualitative research, such as interviews and area observation techniques, is also needed. These methods automatically use micro-levels of socio-spatial analyses.

(2) Conventional data bases should be questioned more critically. Research should be question- rather than data-driven. Decennial census data can only tell part of the story, and register data are incomplete, since many of the more mobile members of the population are not registered. Therefore, both census and register data may be blind to the most recent and dynamic developments taking place in cities, such as illegal housing arrangements. Qualitative and survey research are therefore essential in order to access the stories 'beyond' statistical numbers and then to re-interpret statistics.

(3) Comparisons and a multi-focus approach help to avoid exaggeration of findings. The value of comparative research is twofold. Firstly, comparison makes commonalities with other cases visible and can lead to the identification of general dynamics and features of change. Secondly, comparison helps to identify the specifics of places and processes relevant only to individual cases. The cross-cultural comparative approach used in this project helped to sharpen and qualify our research results by improving the case-study research. Additionally, comparison should encompass different spatial levels. In other words, research should not focus solely on the place of interest but also examine related places in the context of the place of interest.

(4) Combining a range of views, research methods and disciplines helps to limit the effects of positionality. Multiple perspectives and methodological competences play a key role here. Outsiders might have different approaches that overcome the blind spots of national debates. Insider perspectives, on the other hand, understand the history and the dynamics of a place. The interplay of different perspectives, of data related both to aggregate patterns and to individual agency, help to overcome presumptions stemming from particular experiences and approaches.

Resumé and Outlook

The aim of this volume was to say something new about residential change and demographic challenges of the inner city in East Central Europe in the 21st century. The two questions considered here therefore are: what have we learnt about residential change in East Central European inner cities and what research fields and topics will further improve our knowledge on inner-city residential change and demographic processes in cities?

What Did We Learn?

- Residential change is not 'one' process but always consists of a variety of different, sometimes apparently contradictory, processes that are at work in one and the same moment. Change itself includes many facets, such as newly-evolving patterns and the rearrangement of existing ones, as well as inertia that might disappear in the future but might also persist and become subject to a long-lasting, stepwise change that is sometimes difficult to recognize.

- The four case studies reveal the complex dynamics of residential change taking place in the whole city – and our focus area, the inner city, shows only a part of this. It must be emphasized that we assume that some of the processes and dynamics observed in the inner cities are not necessarily restricted to this location, but are also happening in other parts of the cities. Future research might add an important extension to the research presented in this volume by looking at other parts of cities, to see where and how specific processes and dynamics of residential change are evolving.

- Although most of the reported changes are not 'spectacular', they lead to fundamental residential shifts and correlated uses of the urban space. It is, therefore, important to go beyond the most obvious forms of urban change that are found to a greater extent in the capital cities of East Central Europe and to investigate more subtle forms of change in second- and third-order cities, looking at both their differences and their similarities to the respective national metropolises.

- We learnt that residential change can only be explained by the interplay of macro-processes and the specificities of local settings, such as local housing market conditions and logics, socio-demographic structures and specific modes of local governance and planning. The development of inner-city change is neither determined just by location, physical structures and the type of building stock, nor does just the societal framework or the local context explain why an inner city develops the way it does.

- Residential change (as a local phenomenon) is closely connected to demographic shifts (as a macro-process). It shows in a specific way how demographic shifts are socially embedded within specific settings and how they impact on residential behaviour, housing preferences and housing markets. Demographic shifts are therefore a challenge for (local) urban development and, in a more general sense, for the quality of urban life as a response to changing conditions.

- We looked at the two decades from about 1989 until 2009 – the period of post-socialist transition – and found the rearrangement of existing patterns, the development of new ones and a juxtaposition of patterns of different origin. Although we were able to assess the 20-year period with our mixed-method approach, we wish to underline the importance of long-term research in order to detect emerging trends or developments that had started before 1989. Our research is, therefore, both a resumé of the last two decades and an investigation that mirrors a picture at a given moment in time.

- Urban research in East Central Europe after 1990 focused particularly on capitals, rather than on medium-sized cities. In the case of our four second-order cities – Łódź, Gdańsk, Brno and Ostrava – we have filled this gap with respect to residential change and housing demography. There is certainly an imbalance in the empirical evidence available for capital cities, compared with smaller ones, but there is also sometimes the assumption that medium-sized cities will follow similar paths to large cities. However, our research shows that second-order, deindustrializing cities, such as Łódź, face the challenges of population loss and ageing to an extent unknown in the capital cities. It is also apparent that there are 'niches' for distinct small-scale development in the inner cities, for example transitory housing of young people, thanks to the availability of moderately-priced housing – a phenomenon which might be explained by housing market conditions in second-order cities, which are not characterized by excessive growth and demand.

What Future Research Fields Exist?

Our research on residential change in East Central European inner cities does not claim to be exhaustive. Moreover, as is the nature of research, questions emerged that remain unanswered. But it opened up an exciting research field that should be of interest to other urban scholars. In particular, the interplay of macro-processes, such as demographic change, and the specifics of local settings, for example, the peculiarities of housing markets or urban societies, make the analysis of residential change so interesting. The many issues that could not be addressed exhaustively here include:

- systematic analysis of residential change in different parts of the city, together with stronger embedding of inner-city developments into a broader urban framework,
- systematic comparisons of trajectories of residential change in capital cities, on the one hand, and second- and third-order cities, on the other,
- the integration of settlements further down the urban hierarchies into the analyses of urban and demographic change,
- the whole process of suburbanization and its continuous interdependencies with inner-city change, which might hinder the reurbanization of East Central European cities suggested by our research,
- comprehensive analyses of the impacts of demographic processes other than household change, particularly of ageing, on urban change and
- long-term observation of the inner-city processes and trends we analyzed in this volume, first and foremost those that we described as having recently evolved or as incipient.

Within this volume, we have often used words such as *variety, simultaneousness* and *juxtaposition* to describe parallel processes, overlaps and interlinkages between different things happening at the same time. This is the very nature of residential change and one important dimension of living in European cities, irrespective of where they are situated. The manifold impacts that, among others, overarching demographic shifts have on locally specific settings, shape the processes and forms of residential change as locally-based but macro-level contingent issues.

In the introduction, we began with the observation that there is a growing need to consider demographic change in order to analyze and explain residential change as a part of urban development. In our research on four East Central European second-order cities, we found significant evidence for this relationship. We hope that our volume will contribute to expanding this kind of knowledge. However, many questions remain to be answered by future research so that there are many reasons to continue to work on East Central European cities and their development during the 21st century.

Bibliography

Aleš, M. 2001. Vnitřní migrace v České republice v letech 1980–1999. *Demografie*, 43, 187–201.

Altaş, N.E. and Özsoy, A. 1998. Spatial adaptability and flexibility as parameters of user satisfaction for quality housing. *Building and Environment*, 33(5), 315–23.

Altrock, U., Güntner, S., Huning, S. and Peters, D. (eds) 2005. *Zwischen Anpassung und Neuerfindung. Raumplanung und Stadtentwicklung in den Staaten der EU-Osterweiterung*. Cottbus: Brandenburgische TU (Planungsrundschau 11).

Andrusz, G., Harloe, M. and Szelényi, I. (eds) 1996. *Cities after Socialism. Urban and Regional Change and Conflict in Post-Socialist Societies*. Oxford, Cambridge MA.: Blackwell.

Atkinson, R. and Bridge, G. (eds) 2005. *Gentrification in a Global Perspective: The New Urban Colonialism*. London: Routledge (Housing and Society Series).

Atkinson, R. and Rossignolo, C. (eds) 2008a. *The Re-creation of the European City. Governance, Territory & Polycentricity*. Amsterdam: Techne Press.

Atkinson, R. and Rossignolo, C. 2008b. European debates on spatial and urban development and planning, in Atkinson, R. and Rossignolo, C. 2008 (eds), *The Re-creation of the European City. Governance, Policy & Polycentricity*. Amsterdam: Techne Press, 7–16.

Baršová, A. 2002. Problémy bydlení etnických menšin a trendy k rezidenční segregaci v České republice, in Víšek, P. (ed.), *Romové ve městě*. Praha: Socioklub, 5–29.

Barton, A.H. 1955. The concept of property-space in social research, in Lazarsfeld, P.F. and Rosenberg, M. (eds), *The Language of Social Research*. New York: Free Press, 40–53.

Bartoňová, D. 2003. Domácnosti podle výsledků sčítání lidu, domů a bytů v roce 2001. *Demografie*, 45, 268–73.

Bartoňová, D. 2007. Rodiny a domácnosti, in Fialová, L. (ed.), *Populační vývoj České republiky 2001–2006*. Praha: Katedra demografie a geodemografie, Přírodovědecká fakulta Univerzita Karlova, 63–75.

Barvíková, J. 2009. Úsilí o sociální mix, image a stigmatizace panelových sídlišt'. *Demografie*, art. 632. Available at: http:www.demografie.info/?cz_detail_clanku=&artclID=632 [accessed: 28 April 2010].

Bazac-Billaud, L. 1996. Sociální dopad privatizace bytů na sídlištích: příklad Prahy 13, in Olivier, A. and Tuček, M. (eds), *Původní a noví vlastníci: strategie nabývání majetku ve střední a východní Evropě*. Praha: CEFRES (Cahiers du CEFRES, 11), 137–52.

Beauregard, R. 2003. *Voices of Decline. The Postwar Fate of U.S. Cities*. 2nd Edition. New York et al.: Routledge.

Beaverstock, J.V., Smith, R.G. and Taylor, P.J. 1999. A roster of world cities. *Cities*, 16(6), 445–58.

Beaverstock, J.V., Smith, R.G., Taylor, P.J., Walker, D.R.F. and Lorimer, H. 2000. Globalization and world cities: some measurement methodologies. *Applied Geography*, 20(1), 43–63.

Beim, M. and Tölle, A. 2008. Segregationsprozesse durch Altbauverfall und Suburbanisierung in Posen. *disP*, 174(3), 51–65.

Berdahl, D. 2000. Introduction: An Anthropology of Postsocialism, in Berdahl, D., Bunzl, M. and Lampland, M. (eds). *Altering States. Ethnographies of Transition in Eastern Europe and the Former Soviet Union*. Ann Arbor: The University of Michigan Press, 1–14.

Bernardi, L. 2007. *An Introduction to Anthropological Demography*. Rostock: Max Planck Institute for Demographic Research (MPIDR Working Paper, WP-2007-031).

Bernt, M. and Holm, A. 2005. Exploring the substance and style of gentrification: Berlin's 'Prenzlberg', in Atkinson, R. and Bridge, G. (eds), *Gentrification in a Global Context*. London: Routledge, 106–20.

Bernt, M. and Kabisch, S. 2006. Ostdeutsche Grosssiedlungen zwischen Stabilisierung und Niedergang. *disP*, 164(1), 5–15.

Bernt, M. 2008. Die politische Tradition 'Europäische Stadt' und die Schrumpfung. *Berliner Debatte Initial*, 19(4), 102–112.

Beyer, E. 2004. Postsozialismus: Russland, in Oswalt, P. (ed.), *Schrumpfende Städte*, vol. 1. Ostfildern-Ruit: Hatje Cantz, 74–7.

Bierzyński, A. and Węcławowicz, G. 2008. *Łódź Data Analysis Report*. Warsaw: Polish Academy of Sciences, Institute of Geography and Spatial Organization (ConDENSE report, unpublished).

Bijak, J. and Koryś, I. 2006. *Statistics or Reality? International Migration in Poland*. Warsaw: Central European Forum for Migration and Population Research.

Bijak, J., Kicinger, A., Kupiszewski, M. and Śleszyński, P. 2007. *Studium metodologiczne oszacowania rzeczywistej liczby ludności Warszawy*. Warsaw: Central European Forum for Migration and Population Research.

Billari, F.C., Liefbroer, A.C. and Philipov, D. 2006. The postponement of childbearing in Europe: driving forces and implications, in Philipov, D., Liefbroer, A.C. and Billari, F.C. (eds), *Vienna Yearbook of Population Research 2006*. Vienna: Vienna Institute of Demography, 1–17.

Billert, A. 2004. Stadterneuerungsprobleme in Polen. Folgen von fehlenden Marktstrukturen im Wohnungswesen und einem ungenügenden Planungsrecht – ein Praxisbericht und Ausblick. *Städte im Umbruch*, 2, 45–51.

Blacksell, M. and Born, K.M. 2002. Private property restitution: the geographical consequence of official government policies in Central and Eastern Europe. *The Geographical Journal*, 168(2), 178–90.

Blokker, P. 2005. Post-communist modernization, transition studies, and diversity in Europe. *European Journal of Social Theory*, 8, 503–25.

Bodnár, J. 2001. On fragmentation, urban and social, in Gotham, K.F. (ed.), *Critical Perspectives on Urban Redevelopment*, vol. 6. Oxford: JAI Press (Research in Urban Sociology, 6), 173–94.

Bourdieu, P. 1986. The forms of capital, in Richardson, J.G. (ed.), *Handbook of Theory and Research for the Sociology of Education*. New York: Greenwood, 241–58.

Bourne, L.S. and Simmons, J. 2003. Conceptualization and Analysis of Urban Systems: A North American Perspective, in Champion, T. and Hugo, G. (eds), *New Forms of Urbanization. Beyond the Urban-Rural Dichotomy*. Farnham, Burlington: Ashgate, 249–68.

Bouzarovski, S. 2009. Landscapes of flexibility: negotiating the everyday; an introduction. *GeoJournal*, 74, 503–506.

Bouzarovski, S., Haase, A., Hall, R., Steinführer, A., Kabisch, S. and Ogden, P.E. 2010. Household structure, migration trends and residential preferences in inner-city León, Spain: unpacking the demographies of reurbanization. *Urban Geography*, 31(2), 211–35.

Bradshaw, M.J. and Stenning, A.C. 2004a. Introduction: transformation and development, in Bradshaw, M.J. and Stenning, A.C. (eds), *East Central Europe and The Former Soviet Union: The Post Socialist States*. Harlow: Pearson, 1–32.

Bradshaw, M.J. and Stenning, A.C. (eds) 2004b. *East Central Europe and The Former Soviet Union: The Post Socialist States*. Harlow: Pearson/DARG Regional Development Series.

Brednich, R.W. 1982. Zum Stellenwert erzählter Lebensgeschichten in komplexen volkskundlichen Feldprojekten, in Brednich, R.W. (ed.), *Lebenslauf und Lebenszusammenhang. Autobiographische Materialien in der volkskundlichen Forschung*. Freiburg i. Br.: Universität, 46–70.

Bridger, S. and Pine, F. (eds) 1998. *Surviving Post-Socialism. Local Strategies and Regional Responses in Eastern Europe and the Former Soviet Union*. London: Routledge.

Bromley, R.D.F., Tallon, A.R. and Thomas, C.J. 2005. City centre regeneration through residential development: contributing to sustainability. *Urban Studies*, 42, 2407–2429.

Bromley, R.D.F., Tallon, A.R. and Roberts, A.J. 2007. New populations in the British city centre: evidence of social change from the census and household surveys. *Geoforum*, 38, 138–54.

Brophy, P. and Burnett, K. 2003. *A Framework for Community: Development in Weak Market Cities*. Denver: Community Development Partnership Network.

Brown, L.A. and Moore, E.G. 1970. The intra-urban migration process: a perspective. *Geografiska Annaler B*, 52, 1–13.

Brown, S.K. and Bean, F.D. 2005. International migration, in Poston, D.L. and Micklin, M. (eds), *The Handbook of Population*. New York et al.: Kluwer Plenum, 347–82.

Brühl, H., Echter, C., Bodelschwingh, F. and Jekel, G. 2005. *Wohnen in der Innenstadt – eine Renaissance?* Berlin: Deutsches Institut für Urbanistik (Difu-Berichte zur Stadtforschung, 41).

Bujwicka, A. and Michalska-Żyła, A. 2007. Wizje centrum Łodzi w świadomości jego mieszkańców, in Malikowski, M. and Solecki, S. (eds), *Przemiany przestrzenne w dużych miastach Polski i Europy Środkowo-Wschodniej.* Kraków: Nomos, 272–86.

Bunchandranon, C., Howe, G. and Payumo, A.S. 1997. Ageing as an urban experience, in Beall, J. (ed.), *A City for All. Valuing Difference and Working with Diversity.* London, New Jersey: Zed Books, 141–58.

Burawoy, M. 1999. Afterword, in Burawoy, M. and Verdery, K. (eds), *Uncertain Transition. Ethnographies of Change in the Postsocialist World.* Lanham, Oxford: Rowman & Littlefield Publishers, 301–12.

Burawoy, M. and Verdery, K. (eds) 1999. *Uncertain Transition. Ethnographies of Change in the Postsocialist World.* Lanham, Oxford: Rowman & Littlefield Publishers.

Burcin, B. and Kučera, T. 2006. Socio-demographic consequences of the renewal of Prague's historical centre and its functional transformation, in Enyedi, G. and Kovács, Z. (eds), *Social Changes and Social Sustainability in Historical Urban Centres. The Case of Central Europe.* Pécs: Centre for Regional Studies of the Hungarian Academy of Sciences (Discussion Papers Special), 174–88.

Burgers, J. and Musterd, S. 2002. Understanding urban inequality: a model based on existing theories and an empirical illustration. *International Journal of Urban and Regional Research*, 26, 403–13.

Burgess, E.W. 1925/1984. The growth of the city: an introduction to a research project, in Park, R.E., Burgess, E.W. and McKenzie, R.D. (eds), *The City.* Chicago, London: The University of Chicago Press (Midway Reprint, The Heritage of Sociology), 47–62.

Burgess, E.W. 1927. The determination of gradients in the growth of the city. *Publications of the American Sociological Society*, 21, 178–84.

Burjanek, A. 1996. *Problém urbánní deprivace.* Unpublished PhD thesis. Brno: Masarykova univerzita, Fakulta sociálních studií.

Burjanek, A. 1997. Segregace. *Sociologický časopis*, 33(4), 423–34.

Burjanek, A. 2009. Sociálně vyloučené lokality města: názvosloví a charakteristiky, in Ferenčuhová, S., Galčanová, L., Hledíková, M. and Vacková, B. (eds), *Město: proměnlivá ne/samozřejmost.* Brno, Praha: Masarykova univerzita, nakladatelství Pavel Mervart, 51–67.

Bürkner, H.-J. 2008. *Does Demography Overpower the Social Question?* Keynote given at the conference 'Socio-demographic change of European cities and its spatial consequences', Leipzig, 14–16 April 2008.

Burton, E. 2003. Housing for an urban renaissance: implications for social equity. *Housing Studies*, 18, 537–62.

Buzar, S. and Grabkowska, M. 2006. The social reproduction of flexibility in the housing environment: stories from inner-city Gdańsk, in Platje, J., Słodczyk, J. and Filho, W.L. (eds), *Current Issues of Sustainable Development – Priorities and Trends*. Opole: Uniwersytet Opolski (Economic and Environmental Studies 8), 157–75.

Buzar, S., Ogden, P.E. and Hall, R. 2005. Households matter: the quiet demography of urban transformation. *Progress in Human Geography*, 29(4), 413–36.

Buzar, S., Ogden, P.E., Hall, R., Haase, A., Kabisch, S. and Steinführer, A. 2007. Splintering urban populations: emergent landscapes of reurbanisation in four European cities. *Urban Studies*, 44(4), 651–77.

Carmon, N. 2001. Neighbourhood: general, in Smelser, N.J. and Baltes, P.B. (eds), *International Encyclopedia of the Social & Behavioral Sciences*, vol. 15. Oxford: Elsevier, 10490–96.

Čermák, Z. 2005. Migrace a suburbanizační procesy v České republice. *Demografie*, 47, 169–76.

Čermák, Z., Drbohlav, D., Hampl, M., Kučera, T. 1995. *Faktické obyvatelstvo Prahy*. Praha: Přirodovědecká Fakulta Univerzity Karlovy.

Cheshire, P. 1995. A new phase of urban development in Western Europe? The evidence for the 1980s. *Urban Studies*, 32, 1045–63.

Cheshire, P. 2006. Resurgent cities, urban myths and policy hubris: What we need to know. *Urban Studies*, 43, 1231–46.

Cheshire, P. and Gordon, I. (eds) 2006. The resurgent city. *Urban Studies*, 43(8).

Chlupová, D., Polešáková, M. and Rohrerová, L. 2009. Výsledky dotazníkové akce o změnách v obecním bytovém fondu ve vybraných městech (2006, 2007). *Urbanismus a územní rozvoj*, 12(4), 12–27.

Cieśla, S. 1984. Die Wohnverhältnisse und das soziale Milieu der alten Wohnviertel in Polen. *Archiv für Kommunalwissenschaften*, 23, 235–47.

Clark, W.A.V. and Dieleman, F.M. (eds) 1996. *Households and Housing: Choice and Outcomes in the Housing Market*. New Brunswick NJ: Rutgers University, Centre for Urban Policy Research.

Clark, W.A.V., Deurloo, M.C. and Dieleman, F.M. 1984. Housing consumption and residential mobility. *Annals of the Association of American Geographers*, 74(1), 29–43.

Clay, P.L. 1979. *Neighborhood Renewal. Middle Class Resettlement and Incumbent Upgrading in American Neighborhoods*. Lexington Mass. et al.: Heath.

CoE [Council of Europe] 2005. *Recent Demographic Developments in Europe 2004*. Strasbourg: Council of Europe.

Coleman, D. 2004. Why we don't have to believe without doubting in the 'second demographic transition' – some agnostic comments, in *Vienna Yearbook of Population Research 2004*. Vienna: Austrian Academy of Sciences, 11–24.

Coleman, D. 2005. Population Problems and Prospects in Europe. *Genus*, LXI (3–4), 413–64.

Colomb, C. 2007. Unpacking New Labour's 'Urban Renaissance' agenda: towards a socially sustainable reurbanisation of British cities? *Planning, Practice & Research*, 22, 1–24.

Coolen, H. 2008. *The Meaning of Dwelling Features: Conceptual and Methodological Issues.* Amsterdam: IOS Press.

Couch, C. 1990. *Urban Renewal Theory and Practice.* Basingstoke: Macmillan.

Couch, C., Fraser, C. and Percy, S. 2003. *Urban Regeneration in Europe.* Oxford: Blackwell.

Couch, C., Karecha, J., Nuissl, H. and Rink, D. 2005. Decline and sprawl: an evolving type of urban development – observed in Liverpool and Leipzig. *European Planning Studies*, 13, 117–36.

Czepczyński, M. 2008. *Cultural Landscapes of Post-Socialist Cities: Representation of Powers and Needs.* Farnham, Burlington: Ashgate.

Dangschat, J.S. 2000. Segregation, in Häußermann, H. (ed.), *Großstadt – Soziologische Stichworte.* 2nd Edition. Opladen: Leske + Budrich: 209–21.

Dangschat, J.S. 2002. Residentielle Segregation – die andauernde Herausforderung an die Stadtforschung, in Fassmann, H., Kohlbacher, J. and Reeger, U. (eds), *Zuwanderung und Segregation. Europäische Metropolen im Vergleich.* Klagenfurt: Drava, 25–36.

Dangschat, J.S. and Blasius, J. 1987. Social and spatial disparities in Warsaw in 1978: an application of correspondence analysis to a 'socialist city'. *Urban Studies*, 24, 173–91.

Davidson, M. and Lees, L. 2005. New build 'gentrification' and London's riverside renaissance. *Environment and Planning A*, 37, 1165–90.

Dawson, A.H. 1979. Factories and cities in Poland, in French, R.A. and Hamilton, F.E.I. (eds), *The Socialist City. Spatial Structure and Urban Policy.* Chichester et al.: Wiley, 349–86.

Deane, G.D. 1990. Mobility and adjustments: paths to the resolution of residential stress. *Demography*, 27(1), 65–79.

Demszky von der Hagen, A.-M. 2006. *Alltägliche Gesellschaft.* München, Mering: Rainer Hampp.

DG Regio [Directorate General Regional Policy] 2004. *Interim Territorial Cohesion Report (Preliminary Results of ESPON and EU Commission Studies).* Luxembourg: Office for Official Publications of the European Communities.

Dieleman, F.M. 2001. Modelling residential mobility: a review of recent trends in research. *Journal of Housing and the Built Environment*, 16, 249–65.

Donner, C. 2006. *Housing Policies in Central Eastern Europe.* Vienna.

Dorbritz, J. and Philipov, D. 2002. Der Wandel in den Mustern der Familienbildung und der Ehescheidungen in den Reformstaaten Mittel- und Osteuropas – Die Folgen des Austausches der Wirtschafts- und Sozialordnung. *Zeitschrift für Bevölkerungswissenschaft*, 27, 427–63.

Douglas, M.J. 1997. *A Change of System. Housing System Transformation and Neighbourhood Change in Budapest.* Utrecht: Faculteit Ruimtelijke Wetenschappen Universiteit Utrecht (Nederlandse Geografische Studies, 222).

Drbohlav, D. 2000. Die Tschechische Republik und die internationale Migration, in Fassmann, H. and Münz, R. (eds), *Ost-West-Wanderung in Europa*. Wien: Böhlau, 163–81.

Drbohlav, D. and Čermák, Z. 1998. International migrants in Central European cities, in Enyedi, G. (ed.), *Social Change and Urban Restructuring in Central Europe*. Budapest: Akadémiai Kiadó, 87–107.

Droth, W. and Dangschat, J.S. 1985. Räumliche Konsequenzen der Entstehung 'neuer Haushaltstypen', in Friedrichs, J. (ed.), *Die Städte in den 80er Jahren. Demographische, ökonomische und technologische Entwicklungen*. Opladen: Westdeutscher Verlag, 147–80.

Dunford, M. and Smith, A. 2004. Economic restructuring and employment change, in Bradshaw, M. and Stenning, A. (eds), *East Central Europe and the former Soviet Union. The Post-Socialist States*. Harlow: Pearson, 33–58.

Eckardt, F. (ed.) 2006. *Paths of Urban Transformation*. Frankfurt/M.: Peter Lang.

Ein Haus in Europa: Schillerpromenade 27, 12049 Berlin. Zum Wandel der Großstadtkultur am Beispiel eines Berliner Mietshauses. Begleitbuch zur Ausstellung 1997, edited by Bezirksamt Neukölln von Berlin, Abteilung Bildung und Kultur, Kulturamt/Heimatmuseum. Opladen: Leske + Budrich.

Elvers, H.-D., Gross, M. and Heinrichs, H. 2008. The diversity of environmental justice: towards a European approach. *European Societies*, 10(5), 835–56.

Enyedi, G. (ed.) 1998. *Social Change and Urban Restructuring in Central Europe*. Budapest: Akadémiai Kiadó.

Enyedi, G. and Kovács, Z. (eds) 2006. *Social changes and social sustainability in historical urban centres*. Pécs: Centre for Regional Studies of the Hungarian Academy of Sciences (Discussion Papers Special).

EUROCITIES 2007. *Analysis of EUROCITIES Demographic Change Survey 2007*. AGM version. Brussels.

Eurostat 2008. *The Life of Women and Men in Europe – A Statistical Portrait*. Luxembourg: Office for Official Publications of the European Communities.

Eyal, G., Szelényi, I. and Townsley, E. 1998. *Making Capitalism Without Capitalists. Class Formation and Elite Struggles in Post-Communist Central Europe*. London, New York: Verso.

Fialová, L. and Kučera, M. 1997. The main features of population development in the Czech Republic during the transformation of society. *Czech Sociological Review*, 5, 93–111.

Flynn, M. and Oldfield, J. 2006. Trans-national approaches to locally situated concerns: exploring the meanings of post-socialist space. *Journal of Communist Studies and Transition Politics*, 22(1), 3–23.

Fojtík, K. 1963. Dům na předměstí. *Brno v minulosti a dnes*, 5, 45–68.

Forrest, R. and Kearns, A. 2001. Social capital, social cohesion and the neighbourhood. *Urban Studies*, 38(12), 2125–43.

Frank, S. 2005. Eine kurze Geschichte der 'europäischen Stadtpolitik' – erzählt in drei Sequenzen, in Altrock, U., Güntner, S., Huning S. and Peters, D. (eds), *Zwischen Anpassung und Neuerfindung. Raumplanung und Stadtentwicklung in den Staaten der EU-Osterweiterung.* Cottbus: Brandenburgische TU (Planungsrundschau, 11).

Franz, P. 1989. *Stadtteilentwicklung von unten. Zur Dynamik und Beeinflußbarkeit ungeplanter Veränderungsprozesse auf Stadtteilebene.* Basel et al.: Birkhäuser (Stadtforschung aktuell, 21).

Frey, W.H. and Kobrin, F.E. 1982. Changing families and changing mobility: their impact on the central city. *Demography,* 19(3), 261–77.

Friedrichs, J. (ed.) 1978. *Stadtentwicklungen in kapitalistischen und sozialistischen Ländern.* Reinbek bei Hamburg: Rowohlt (Rowohlts deutsche Enzyklopädie, 378).

Friedrichs, J. 1993. A theory of urban decline: economy, demography and political elites. *Urban Studies,* 30, 907–17.

Friedrichs, J. and Kahl, A. 1991. Strukturwandel in der ehemaligen DDR – Konsequenzen für den Städtebau. *Archiv für Kommunalwissenschaften,* 30, 169–97.

Frištenská, H. 2002. Romové ve městě – společné bydlení, in Víšek, P. (ed.), *Romové ve městě.* Praha: Socioklub, 49–68.

Galster, G. 2001. On the nature of neighbourhood. *Urban Studies,* 38(12), 2111–24.

Gawryszewski, A., Korcelli, P. and Nowosielska, E. 1998. *Funkcje metropolitalne Warszawy.* Warszawa: IGiPZ PAN (Zeszyty, 53).

Gentile, M. and Borén, T. 2007. Metropolitan processes in post-communist states: an introduction. *Geografiska Annaler B,* 89(2), 95–110.

Giddens, A. 2009. *Sociology.* 6th Edition, revised and updated with Philip W. Sutton. Cambridge, Malden: Polity Press.

Glick, P.C. 1947. The family life cycle. *American Journal of Sociology,* 12, 164–74.

Göb, R. 1977. Die schrumpfende Stadt. *Archiv für Kommunalwissenschaften,* 16, 149–77.

Gober, P. 1990. The urban demographic landscape, in Myers, D. (ed.), *Housing Demography. Linking Demographic Structure and Housing Markets.* Madison, London: University of Wisconsin Press, 232–48.

Gober, P. 1992. Urban housing demography. *Progress in Human Geography,* 16, 171–89.

Gowan, P. 1995. Neo-liberal theory and practice for Eastern Europe. *The New Left Review,* 213, 3–60.

Grabher, G. and Stark, D. 1997. Organizing diversity: evolutionary theory, network analysis and postsocialism. *Regional Studies,* 31(5), 533–44.

Grabkowska, M. 2007. Property tenure changes in inner-city Gdansk after 1989 in the view of the second demographic transition, in Komar, B. and Kucharczyk-Brus, B. (eds), *Housing and environmental conditions in post-communist countries.* Gliwice: Wydawnictwo Politechniki Śląskiej, 133–45.

Grabkowska, M. and Sagan, I. 2008. *Gdańsk Data Analysis Report*, Gdańsk: Institute of Geography of the University of Gdańsk (conDENSE report, unpublished).

Greene, J.C. 2008. Is mixed methods social inquiry a distinctive methodology? *Journal of Mixed Methods Research*, 2(1), 7–22.

Grimm, F. 1994. *Zentrensysteme als Träger der Raumentwicklung in Mittel- und Osteuropa.* Leipzig: Institut für Länderkunde (Beiträge zur Regionalen Geographie, 37).

Grossmann, K. 2005. 'Abriss Ost.' *Informationen zur modernen Stadtgeschichte No 1*, 47–51.

Grossmann, K. 2007. *Am Ende des Wachstumsparadigmas? Zum Wandel von Deutungsmustern in der Stadtentwicklung. Der Fall Chemnitz.* Bielefeld: Transcript.

Grossmann, K. 2010. Die symbolische Differenzierung der Stadt, in Altrock, U., Huning, S., Kuder, T., Nuissl, H. and Peters, D. (eds), *Symbolishe Orte. Planerische (De-)Konstruktionen.* Berlin (Planungsrundschau, 19), 23–41.

Grossmann, K., Haase, A., Rink, D., Steinführer, A. and Arndt, T. 2008a. *Segregation under Conditions of Housing Markets with Supply-Surplus: The Case of Leipzig.* Paper given at the First ISA Forum of Sociology 'Sociological Research and Public Debate'. Barcelona, 5–8 September 2008.

Grossmann, K., Haase, A., Rink, D. and Steinführer, A. 2008b. Urban shrinkage in East Central Europe? Benefits and limits of a cross-national transfer of research approaches, in Nowak, M. and Nowosielski, M. (eds), *Declining Cities/Developing Cities. Polish and German Perspectives.* Poznań: Instytut Zachodni, 77–99.

Grotowska-Leder, J. 2002. *Wielkomiejska bieda. Od epizodu do underclass.* Lodz: UL Press.

Grundy, E. 1999. Intergenerational perspectives on family and household change in mid- and later life in England and Wales, in McRae, S. (ed.), *Changing Britain. Families and Households in the 1990s.* Oxford: Oxford University Press, 201–28.

Guest, A. and Wierzbicki, S. 1999. Social ties at the neighbourhood level: two decades of GSS evidence. *Urban Affairs Review*, 35, 92–111.

Gurvitch, G. 1949. Microsociology and sociometry. *Sociometry*, 12(1/3), 1–31.

GUS [Główny Urząd Statystyczny] 2004a. *Miasta wojewódzkie.* Podstawowe dane statystyczne, 7. Warszawa.

GUS [Główny Urząd Statystyczny] 2006. *Rocznik demograficzny. Demographic Yearbook of Poland.* Warszawa.

Haase, A. and Steinführer, A. 2009. *New Life in Old Houses. Silent Residential Change in the Postsocialist Inner Cities of Lodz and Brno.* Available at: http://www.evropskemesto.cz/cms/index.php?option=com_content&task=view&id=565&Itemid=190 [accessed: 3 November 2009].

Haase, A., Kabisch, S. and Steinführer, A. 2005. Reurbanisation of inner-city areas in Europe. Scrutinising a concept of urban development with reference to demographic and household changes, in Sagan, I. and Smith, D.M. (eds), *Society, Economy, Environment – Towards the Sustainable City.* Gdańsk, Poznań: Bogucki Wydawnictwo Naukowe, 77–93.

Haase, A., Steinführer, A., Kabisch, S. and Gierczak, D. 2007. How inner-city housing and demographic change are intertwined in East-Central European cities. Comparative analyses in Polish and Czech cities for the transition period, in Komar, B. and Kucharczyk-Brus, B. (eds), *Housing and Environmental Conditions in Post-Communist Countries.* Gliwice: Wydawnictwo Politechniki Śląskiej, 148–74.

Haase, A., Grossmann, K., Kabisch, S. and Steinführer, A. 2008. Städte im demographischen Wandel. Perpektivenwechsel für Ostmitteleuropa. *Osteuropa*, 58(1), 77–90.

Haase, A., Maas, A., Steinführer, A. and Kabisch, S. 2009. From long-term decline to new diversity: socio-demographic change in Polish and Czech inner cities. *Journal of Urban Regeneration and Renewal*, 3(1), 31–45.

Haase, A., Kabisch, S., Steinführer, A., Bouzarovski, S., Hall, R. and Ogden, P. 2010. Emergent spaces of reurbanisation: exploring the demographic dimension of inner-city residential change in a European setting. *Population, Space and Place*, 16(5), 443–463.

Haišman, T. 1999. Romové v Československu v letech 1945–1967. Vývoj institucionálního zájmu a jeho dopady, in *Romové v České republice (1945–1998).* Praha: Socioklub, 137–83.

Hajská, M. and Poduška, O. 2004. *Sociálně vyloučená lokalita Bohdálek. Zpráva z výzkumu.* Ostrava/Praha.

Hall, R. and Ogden, P.E. 2003. The rise of living alone in inner London: trends among the population of working age. *Environment and Planning A*, 35, 871–88.

Hall, R., Ogden, P.E. and Hill, C. 1999. Living alone: evidence from England and Wales and France for the last two decades, in McRae, S. (ed.), *Changing Britain. Families and Households in the 1990s.* Oxford: Oxford University Press, 265–96.

Hamilton, F.E.I. 1979a. Urbanization in socialist Eastern Europe: the macro-environment of internal city structure, in French, R.A. and Hamilton, F.E.I. (eds), *The Socialist City. Spatial Structure and Urban Policy.* Chichester et al.: Wiley, 167–94.

Hamilton, F.E.I. 1979b. Spatial structure in East European cities, in French, R.A. and Hamilton, F.E.I. (eds), *The Socialist City. Spatial Structure and Urban Policy.* Chichester et al.: Wiley, 195–262.

Hamilton, F.E.I. 1999. Transformation and space in Central and Eastern Europe. *The Geographical Journal*, 165(2), 135–44.

Hamilton, F.E.I. and Burnett, A.D. 1979. Social processes and residential structure, in French, R.A. and Hamilton, F.E.I. (eds), *The Socialist City. Spatial Structure and Urban Policy.* Chichester et al.: Wiley, 263–304.

Hamilton, F.E.I., Dimitrovska-Andrews, K. and Pichler-Milanović, N. (eds) 2005. *Transformation of Cities in Central and Eastern Europe: Towards Globalization.* Tokyo, New York: United Nations University Press.

Hamnett, C. 2001. Social segregation and social polarisation, in Paddison, R. (ed.), *Handbook of Urban Studies.* London: Sage, 162–76.

Hann, C. (ed.) 2002. *Postsocialism. Ideas, Ideologies and Practices in Eurasia.* London, New York: Routledge.

Harkness, J.A. and Schoua-Glusberg, A. 1998. Questionnaires in translation, in Harkness, J.A. (ed.), *Cross-Cultural Survey Equivalence.* Mannheim: Zentrum für Umfragen, Methoden und Analysen (ZUMA-Nachrichten Spezial, 3), 87–127.

Harloe, M. 1996. Cities in transition, in Andrusz, G., Harloe, M. and Szelényi, I. (eds), *Cities after Socialism. Urban and Regional Change and Conflict in Post-Socialist Societies.* Oxford, Cambridge MA: Blackwell, 1–29.

Harth, A. 1994. Lebenslagen und Wohnmilieus, in Herlyn, U. and Hunger, B. (eds), *Ostdeutsche Wohnmilieus im Wandel: eine Untersuchung ausgewählter Stadtgebiete als sozialplanerischer Beitrag zur Stadterneuerung.* Basel et al.: Birkhäuser (Stadtforschung aktuell, 47), 47–212.

Hassenpflug, D. 2002. Die europäische Stadt als Erinnerung, Leitbild und Fiktion, in Hassenpflug, D. (ed.), *Die Europäische Stadt. Mythos und Wirklichkeit.* Münster: Lit, 11–48.

Hastings, A. 2004. Stigma and social housing estates: beyond pathological explanations. *Journal of Housing and the Built Environment*, 19(3), 233–54.

Hatt, P. 1946. The concept of natural area. *American Sociological Review*, 11, 423–7.

Häußermann, H. and Siebel, W. 2004. *Stadtsoziologie.* Frankfurt/M.: Campus.

Haviland, W.A. 2003. *Anthropology.* 10th Edition. Belmont CA: Wadsworth/ Thomson.

Hegedüs, J. and Tosics, I. 1998. Towards new models of the housing system, in Enyedi, G. (ed.), *Social Change and Urban Restructuring in Central Europe.* Budapest: Akadémiai Kiadó, 137–67.

Helbrecht, I. 1996. Die Wiederkehr der Innenstädte. Zur Rolle von Kultur, Kapital und Konsum in der Gentrification. *Geographische Zeitschrift*, 84, 1–15.

Hemmnet, J. 2003. Introduction to special issue: ethnographies of postsocialism, *Anthropology of Eastern Europe Review*, 21(2), 3–6.

Herfert, G. 2007. Regionale Polarisierung der demographischen Entwicklung in Ostdeutschland – Gleichwertigkeit der Lebensverhältnisse? *Raumforschung und Raumordnung*, 65, 435–55.

Herlyn, U. 1990. *Leben in der Stadt. Lebens- und Familienphasen in städtischen Räumen.* Opladen: Leske + Budrich.

Hirschman, A. O. 1970. *Exit, Voice, and Loyalty. Responses to Decline in Firms, Organizations, and States.* Cambridge Mass.: Harvard University Press.

Hirt, S. 2005. Planning the post-communist city: experiences from Sofia. *International Planning Studies*, 10(3/4), 219–40.

Hirt, S. and Stanilov, K. 2007. The perils of post-socialist transformation: residential development in Sofia, in Stanilov, K. (ed.), *The Post-Socialist City: Urban Form and Space Transformations in Central and Eastern Europe after Socialism*. Dordrecht: Springer (GeoJournal Library, 92), 215–44.

Hörschelmann, K. 2002. History after the end: post-socialist difference in a (post)modern world. *Transactions of the Institute of British Geographers*, 27(1), 52–66.

Housing Statistics in the European Union 2004. Edited by the National Board of Housing, Building and Planning, Sweden and the Ministry for Regional Development of the Czech Republic. Karlskrona: Boverket, 2005.

Howell, N. 1986. Demographic anthropology. *Annual Review of Anthropology*, 15, 219–46.

Huinink, J. 2001. *The Macro-Micro-Link in Demography – Explanations of Demographic Change*. Paper given at the conference 'The Second Demographic Transition in Europe'. Bad Herrenalb, 23–28 June 2001.

Huinink, J. 2007. *Wie neu sind die 'neuen' Lebensformen?* Paper given at the workshop 'Wandel der Lebensformen in Deutschland – Ausmaß, Ursachen und Konsequenzen im sozialpolitischen Kontext'. Rostock, 22–23 March 2007.

Huinink, J. and Wagner, M. 1998. Individualisierung und die Pluralisierung von Lebensformen, in Friedrichs, J. (ed.), *Die Individualisierungs-These*. Opladen: Leske + Budrich, 85–106.

Hummel, D. and Lux, A. 2007. Population decline and infrastructure: The case of the German water supply system. *Vienna Yearbook of Population Research 2007*, 167–91.

Humphrey, C. 2002. Does the category 'postsocialist' still make sense? in Hann, C. (ed.), *Postsocialism: Ideas, Ideologies and Practices in Eurasia*. London, New York: Routledge, 12–15.

Iglicka, K. 2002. *Poland's Post-War Dynamic of Migration*. Aldershot: Ashgate.

Jakóbczyk-Gryszkiewicz, J. 1997. Demographic characteristics of Łódź, in Liszewski, S. and Young, C. (eds), *A Comparative Study of Łódź and Manchester: Geographies of European Cities in Transition*. Łódź: University of Łódź, 111–24.

Jałowiecki, B. 1980. *Człowiek w przestrzeni miasta*. Katowice: Śląski Instytut Naukowy.

Jałowiecki, B. and Łukowski, W. 2007. *Gettoizacja polskiej przestrzeni miejskiej*. Warszawa: SWPS and Scholar.

Jarosz, M. (ed.) 2008. *Wykluczeni. Wymiar społeczny, materialny i etniczny*. Warszawa: Instytut Studiów Politycznych PAN.

Jarvis, H., Pratt, A.C. and Cheng Chong, W. 2001. *The Secret Life of Cities: The Social Reproduction of Everyday Life*. Harlow: Prentice Hall.

Jędrzejko, M. 2008. Polskie favele? in Jarosz, M. (ed.), *Wykluczeni. Wymiar społeczny, materialny i etniczny*. Warszawa: Instytut Studiów Politycznych PAN, 177–210.

Jenks, M., Burton, E. and Williams, K. (eds) 1996. *The Compact City: A Sustainable Urban Form?* London: Taylor & Francis.

John, V. 1998. *Socializace romské mládeže v Ostravě.* Ostrava: DP PedF Ostravské univerzity.

Johnson, T.P. 1998. Approaches to equivalence in cross-cultural and cross-national survey research, in Harkness, J.A. (ed.). *Cross-Cultural Survey Equivalence.* Mannheim: Zentrum für Umfragen, Methoden und Analysen (ZUMA-Nachrichten Spezial, 3), 1–40.

Kabisch, S., Steinführer, A., Haase, A., Grossmann, K., Peter, A. and Maas, A. 2008. *Demographic Change and Housing in European Cities.* Final report for EUROCITIES. Leipzig, Brussels. [Online]. Available at: http://www. eurocities.eu/uploads/load.php?file=housing_final-MROD.pdf [accessed: 12 October 2009].

Kaczmarek, S. 1997. Spatial differentiation of housing conditions and urban landscape in Łódź, in Liszewski, S. and Young, C. (eds), *A Comparative Study of Łódź and Manchester: Geographies of European Cities in Transition.* Łódź: University of Łódź, 175–85.

Kaczmarek, J. 2000. Świat życia człowieka w mieście postsocjalistycznym, in Jażdżewska, J. (ed.), *Miasto socjalistyczne. Organizacja przestrzeni miejskiej i jej przemiany.* Łódź: UL, Komisja geografii osadnictwa i ludności PTU, Polskie Towarzystwo Naukowe (XIII Konwersatorium Wiedzy o Mieście), 53–60.

Kährik, A. 2002. Changing social divisions in the housing market of Tallinn, Estonia. *Housing, Theory and Society*, 19(1), 48–56.

Kährik, A. 2006. *Socio-spatial residential segregation in post-socialist cities: The case of Tallinn, Estonia.* PhD thesis. Tartu: University of Tartu.

Kaltenberg-Kwiatkowska, E. 2004. Urban private spaces – perception and evaluation of new and old housing, in Sagan, I. and Czepczyński, M. (eds), *Featuring the Quality of Urban Life in Contemporary Cities of Eastern and Western Europe.* Gdańsk, Poznań: Bogucki Wydawnictwo Naukowe, 47–59.

Kames, M. 2009: *Das Wintergartenhochhaus Leipzig. Architektur- und wohnsoziologische Studie einer Hausbiographie.* Unpublished diploma thesis. Leipzig: Universität Leipzig.

Karsten, L. 2003. Family gentrifiers: challenging the city as a place simultaneously to build a career and to raise children. *Urban Studies*, 40, 2573-85.

Kawka, R. 2007. Regional disparities in the GDR – do they still matter? in Lentz, S. (ed.), *Restructuring Eastern Germany.* Berlin, Heidelberg: Springer (German Annual of Spatial Research and Policy), 41–55.

Kazepov, Y. 2004. *Cities of Europe. Changing Contexts, Local Arrangements, and the Challenge to Urban Cohesion.* Oxford, Cambridge MA: Blackwell.

Keely, C. 2001. Replacement migration: the wave of the future? *International Migration*, 39(6), 103–10.

Keilman, N. 1995. Household concepts and household definitions in Western Europe: different levels but similar trends in household developments, in van Imhoff, E., Kuijsten, A., Hooimeijer, P. and van Wissen, L. (eds), *Household Demography and Household Modelling*. New York, London: Plenum Press (The Plenum Series on Demographic Methods and Population Analysis), 111–35.

Keim, K.-D. 1979. *Milieu in der Stadt: ein Konzept zur Analyse älterer Wohnquartiere*. Stuttgart et al.: Kohlhammer (Schriften des Deutschen Instituts für Urbanistik, 63).

Kelle, U. 2001. Soziologische Erklärungen zwischen Mikro- und Makroebene und die Integration qualitative und quantitativer Methoden, in Fielding, N. and Schreier, M. (eds), *Qualitative und quantitativer Forschung: Übereinstimmungen und Divergenzen. Forum Qualitative Sozialforschung* [Online], 2(1). Available at: http://www.qualitative-research.net/fqs/fqs.htm [acessed: 21 July 2010].

Kelle, U. 2008. *Die Integration qualitativer und quantitativer Methoden in der empirischen Sozialforschung. Theoretische Grundlagen und methodologische Konzepte*. Wiesbaden: VS Verlag für Sozialwissenschaften.

Kelle, U. and Kluge, S. (eds) 2001. *Integration qualitativer und quantitativer Verfahren in der Lebenslaufforschung*. Weinheim: Juventa (Statuspassagen und Lebenslauf, 4).

Kendig, H.L. 1990. A life course perspective on housing attainment, in Myers, D. (ed.), *Housing Demography. Linking Demographic Structure and Housing Markets*. Madison, London: University of Wisconsin Press, 133–56.

Kenyon, E.L. 1997. Seasonal sub-communities: the impact of student households on residential communities. *The British Journal of Sociology*, 48(2), 286–301.

Kenyon, L. 1999. A home from home. Students' transitional experience of home, in Chapman, T. and Hockey, J. (eds), *Ideal homes? Social change and domestic life*. London: Routledge, 84–95.

Kertzer, D.I. 1991. Household history and sociological theory. *Annual Review of Sociology*, 17, 155–79.

Kirschenbaum, A. 1983. Sources of neighbourhood residential change: a micro-level analysis. *Social Indicators Research*, 12, 183–98.

Kiss, É. 2004. Spatial impacts of post-socialist industrial transformation in the major Hungarian cities. *European Urban and Regional Studies*, 11, 81–7.

Kluge, S. 2000. Empirisch begründete Typenbildung in der qualitativen Sozialforschung. *Forum Qualitative Sozialforschung* [Online], 1(1). Available at: http://www.qualitative-research.net/index.php/fqs/article/viewArticle/1124/2497 [accessed: 21 July 2010].

Kluge, S. and Kelle, U. 1999. *Vom Einzelfall zum Typus: Fallvergleich und Fallkontrastierung in der qualitativen Sozialforschung*. Opladen: Leske + Budrich.

Klusáček, P., Martinát, S., Vaishar, A. and Zapletalová, J. 2007. *Brno and Ostrava Data Analysis Report*. Ostrava, Brno: Institute of Geonics, Academy of Sciences of the Czech Republic (conDENSE report, unpublished).

Kohler, H.P., Billari, F.C. and Ortega, J.A. 2002. The emergence of lowest-low fertility in Europe during the 1990s. *Population and Development Review*, 28, 641–80.

Kojder, A. (ed.) 2007. *Jedna Polska? Dawne i nowe zróżnicowania społeczne.* Kraków: WAM.

Konietzka, D. and Kreyenfeld, M. (eds) 2007. *Ein Leben ohne Kinder. Kinderlosigkeit in Deutschland.* Wiesbaden: VS Verlag für Sozialwissenschaften.

Korcelli, P. 1996. Die Städte Polens im Wandel – ihre demographischen und ökonomischen Determinanten, in Brade, I. and Grimm, F. (eds), *Städtesysteme und Regionalentwicklung in Mittel- und Osteuropa: Rußland, Ukraine, Polen.* Leipzig: Institut für Länderkunde, 148–65.

Korcelli, P. 1997. The urban system of Poland in an era of increasing inter urban competition. *Geographia Polonica*, 69, 45–54.

Kornai, J. 1992. *The Socialist System. The Political Economy of Communism.* Princeton NJ: Princeton University Press.

Kostelecký, T. 2000. Housing and its influence on the development of social inequalities in the post-communist Czech Republic. *Czech Sociological Review*, 8, 177–93.

Kostelecký, T., Nedomová, A. and Vajdová, Z. 1997. *The Housing Market and Its Impact on Social Inequality. A Study of Prague and Brno (Czech Republic).* Wien: Institut für die Wissenschaften vom Menschen (SOCO Project Paper, 54).

Kostinskiy, G. 2001. Post-socialist cities in flux, in Paddison, R. (ed.), *Handbook of Urban Studies.* London et al.: Sage, 451–65.

Kotowska, I.E. 1999a. Rodziny i gospodarstwa domowe, in Kotowska, I.E. (ed.), *Przemiany demograficzne w Polsce w latach 90. w świetle koncepcji drugiego przejścia demograficznego.* Warszawa: Oficyna Wydawnicza Szkoły Głównej Handlowej, 211–21.

Kotowska, I.E. (ed.) 1999b. *Przemiany demograficzne w Polsce w latach 90. w świetle koncepcji drugiego przejścia demograficznego.* Warszawa: Oficyna Wydawnicza Szkoły Głównej Handlowej.

Kotowska, I.E. 2001. Poland – Demographic changes since 1989. *Der Donauraum*, 41(4), 50–73.

Kotowska, I.E., Jóźwiak, J., Matysiak, A. and Baranowska, A. 2008. Poland: Fertility decline as a response to profound societal and labour market changes? *Demographic Research* [Online], 19(22), 795–854. Available at: http://www. demographic-research.org [accessed: 12 October 2009].

Kotus, J. 2006. Changes in the spatial structure of a large Polish city – the case of Poznań. *Cities*, 23(5), 364–81.

Kovács, Z. 1994. A city at the crossroads: social and economic transformation in Budapest. *Urban Studies*, 31, 1081–96.

Kovács, Z. 2009. Social and economic transformation of historical neighbourhoods in Budapest. *Tijdschrift voor Economische en Sociale Geografie*, 100(4), 399–416.

Kovács, Z. and Wießner, R. (eds) 1997. *Prozesse und Perspektiven der Stadtentwicklung in Ostmitteleuropa.* Passau: L.I.S.

Kovács, Z. and Wießner, R. 1999. *Stadt- und Wohnungsmarktentwicklung in Budapest. Zur Entwicklung der innerstädtischen Wohnquartiere im Transformationsprozeß.* Leipzig: Institut für Länderkunde (Beiträge zur regionalen Geographie, 48).

Kovács, Z. and Wießner, R. 2004. Budapest – restructuring a European metropolis. *Europa regional*, 12(1), 22–31.

Kraler, A. and Iglicka, K. 2002. Labour migration in Central and Eastern European countries, in Laczko, F., Stacher, I. and Klekowski von Koppenfels, A. (eds), *New Challenges for Migration Policy in Central and Eastern Europe.* The Hague: Asser Press, 27–58.

Kretschmerová, T. and Šimek, M. 2004. Projekce obyvatelstva České republiky do roku 2050. *Demografie*, 46, 91–9.

Kuča, K. 2000. *Brno: vývoj města, předměstí a připojených vesnic.* Praha, Brno: Baset.

Kučera, M. and Fialová, L. 1996. *Demografické chování obyvatelstva České republiky během přeměny společnosti po roce 1989.* Praha: Sociologický ústav AV ČR.

Kučera, T., Kučerová, O.V., Opara, O.B. and Schaich, E. (eds) 2000. *New Demographic Faces of Europe. The Changing Population Dynamics in Countries of Central and Eastern Europe.* Berlin et al.: Springer.

Kühn, M. and Liebmann, H. 2009. *Regenerierung der Städte. Strategien der Politik und Planung im Schrumpfungskontext.* Wiesbaden: VS Verlag für Sozialwissenschaften.

Kuijsten, A.C. 1996. Changing family patterns in Europe: a case of divergence? *European Journal of Population*, 12, 115–43.

Kujath, H.J. 1988. Reurbanisierung? – Zur Organisation von Wohnen und Leben am Ende des städtischen Wachstums. *Leviathan*, 16, 23–43.

Kupiszewski, M. and Bijak, J. 2006. *Ocena prognozy ludności GUS 2003 z perspektywy aglomeracji Warszawskiej.* Warsaw: Central European Forum for Migration and Population Research (CEFMR Working Paper, 1/2006).

Kupiszewski, M., Durham, H. and Rees, P. 1998. Internal migration and urban change in Poland. *European Journal of Population*, 14, 265–90.

Kurek, S. 2008. *Typologia starzania się ludności Polski w ujęciu przestrzennym.* Kraków: Wydawnictwo Naukowe Akademii Pedagogicznej.

Ladányi, J. 1993. Patterns of residential segregation and the gypsy minority in Budapest. *International Journal of Urban and Regional Research*, 17, 30–41.

Lambert, C. and Boddy, M. 2002. *Transforming City Centres: Trends in Re-urbanisation and New Housing Development in UK Cities.* Paper presented at the Urban Affairs Association Annual Conference, Boston.

Lane, D. (ed.) 2007. *Transformation of State Socialism: System Change, Capitalism, or Something Else?* Houndmills, Basingstoke et al.: Palgrave.

Landale, N.S. and Guest, A.M. 1985. Constraints, satisfaction and residential mobility: Speare's model reconsidered. *Demography*, 22, 199–222.

Langhamrová, J. and Fiala, T. 2003. Kolik je vlastně Romů v České republice? *Demografie*, 45(1), 23–32.

Laslett, P. 1974. *Household and Family in Past Time*. London: Cambridge University Press.

Lazarsfeld, P. F. 1937. Some remarks on the typological procedures in social research. *Zeitschrift für Sozialforschung*, VI, 119–39.

Lee, G.S., Schmidt-Dengler, P., Felderer, B. and Helmenstein, C. 2001. Austrian Demography and Housing Demand: Is there a connection. *Empirica*, 28, 259–76.

Lees, L. 2008. Gentrification and social mixing: towards an inclusive urban renaissance? *Urban Studies*, 45, 2449–70.

Lees, L. and Ley, D. 2008. Introduction to special issue on gentrification and public policy. *Urban Studies*, 45, 2379–84.

Le Galès, P. 2002. *European Cities. Social Conflicts and Governance*. Oxford: Oxford University Press.

Lesthaeghe, R. 1995. The second demographic transition in western countries: an interpretation, in Mason, K.O. and Jensen, A.-M. (eds), *Gender and Family Change in Industrialized Countries*. Oxford: Clarendon Press (International Studies in Demography), 17–62.

Lesthaeghe, R. 2007. *Population Development in Europe – Causes and Consequences of the Second Demographic Transition*. Keynote given at the International Conference Demographic Change, Berlin, 2 February 2007. [Online]. Available at: http://www.arl-net.org/pdf/veranst/demographic_change/Lesthaege.pdf [accessed: 4 October 2009].

Lesthaeghe, R. and Surkyn, J. 2002. New forms of household formation in Central and Eastern Europe: are they related to newly emerging value orientations? *UNECE Economic Survey of Europe*, 1, 197–216.

Lesthacghe, R. and van de Kaa, D.J. 1986. Twee demografische transities? in Lesthaeghe, R. and van de Kaa, D.J. (eds), *Bevolking: groei en krimp*. Deventer: Van Loghum Slaterus, 9–24.

Lever, W.F. 1993. Reurbanisation – The Policy Implications. *Urban Studies*, 30, 267–84.

Levin, E., Montagnoli, A. and Wright, R.E. 2009. Demographic change and the housing market: evidence from a comparison of Scotland and England. *Urban Studies*, 46(1), 27–43.

Ley, D. 1996. *The New Middle Class and the Remaking of the Central City*. Oxford, New York: Oxford University Press (Oxford Geographical and Environmental Studies).

Ley, D. 2003. Artists, aestheticisation and the field of gentrification. *Urban Studies*, 40(12), 2527–44.

Littlewood, A. and Munro, M. 1997. Moving and improving: strategies for attaining housing equilibrium. *Urban Studies*, 34, 1771–87.

Liszewski, S. 1997. The origins and stages of development of industrial Łódź and the Łódź urban region, in Liszewski, S. and Young, C. (eds), *A Comparative*

Study of Łódź and Manchester: Geographies of European Cities in Transition. Łódź: Łódź University Press, 11–34.

Loomis, C.P. and Hamilton, C.H. 1936. Family life cycle analysis. *Social Forces*, 15, 225–31.

Lowe, S. 2003. Housing in post-communist Europe – issues and agendas, in Lowe, S. and Tsenkova, S. (eds), *Housing Change in East and Central Europe. Integration or Fragmentation?* Aldershot, Burlington: Ashgate, XIII-XIX.

Lowe, S. and Tsenkova, S. (eds) 2003. *Housing Change in East and Central Europe. Integration or Fragmentation?* Aldershot, Burlington: Ashgate.

Lupton, R. and Power, A. 2004. *What We Know About Neighbourhood Change: A Literature Review.* London (LSE STICERD Research Paper, CASEreport, 27). [Online]. Available at: http://sticerd.lse.ac.uk/dps/case/cr/CASEreport27. pdf [accessed: 9 October 2009].

Lux, M. 2000. *The Housing Policy Changes and Housing Expenditures in the Czech Republic.* Praha: Sociogický ústav AV ČR.

Lux, M. (ed.) 2003. *Housing Policy: An End or a New Beginning?* Budapest: Local Government and Public Service Reform Initiative, Open Society Institute.

Lux, M. 2004. Housing the poor in the Czech Republic: Prague, Brno and Ostrava, in Fearn, J. (ed.), *Too Poor to Move, Too Poor to Stay.* Budapest: LGI – OSI, 23–66.

MacGregor, S. 1990. The inner-city battlefield: politics, ideology, and social relation, in MacGregor, S. and Pimlott, B. (eds), *Tackling the Inner City. The 1980s Reviewed, Prospects for the 1990s.* Oxford: Clarendon Press, 64–92.

Maier, K. 2003. Sídliště: problém a multikriteriální analýza jako součást přípravy k jeho řešení. *Sociologický časopis*, 39, 653–66.

Majer, A. 1988. Lodz – Aufstieg und Alltagsleben einer Industriestadt, in Strubelt, W. and Frackiewicz, L. (eds), *Soziale Probleme von Industriestädten.* Bonn: Bundesforschungsanstalt für Landeskunde und Raumordnung, 80–88.

Malina, J. 2009. Metoda emické deskripce, in *Antropologický slovník.* Brno: Akademické nakladatelství CERM v Brně. [Online]. Available at: http://is.muni.cz/do/1431/UAntrBiol/el/antropos/slovnik.html [accessed: 17 January 2010].

Mandič, S. 2001. Residential mobility versus 'in-place' adjustments in Slovenia: viewpoint from a society 'in transition'. *Housing Studies*, 16(1), 53–73.

Mangen, S.P. 2004. *Social Exclusion and Inner City Europe. Regulating Urban Regeneration.* Houndmills, Basingstoke: Palgrave Macmillan.

Mankiw, N.G. and Weil, D.N. 1989. The baby boom, the baby bust, and the housing market. *Regional Science and Urban Economics*, 19, 235–58.

Marcińczak, S. 2007. The socio-spatial structure of post-socialist Łódź, Poland. Results of national census 2002. *Bulletin of Geography* (Socio-Economic Series), 8, 65–82.

Marcinowicz, D. 2006. Zachowania studentów na rynku wynajmu mieszkań w Poznaniu, in Gaczek, W.M., Kaczmarek, M. and Marcinowicz, D. (eds),

Poznański ośrodek akademicki, próba określenia wpływu studentów na rozwój miasta akademicki. Poznań: Bogucki Wydawnicto Naukowe, 49–55.

Marcuse, P. 1991. Die Zukunft der 'sozialistischen' Städte. *Berliner Journal für Soziologie*, 1, 203–10.

Marcuse, P. and van Kempen, R. 2000. Conclusion: a changed spatial order, in Marcuse, P. and van Kempen, R. (eds), *Globalizing Cities: A New Spatial Order?* Oxford: Blackwell (Studies in Urban and Social Change), 249–75.

Mareš, P. 2000. Chudoba, marginalizace, sociální vyloučení. *Sociologický časopis*, 3, 285–97.

Markowski, T. and Stawasz, D. (eds) 2007. *Rewitalizacja a rozwój funkcji metropolitalnych miasta Łodzi.* Łódź: Wydawnictwo Uniwersytetu Łódzkiego.

Massey, D.S. and Denton, N.A. 1988. The dimensions of residential segregation. *Social Forces*, 67, 281–315.

Massey, D.S. and Denton, N.A. 1998. *American Apartheid: Segregation and the Making of the American Underclass*, Cambridge, MA: Harvard University Press.

Matějů, P., Večerník, J. and Jeřábek, H. 1979. Social structure, spatial structure and problems of urban research: the example of Prague. *International Journal of Urban and Regional Research*, 3, 181–202.

Matthiesen, U. 1998. Milieus in Transformationen. Positionen und Anschlüsse, in Matthiesen, U. (ed.), *Die Räume der Milieus. Neue Tendenzen in der sozial- und raumwissenschaftlichen Milieuforschung, in der Stadt- und Raumplanung.* Berlin: Edition Sigma, 17–79.

Matysiak, A. and Nowok, B. 2006. *Stochastic forecast of the population of Poland, 2005–2050.* Rostock: Max Planck Institute for Demographic Research (MPIDR Working Paper, WP-2006-026).

McAuley, W.J. and Nutty, C.L. 1982. Residential preferences and moving behavior. *Journal of Marriage and the Family*, 44(2), 301–09.

Metzger, J.T. 2000. Planned abandonment: the neighborhood life-cycle theory and national urban policy. *Housing Policy Debate*, 11(1), 7–40.

Mikulík, O. and Vaishar, A. 1996. Residential environment and territorially functional structure of the Brno city in the period of transformation. *Geografie – Sborník České geografické společnosti*, 101, 128–42.

MMB [Magistrát města Brna] 2008. *Generel bydlení města Brna. Analytické údaje o bytovém fondu a problematice bydlení.* Brno.

Morgan, D.L. 2007. Paradigms lost and pragmatism regained: methodological implications of combining qualitative and quantitative methods. *Journal of Mixed Methods Research*, 1, 48–76.

Morris, E.W. and Winter, M. 1975. A theory of family housing adjustment. *Journal of Marriage and the Family*, 37(1), 79–88.

Moss, T. 2008. 'Cold spots' of urban infrastructure: 'shrinking' processes in eastern Germany and the modern infrastructural ideal. *International Journal of Urban and Regional Research*, 32(2), 36–51.

Mueller, E.J. 2005. Neighbourhood, in Caves, R.W. (ed.), *Encyclopedia of the City*. London and New York: Routlege, 326–7.

Mulder, C.H. 1996. Housing choice. Assumptions and approaches. *Netherlands Journal of Housing and the Built Environment*, 11, 209–32.

Mulder, C.H. 2006. Population and housing: a two-sided relationship. *Demographic Research* [Online], 15, art. 13, 401–12. Available at: http://www.demographic-research.org [accessed: 4 October 2009].

Mulder, C.H. 2007. The family context and residential choice: a challenge for new research. *Population, Space and Place*, 13(4), 265–78.

Mulder, C.H. and Dieleman, F.M. 2002. Living arrangements and housing arrangements: introduction to the special issue. *Journal of Housing and the Built Environment*, 17, 209–13.

Mulder, C.H. and Hooimeijer, P. 1999. Residential relocations in the life course, in van Wissen, L.J.G. and Dykstra, P.A. (eds), *Population Issues: An Interdisciplinary Focus*. New York: Plenum, 159–86.

Mulder, C.H. and Manting, D. 1994. Strategies of nest-leavers: 'settling down' versus flexibility. *European Sociological Review*, 10(2), 155–72.

Mulíček, O. and Seidenglanz, D. 2004. *Aktuální demografické procesy ve městě Brně a jejich prostorová diferenciace. Díl I: Aktuální demografické procesy.* Brno: Masarykova univerzita, Výzkumné centrum regionálního rozvoje.

Mullins, P. 1995. Households, consumerism and metropolitan development, in Troy, P. (ed.), *Australian Cities. Issues, Strategies and Policies for Urban Australia in the 1990s*. Cambridge: Cambridge University Press, 87–109.

Murphy, M.J. and Sullivan, O. 1985. Housing tenure and family formation in contemporary Britain. *European Sociological Review*, 1(3), 230–43.

Murzyn, M. 2004. From neglected to trendy. The process of urban revitalization in the Kazimierz district in Cracow, in Sagan, I. and Czepczyński, M. (eds), *Featuring the Quality of Urban Life in Contemporary Cities of Eastern and Western Europe*. Gdańsk, Poznań: Bogucki Wydawnictwo Naukowe, 257–74.

Murzyn, M.A. 2006. 'Winners' and 'losers' in the game: the social dimension of urban regeneration in the Kazimierz Quarter in Krakow, in Enyedi, G. and Kovács, Z. (eds), *Social Changes and Social Sustainability in Historical Urban Centres. The Case of Central Europe*. Pécs: Centre for Regional Studies of the Hungarian Academy of Sciences (Discussion Papers Special), 81–106.

Musil, J. 1967. *Sociologie soudobého města*. Praha: Svoboda.

Musil, J. 1985. *Lidé a sídliště*. Praha: Svoboda.

Musil, J. 1992a. Recent changes in the housing system and policy in Czechoslovakia: an institutional approach, in Turner, B., Hegedüs, J. and Tosics, I. (eds), *The Reform of Housing in Eastern Europe and the Soviet Union*. London et al.: Routlege, 62–70.

Musil, J. 1992b. Změny městských systémů v postkomunistických společnostech střední Evropy. *Sociologický časopis*, 28(4), 451–62.

Musil, J. 1993. Changing urban systems in post-communist societies in Central Europe: analysis and prediction. *Urban Studies*, 30, 899–905.

Musil, J. 2002. Urbanizace českých zemí a socialismus, in Horská, P., Maur, E. and Musil, J., *Zrod velkoměsta. Urbanizace českých zemí a Evropa.* Praha, Litomyšl: Paseka, 237–97.

Musil, J. 2004. Poznámky o české sociologii za komunistického režimu. *Sociologický časopis*, 40(5), 573–95.

Myers, D. 1990a. Introduction: the emerging concept of housing demography, in Myers, D. (ed.), *Housing Demography. Linking Demographic Structure and Housing Markets.* Madison, London: University of Wisconsin Press, 3–31.

Myers, D. (ed.) 1990b. *Housing Demography. Linking Demographic Structure and Housing Markets.* Madison, London: University of Wisconsin Press.

Mykhnenko, V. and Turok, I. 2007. *Shrinking Cities: East European Urban Trajectories 1960–2005.* Glasgow: Centre for Public Policy for Regions. [Online]. Available at: www.cppr.ac.uk/centres/cppr/publications [accessed: 4 October 2009].

Mykhnenko, V. and Turok, I. 2008. East European cities – patterns of growth and decline. *International Planning Studies*, 13, 311–42.

Mynarska, M. and Bernardi, L. 2007. *Meanings and Attitudes Attached to Cohabitation in Poland. Qualitative Analyses of the Slow Diffusion of Cohabitation among the Young Generation.* Rostock: Max Planck Institute for Demographic Research (MPIDR Working Paper, WP-2007-006).

Navrátil, P. and Šimíková, I. 2002. *Hodnocení projektů zaměřených na snižování rizika sociálního vyloučení romské populace. Část 1. Typologie projektů.* Brno: Výzkumný ústav práce a sociálních věcí.

Navrátil, P. et al. 2003. *Romové v české společnosti.* Praha.

Nečas, C. 1997. *Historický kalendář. Dějiny českých Romů v datech.* Olomouc: Vydavatelství Univerzity Palackého.

Nedović-Budić, Z. and Tsenkova, S., with Marcuse, P. 2006. The urban mosaic of post-socialist Europe: introduction, in Nedović-Budić, Z. and Tsenkova, S. (eds), *The Urban Mosaic of Post-Socialist Europe: Space, Institutions and Policy.* Heidelberg: Physica, 3–20.

Ninomiya, H. 2001. Sociability in history, in Smelser, N.J. and Baltes, P.B. (eds), *International Encyclopedia of the Social & Behavioral Sciences*, vol. 21. Oxford: Elsevier, 14208–12.

Nowak-Lewandowska, R. 2006. Emigro, ergo sum: Die Emigration der Polen und ihre Folgen. *Osteuropa*, 56(11–12), 167–78.

Ogden, P.E. and Hall, R. 2000. Households, reurbanisation and the rise of living alone in the principal French cities, 1975–90. *Urban Studies*, 37(2), 367–90.

Ogden, P.E. and Hall, R. 2004. The second demographic transition, new household forms and the urban population of France during the 1990s. *Transactions of the Institute of British Geographers*, 29, 88–105.

Ogden, P.E. and Schnoebelen, F. 2005. The rise of the small household: demographic change and household structure in Paris. *Population, Space and Place*, 11, 251–68.

Okólski, M. 2006. Płodność i rodzina w okresie transformacji, in Wasilewski, J. (ed.), *Współczesne społeczeństwo polskie*. Warszawa: Wydawnictwo Naukowe SCHOLAR, 103–43.

Orbell, J.M. and Uno, T. 1972. A theory of neighborhood problem solving: political action vs. residential mobility. *The American Political Science Review*, 66, 471–89.

Oswalt, P. (ed.) 2005. *Shrinking Cities 1: International Research*. Ostfildern: Hatje Cantz.

Ouředníček, M. (ed.) 2006. *Sociální geografie pražského městského regionu*. Praha: Univerzita Karlova, Katedra Sociální geografie a regionálního rozvoje.

Outhwaite, W. and Ray, L. 2005. *Social Theory and Postcommunism*. Oxford: Blackwell.

Oxley, M. 2001. Meaning, science, context and confusion in comparative housing research. *Journal of Housing and the Built Environment*, 16, 89–106.

Park, R.E. 1915. The city. Suggestions for the investigation of human behavior in the city environment. *American Journal of Sociology*, 20, 577–612.

Park, R.E. 1967. The urban community as a spatial pattern and a moral order, in Park, R.E., *On Social Control and Collective Behaviour: Selected Papers*, edited by Ralph H. Turner. Chicago, London: University of Chicago Press (The Heritage of Sociology), 55–68 (first published in 1925).

Park, R.E., Burgess, E.W. and McKenzie, R.D. 1984. *The City*. Chicago, London: The University of Chicago Press (Midway Reprint, The Heritage of Sociology) (essay first published in 1925).

Parysek, J.J. 2005. Development of Polish towns and cities and factors affecting this process at the turn of the century. *Geographia Polonica*, 78, 99–115.

Parysek, J.J. and Mierzejewska, L. 2006. City profile Poznań. *Cities*, 23, 291–305.

Pavelčíková, N. 1999. *Romské obyvatelstvo na Ostravsku (1945–1975)*. Ostrava: Ostravská univerzita, Filozofická fakulta.

Pavelčíková, N. 2008. Jak Romové šli do města. Proměny romské kultury v kontaktu s městským prostředím, in *Miasto w obrazie, legendzie, opowieści...* Wrocław, Kraków: Polskie Towarzystwo Ludoznawcze, 209–16.

Peter, A. 2009. *Quartiere auf Zeit. Lebensqualität im Alter in schrumpfenden Städten*. Wiesbaden: VS Verlag für Sozialwissenschaften.

Petsimeris, P. 1998. Urban decline and the new social and ethnic divisions in the core cities of the Italian triangle. *Urban Studies*, 35(3), 449–65.

Petsimeris, P. 2005. Out of squalor and towards another urban renaissance? Gentrification and neighbourhood transformations in Southern Europe, in Atkinson, R. and Bridge, G. (eds), *Gentrification in a Global Perspective: The New Urban Colonialism*. London: Routledge, 240–55.

Philipov, D. and Dorbritz, J. 2003. *Demographic Consequences of Economic Transition in Countries of Central and Eastern Europe*. Strasbourg: Council of Europe (Population Studies, 23).

Pichler-Milanović, N. 2001. Urban housing markets in Central and Eastern Europe: convergence, divergence or policy 'collapse'. *European Journal of Housing Policy*, 1, 145–87.

Pichler-Milanovic, N. and Dimitrovska Andrews, K. 2005. Conclusions, in Hamilton F.E.I., Dimitrovska Andrews, K. and Pichler-Milanovic, N. (eds), *Transformation of Cities in Central and Eastern Europe: Towards Globalization*. Tokyo: United Nations University Press, 465–87.

Pickel, A. 2002. Transfomation theory: scientific or political? *Communist and Post-Communist Studies*, 35(1), 105–14.

Pickles, J. and Smith, A. (eds) 1998. *Theorising Transition. The Political Economy of Post-Communist Transformations*. London, New York: Routledge.

Pickvance, C.G. 2002. State socialism, post-socialism and their urban patterns: theorising the Central and Eastern European experience, in Eade, J. and Mele, C. (eds), *Understanding the City: Contemporary and Future Perspectives*. Oxford: Blackwell, 183–203.

Pine, F. and Bridger, S. 1998. Introduction: transitions to post-socialism and cultures of survival, in Pine, F. and Bridger, S. (eds), *Surviving post-socialism*. London, New York: Routledge, 1–15.

Polanska, D. 2008. Decline and revitalization in post-communist urban context: a case of the Polish city – Gdansk. *Communist and Post-Communist Studies*, 41, 359–74.

Porter, L. and Shaw, K. (eds) 2008. *Whose Urban Renaissance? An International Comparison of Urban Regeneration Strategies*. Oxon: Routledge.

Pospíšilová, J. 2009. Sociálně vyloučené skupiny obyvatelstva, in Vaishar, A., Klusáček, P., Krejčí, T., Martinát, S., Pavelčíková, N., Pospíšilová, J. and Zapletalová, J. (eds), *Současný vývoj vnitřních částí Brna a Ostravy*. Brno, Ostrava: Ústav geoniky AV ČR v.v.i. (Studia Geographica, 100).

Priemus, H. and Mandič, S. 2000. Rental housing in Central and Eastern Europe as no man's land. *Journal of Housing and the Built Environment*, 15, 205–15.

Raban, P. 1986. *Housing Policy in Czechoslovakia*. Prague: Orbis Press Agency.

Rabušic, L. 2001. Value change and demographic behaviour in the Czech Republic. *Czech Sociological Review*, 9, 99–122.

Raymer, J. and Willekens, F. 2008. *International migration in Europe: data, models and estimates*. Chichester et al.: Wiley.

Reed, R. 2001. *The Increasing Importance of Housing Demography in the 21st Century*. Paper presented at the RICS 'Cutting Edge 2001' Conference, Oxford Brookes University, 5–7 September 2001.

Reiter, R. 2008. The 'European City' in the European Union, in Atkinson, R. and Rossignolo, C. (eds), *The Re-Creation of the European City. Governance, Territory & Polycentricity*, Amsterdam: Techne Press, 17–38.

Richter, P. 2006. *Der Plattenbau als Krisengebiet. Die architektonische und politische Transformation industriell errichteter Wohngebäude aus der DDR am Beispiel der Stadt Leinefelde*. Unpublished PhD thesis. Hamburg: Universität Hamburg, FB Kulturgeschichte und Kulturkunde.

Riemer, S. 1943. Sociological theory of home adjustment. *American Sociological Review*, 8(3), 272–8.

Rietdorf, W., Liebmann, H. and Schmigotzki, B. 2001. *Weiterentwicklung großer Neubaugebiete in Ostmitteleuropa als Bestandteil einer ausgeglichenen, nachhaltigen Siedlungsstruktur- und Stadtentwicklung. Internationale Vergleichsprojekte für das Planspiel Leipzig-Grünau*. Erkner: Institut für Regionalentwicklung und Strukturplanung.

Riley, R., Niżnik, A.M. and Burdack, J. 1999. Łódź. Transformation einer altindustriellen Stadt in der postsozialistischen Periode. *Europa Regional*, 7(1), 22–31.

Rink, D. 1997. Zur Segregation in ostdeutschen Großstädten, in Kabisch, S., Kindler, A. and Rink, D. (eds), *Sozialatlas der Stadt Leipzig*. Leipzig: UFZ – Umweltforschungszentrum Leipzig-Halle, 26–46.

Roberts, P. and Sykes, H. (eds) 2000. *Urban Regeneration. A Handbook*. London et al.: Sage.

Robson, B., Lymperopoulou, K. and Rae, A. 2009. *Understanding the Different Roles of Deprived Neighbourhoods: A Typology*. [Online]. Available at: http://www.communities.gov.uk/documents/communities/pdf/1152906.pdf [accessed: 17 January 2010].

Rossi, P.H. 1980. *Why Families Move*. Beverly Hills, London: Sage (first edition in 1955).

Ruoppila, S. 2006. *Residential Differentiation, Housing Policy and Urban Planning in the Transformation from State Socialism to a Market Economy. The Case of Tallinn*. Helsinki: Helsinkie University of Technology, Centre for Urban and Regional Studies (Centre for Urban and Regional Studies Publications, A 33), 64 pp. Available at: http://ethesis.helsinki.fi/julkaisut/val/sospo/vk/ruoppila/resident.pdf [accessed: 28 January 2010].

Růžička, M. 2006. Geografie sociální exkluze, in Ferenčuhová, S., Šuleřová, M. and Vacková, B. (eds), *Město*. Brno: Fakulta sociálních studií Masarykovy univerzity (Sociální studia 2/2006), 117–32.

Rychtaříková, J. 1999. Is Eastern Europe experiencing a second demographic transition? *Acta Universitatis Carolinae – Geographica*, 34, 19–44.

Rychtaříková, J. 2000. Demographic transition or demographic shock in recent population development in the Czech Republic? *Acta Universitatis Carolinae – Geographica*, 35, 89–102.

Rychtaříková, J. 2008. Different risks of population ageing: EU old and new members, in Dostál, P. (ed.), *Evolution of Geographical Systems and Risk Processes in the Global Context*. Prague: Charles University, Faculty of Science, Nakladatelství P3K, 101–9.

Sagan, I. 1995. The policy of sustainable development in post-socialist cities, in Sagan, I. and Smith, D.M. (eds), *Society, Economy, Environment – Towards the Sustainable City*. Gdańsk, Poznań: Bogucki Wydawnictwo Naukowe, 45–58.

Sagan, I. 2000a. *Miasto. Scena konfliktów i współpracy. Rozwój miast w świetle koncepcji reżimu miejskiego.* Gdańsk: Wydawnictwo Uniwersytetu Gdańskiego.

Sagan, I. 2000b. Społeczny i rynkowy wymiar miejsca w mieście socjalistycznym i postsocjalistycznym, in Jażdżewska, J. (ed.), *Miasto socjalistyczne. Organizacja przestrzeni miejskiej i jej przemiany.* Łódź: UL, Komisja geografii osadnictwa i ludności PTU, Polskie Towarzystwo Naukowe (XIII Konwersatorium Wiedzy o Mieście), 67–72.

Sagan, I. 2000c. Place as a social and market good in socialist and post-socialist city, in Department of Urban and Tourism Geography of Lodz University (ed.), *Post-socialist city. The XIII Seminar of the Knowledge of the City.* Lodz: Lodz University 67–71.

Sailer, U. 2001. Residential mobility during transformation: Hungarian cities in the 1990s, in Meusburger, P. and Jöns, H. (eds), *Transformations in Hungary: Essays in Economy and Society.* Heidelberg: Physica, 329–54.

Sailer-Fliege, U. 1999. Characteristics of post-socialist urban transformation in East Central Europe. *GeoJournal*, 49, 7–16.

Sakwa, R. 1999. *Postcommunism.* Buckingham, Philadelphia: Open University Press.

Sampson, R.J. 1988. Local friendship ties and community attachment in mass society: a multilevel systemic model. *American Sociological Review*, 53(5), 766–79.

Schäfers, B. (ed.) 2001. *Grundbegriffe der Soziologie.* Opladen: Leske + Budrich.

Schögel, K. 2006. The Comeback of the European Cities. *International Review of Sociology*, 16(2), 471-85.

Schneider, N.F., Rosenkranz, D. and Limmer, R. 1998. *Nichtkonventionelle Lebensformen. Entstehung, Entwicklung, Konsequenzen.* Opladen: Leske + Budrich.

Schneiderová, A., Solanský, O. and Pobořil, M. 2002. *Sociálně demografická analýza města Ostravy v návaznosti na rozvoj sociálních služeb. II. etapa. Díl II: Sociodemografická struktura Ostravy – současný stav a očekávaný vývoj.* Ostrava: Ostravská univerzita.

Schnur, O. 2008. *Neighbourhood Trek. Vom Chicago Loop nach Bochum-Hamme – Quartiersforschungskonzepte im Überblick.* Berlin: Geographisches Institut, Humboldt-Universität (Arbeitsberichte, 145).

Schütz, A. 1962. *Collected Papers I: The Problem of Social Reality.* 3rd Edition. The Hague: Nijhoff.

Schwirian, K.P. 1983. Models of neighborhood change. *Annual Review of Sociology*, 9, 83–102.

Segert, D. (ed.) 2007. *Postsozialismus. Hinterlassenschaften des Staatssozialismus und neue Kapitalismen in Europa.* Wien: Braumüller.

Seo, J.K. 2002. Re-urbanisation in regenerated areas of Manchester and Glasgow. New residents and the problems of sustainability. *Cities*, 19, 113–21.

Siebel, W. 2004. *Die europäische Stadt.* Frankfurt/M.: Suhrkamp.

Sidiropulu Janků, K. 2008. *Living in the Ostrava inner city. Socio-spatial consequences of demographic change in East Central European cities. The case of Ostrava, Czech Republic.* Ostrava (conDENSE report, unpublished).

Šiklová, J. 1999. Romové a nevládní, neziskové romské a proromské občanské organizace přispívající k integraci tohoto etnika, in *Romové v České republice (1945–1998)*, Praha: Socioklub, 271–89.

Silva, E.B. and Smart, C. 1999. The 'new' practices and policies of family life, in Silva, E.B. and Smart, C. (eds), *The New Family?* London et al.: Sage, 1–12.

Šimíková, I. and Navrátil, P. 2002. *Hodnocení projektů zaměřených na snižování rizika sociálního vyloučení romské populace. Typologie projektů.* Brno: VÚSPV.

Simmel, G. and Hughes, E.C. 1949. The sociology of sociability. *The American Journal of Sociology*, 55(3), 254–61.

SINGLETOWN 2008. Foges, C. (ed.). *SINGLETOWN Paper.* Part of the Venice Architecture Biennale 2008 'Out There, Architecture Beyond Building', 14 September–23 November.

Škrabal, J. and Šimek, M. 2007. Jak je to s počtem obyvatel v obcích? *Veřejná správa, Příloha*, 18(10), I–IV.

Slany, K. (ed.). 2005. *International migration: A multidimensional analysis.* Cracow: AGH University of Science and Technology.

Slater, T. 2008. 'A literal necessity to be re-placed': a rejoinder to the gentrification debate. *International Journal of Urban and Regional Research*, 32(1), 212–23.

Slater, T., Curran, W. and Lees, L. 2004. Gentrification research: new directions and critical scholarship. *Environment and Planning A*, 36, 1141–50.

Śleszyński, P. 2004a. Regionalne różnice pomiędzy liczbą ludności według narodowego spisu powszechnego w 2002 roku i szacowaną na podstawie ewidencji bieżącej. *Studia Demograficzne*, 145, 93–102.

Śleszyński, P. 2004b. Różnice liczby ludności wykazane w NSP 2002, Suplement. *Studia Demograficzne*, 146, 104–9.

Śleszyński, P., Bański, J., Degórski, M., Komornicki, T. and Więckowski M. 2007. *Stan zaawansowania planowania przestrzennego w gminach.* Warszawa:IGiPZ PAN (Prace Geograficzne, 211).

Šmídová, O. 1996. Vlastnictví a kvazi-vlastnictví bytů za socialismu a jejich postsocialistická mutace, in Olivier, A. and Tuček, M. (eds), *Původní a noví vlastníci. Strategie nabývání majetku ve střední a východní Evropě.* Praha: CEFRES (Cahiers du CEFRES, 11), 115–36.

Šmídová, O. 1999. Co vyprávějí naše byty, in Konopásek, Z. (ed.), *Otevřená minulost. Autobiografická sociologie státního socialismu*, Praha: Karolinum, 171–203.

Smigiel, C. (ed.) 2009. *Gated and Guarded Housing in Eastern Europe.* Leipzig: Leibniz-Institut für Länderkunde (forum ifl, 11/2009).

Smith, D.M. (ed.) 1995. *Lodz. Geographical Studies of a Polish city.* London: Department of Geography. Queen Mary and Westfield College (Research Papers in Geography, 8).

Smith, D.M. 1996. The socialist city, in Andrusz, G., Harloe, M. and Szelenyi, I. (eds). *Cities after Socialism. Urban and Regional Change and Conflict in Post-Socialist Societies.* Oxford: Blackwell, 70–99.

Smith, D.P. and Butler, T. 2007. Conceptualising the sociospatial diversity of gentrification: 'to boldly go' into contemporary gentrified spaces, the 'final frontier'? *Environment and Planning A*, 39, 2–9.

Smith, D.P. and Holt, L. 2007. Studentification and 'apprentice' gentrifiers within Britain's provincial towns and cities: extending the meaning of gentrification. *Environment and Planning A*, 39, 142–61.

Šnejdová, I. 2006. Změny ve vzdělanostní struktuře obyvatelstva Pražského městského regionu, in Ouředníček, M. (ed.), *Sociální geografie Pražského městského regionu.* Praha: Univerzita Karlova, Katedra Sociální geografie a regionálního rozvoje, 115–27.

Sobotka, T. 2008. The rising importance of migrants for childbearing in Europe. Overview Chapter 7, in Frejka, T., Sobotka, T., Hoem, J.M. and Toulemon, L. (eds), Childbearing Trends and Policies in Europe. *Demographic Research* [Online], Special Collection 7, vol. 19, 225–48. Available at: http://www. demographic-research.org [accessed: 12 October 2009].

Sobotka, T., Zeman, K. and Kantorová, V. 2003. Demographic shifts in the Czech Republic after 1989: a second demographic transition view. *European Journal of Population*, 19, 249–77.

Souralová, A. 2008. Lokalita Brno, in Kašparová, I., Ripka, Š. and Sidiropulu Janků, K. (eds), *Dlouhodobý monitoring situace romských komunit v České republice – moravské lokality.* Praha: Úřad vlády České republiky, Brno: Fakulta sociální studií, MU v Brně.

Sowa, K. 1988. Kraków – Bestand und Veränderung. Einige Bemerkungen über die Umwandlungen der Stadt, in Strubelt, W. and Frackiewicz, L. (eds), *Soziale Probleme von Industriestädten.* Bonn: Bundesforschungsanstalt für Landeskunde und Raumordnung, 63–79.

Speare, A.J. 1970. Home ownership, life cycle stage and residential mobility. *Demography*, 7(4), 449–58.

Spiegel, E. 1986. *Neue Haushaltstypen. Entstehungsbedingungen, Lebenssituation, Wohn- und Standortverhältnisse.* Frankfurt/M., New York: Campus (Campus-Forschung, 503).

Standl, H. and Krupickaite, D. 2004. Gentrification in Vilnius (Lithuania) – the example of Užupis, *Europa Regional*, 12(1), 42–51.

Stanilov, K. 2007a. Housing trends in Central and Eastern European cities during and after the period of transition, in Stanilov, K. (ed.), *The Post-Socialist City: Urban Form and Space Transformations in Central and Eastern Europe after Socialism.* Dordrecht: Springer (GeoJournal Library, 92), 173–90.

Stanilov, K. 2007b. Political reform, economic development, and regional growth in post-socialist Europe, in Stanilov, K. (ed.), *The Post-Socialist City. Urban Form and Space Transformations in Central and Eastern Europe after Socialism.* Dordrecht: Springer (GeoJournal Library; 92), 21–34.

Stanilov, K. 2007c. Taking stock of post-socialist urban development: a recapitulation, in Stanilov, K. 2007 (ed.), *The Post-Socialist City: Urban Form and Space Transformations in Central and Eastern Europe after Socialism.* Dordrecht: Springer (GeoJournal Library, 92), 3–17.

Staniszkis, J. 2001. *Postkomunizm. Próba opisu.* Słowo/obraz terytoria, Gdańsk.

Stark, D. and Bruszt, L. 1998. *Postsocialist Pathways. Transforming Politics and Property in East Central Europe.* Cambridge: Cambridge University Press.

Steinführer, A. 2001. Wandel und Persistenz innerstädtischer Segregationsmuster in Ostmitteleuropa. *Europa Regional*, 9(4), 212–22.

Steinführer, A. 2003. Sociálně prostorové struktury mezi setrvalostí a změnou. Historický a současný pohled na Brno. *Sociologický časopis*, 39(2), 169–92.

Steinführer, A. 2004a. Privatisierung auf Umwegen. Wohnungspolitik, Eigentumsstrukturen und Segmentation der Wohnungsmärkte in der Tschechischen Republik. *Geographische Zeitschrift*, 92 (1/2), 76–92.

Steinführer, A. 2004b. *Wohnstandortentscheidungen und städtische Transformation. Vergleichende Fallstudien in Ostdeutschland und Tschechien.* Wiesbaden: VS Verlag für Sozialwissenschaften (Stadtforschung aktuell, 99).

Steinführer, A. 2005. Comparative case studies in cross-national housing research, in Vestbro, D.U., Hürol, Y. and Wilkinson, N. (eds), *Methodologies in Housing Research.* Gateshead: Urban International Press, 91–107.

Steinführer, A. 2006. The urban transition of inner-city areas reconsidered (a German-Czech comparison). *Moravian Geographical Reports*, 14(1), 3–16.

Steinführer, A. 2011. Konstruktionen des 'demographischen Wandels' in der Tschechischen Republik 1990–2008. Oder: Von der Unmöglichkeit eines neutralen Konzepts, in Overath, P. (ed.), *Die vergangene Zukunft Europas. Bevölkerungsforschung und -prognosen im 20. und 21. Jahrhundert.* Köln, Wien, Weimar: Böhlau (forthcoming).

Steinführer, A. and Haase, A. 2007. Demographic change as a future challenge for cities in East Central Europe. *Geografiska Annaler B*, 89(2), 183–95.

Steinführer, A. and Haase, A. 2009a. Demografischer Wandel und städtische Schrumpfung in Ostmitteleuropa nach 1989, in Bohn, T. (ed.), *Von der 'europäischen Stadt' zur 'sozialistischen Stadt' und zurück? Urbane Transformationen im östlichen Europa des 20. Jahrhunderts.* München: Oldenbourg (Bad Wiesseer Tagungen des Collegium Carolinum, 29), 397–417.

Steinführer, A. and Haase, A. 2009b. Flexible–inflexible: socio-demographic, spatial and temporal dimensions of flat sharing in Leipzig (Germany). *GeoJournal*, 74(6), 567–87.

Steinführer, A., Pospíšilová, J. and Grohmannová, J. 2009. Ne/nápadné proměny vnitřního města v postsocialistickém období, in Ferenčuhová, S., Hledíková, M., Galčanová, L. and Vacková, B. (eds), *Město: proměnlivá ne/samozřejmost.* Brno, Praha: Masarykova univerzita, nakladatelství Pavel Mervart, 129–52.

Steinführer, A., Bierzyński, A., Grossmann, K., Haase, A., Klusáček, P. and Kabisch, S. 2010. Population decline in Polish and Czech cities during post-socialism? Looking behind the official statistics. *Urban Studies*, 47(11), 2325–46.

Stenning, A.C. 2004. Urban change and the localities, in Bradshaw, M. and Stenning, A.C. (eds), *East Central Europe and the Former Soviet Union: The Post-Socialist States*. Harlow, Essex: Pearson, 87–108.

Stenning, A.C. 2005a. Post-socialism and the changing geographies of the everyday in Poland. *Transactions of the Institute of British Geographers*, 30(1), 113–27.

Stenning, A.C. 2005b. Out there and in here: studying Eastern Europe in the west, *Area*, 37(4), 378–83.

Stenning, A.C. and Bradshaw, M.J. 2004. Conclusions: facing the future? in Bradshaw, M.J. and Stenning, A.C. (eds), *The Post Socialist Economies of East Central Europe and The Former Soviet Union.* Harlow: Pearson/DARG Regional Development Series, 247–256.

Stenning, A.C. and Hörschelmann, K. 2008. History, geography and difference in the post-socialist world: or, do we still need post-socialism? *Antipode*, 40(2), 312–35.

Storper, M. and Manville, M. 2006. Behaviour, preferences and cities: urban theory and urban resurgence. *Urban Studies*, 43, 1247–74.

Strassmann, W.P. 2001. Residential mobility: contrasting approaches in Europe and the United States. *Housing Studies*, 16, 7–20.

Strzelecki, Z. (ed.) 2003. *Problemy demograficzne Polski przed wejściem do Unii Europejskiej*. Warszawa: Polskie Wydawnictwo Ekonomiczne.

Sunega, P., Čermák, D. and Vajdová, Z. 2002. *Dráhy bydlení v ČR 1960–2001. Minulá, současná a budoucí stěhování občanů ČR ve výzkumu postoje k bydlení.* Praha: Sociologický ústav AV ČR (Pracovní texty/Working papers, 02/5).

Surkyn, J. and Lesthaeghe, R. 2004. Wertorientierungen und die ‚second demographic transition' in Nord-, West- und Südeuropa: Eine aktuelle Bestandsaufnahme. *Zeitschrift für Bevölkerungswissenschaft*, 29, 63–98.

Sýkora, L. 1993. City in transition: the role of rent gaps in Prague's revitalization. *Tijdschrift voor Economische en Sociale Geografie*, 84, 281–93.

Sýkora, L. 1994. Local urban restructuring as a mirror of globalisation processes: Prague in the 1990s. *Urban Studies*, 31, 1149–66.

Sýkora, L. 1996a. Economic and social restructuring and gentrification in Prague. *Acta Facultatis Rerum Naturalium Universitatis Comenianae – Geographica*, 37, 71–81.

Sýkora, L. 1996b. The Czech Republic, in Balchin, P. (ed.), *Housing Policy in Europe.* London, New York: Routledge, 272–88.

Sýkora, L. 1999. Processes of socio-spatial differentiation in post-communist Prague. *Housing Studies*, 14, 679–701.

Sýkora, L. 2000a. *Proměny vnitřní prostorové struktury postkomunistické Prahy.* Online version of the introduction of the professorial dissertation. Praha: Přírodovědecká fakulta Univerzity Karlovy. Available at: www.natur.cuni. cz/~sykora/text/hp.htm [accessed: 30 April 2001].

Sýkora, L. 2000b. Post-communist city, in Jażdżewska, I. (ed.), *Miasto socjalistyczne. Organizacja przestrzeni miejskiej i jej przemiany*. Łódź: UL, Komisja geografii osadnictwa i ludności PTU, Polskie Towarzystwo Naukowe (XIII Konwersatorium Wiedzy o Mieście), 41–45.

Sýkora, L. 2003. Between the state and the market: local government and housing in the Czech Republic, in Lux, M. (ed.), *Housing Policy: An End or a New Beginning?* Budapest: Local Government and Public Service Reform Initiative, Open Society Institute, 47–116.

Sýkora, L. 2005. Gentrification in post-communist cities, in Atkinson, R. and Bridge, G. (eds), *Gentrification in a Global Perspective. The New Urban Colonialism*. London: Routledge, 91–105.

Sýkora, L. 2007. Office development and post-communist city formation, in Stanilov, K. (ed.), *The Post-Socialist City: Urban Form and Space Transformations in Central and Eastern Europe after Socialism*. Dordrecht: Springer (GeoJournal Library, 92), 117–45.

Sýkora, L. 2008. Revolutionary change, evolutionary adaptation and new path dependencies: socialism, capitalism and transformations in urban spatial organizations, in Strubelt, W. and Gorzelak, G. (eds), *City and Region. Papers in Honour of Jiří Musil*. Opladen, Farmington Hills Mich.: Budrich UniPress, 283–95.

Sýkora, L. 2009a. New socio-spatial formations: Places of residential segregation and separation in Czechia. *Tijdschrift voor Economische en Sociale Geografie*, 100(4), 417–35.

Sýkora, L. 2009b. *Nové sociálně-prostorové formace*. Presentation at the conference 'Československé město včera a dnes: každodennost – reprezentace – výzkum', Brno, 13–14 November 2009.

Sýkora, L., Kamenický, J. and Hauptmann, P. 2000. Changes in the spatial structure of Prague and Brno in the 1990s. *Acta Universitatis Carolinae – Geographica*, 35, 61–76.

Sýkora, L. and Bouzarovski, S. 2011. Multiple transformations: conceptualising post-communist urban transition. *Urban Studies* (forthcoming).

Szafrańska, E. 2009a. *Socio-spatial structure of large housing estates in post-socialist Lodz*. Paper presented at the International Symposium 'Revitalising Built Environments: Requalifying Old Places for New Uses', Istanbul, 12–16 October 2009.

Szafrańska, E. 2009b. The attractiveness of residential blocks in the opinion of residents on the example of a residential Shark North in Lodz, in Jazdzewski, Y. (ed.), *Large and Medium-Sized Polish Cities in Transition*. Lodz, Lodz University.

Szczepański, M. and Ślęzak-Tazbir, W. 2007. Between fear and admiration. Social and spatial ghettoes in an old industrial region. *Polish Sociological Review*, 159(3), 299–320.

Szczepański, M. and Ślęzak-Tazbir, W. 2008. Between fear and admiration. Social and spatial ghettos in an old industrial region, in Strubelt, W. and Gorzelak, G.

(eds), *City and region. Papers In honour of Jiří Musil.* Opladen, Farmington Hills Mich.: Budrach UniPress, 297–327.

Szelenyi, I. 1983. *Urban Inequalities under State Socialism.* Oxford: Oxford University Press.

Szelenyi, I. 1996. Cities under socialism – and after, in Andrusz, G., Harloe, M. and Szelenyi, I. (eds), *Cities after Socialism. Urban and Regional Change and Conflict in Post-Socialist Societies.* Oxford, Cambridge MA: Blackwell, 286–317.

Taylor, P.J. and Walker, D.R.F. 2001. World cities. A first multivariante analysis of their service complexes. *Urban Studies*, 38(1), 23–47.

Telgarsky, J.P. and Struyk, R.J. 1991. *Toward a Market-Oriented Housing Sector in Eastern Europe.* Washington, D.C.: The Urban Institute Press.

Temelová, J. 2007. Flagship developments and the physical upgrading of the post-socialist inner city: the golden angel project in Prague. *Geografiska Annaler B*, 89, 169–81.

Tews, H.P. 1993. Neue und alte Aspekte des Strukturwandels des Alters, in Naegele G. and Tews, H.P. (eds), *Lebenslagen im Strukturwandel des Alters. Alternde Gesellschaft – Folgen für die Politik.* Opladen: Westdeutscher Verlag, 15–42.

Timmermans, H., Molin, E. and van Noortwijk, L. 1994. Housing choice processes: stated versus revealed modeling approaches. *Netherlands Journal of Housing and the Built Environment*, 9, 215–27.

Tosics, I. 2005. City development in Central and Eastern Europe since 1990: the impact of internal forces, in Hamilton F.E.I., Dimitrovska Andrews, K. and Pichler-Milanović, N. (eds), *Transformation of Cities in Central and Eastern Europe: towards Globalization.* Tokyo, New York: The United Nations University Press, 44–78.

Tsenkova, S. 2003. Housing policy matters: the reform path in Central and Eastern Europe, in Lowe, S. and Tsenkova, S. (eds), *Housing Change in East and Central Europe. Integration or Fragmentation?* Aldershot, Burlington: Ashgate, 193–204.

Tsenkova, S. 2006. Beyond transitions. Understanding urban change in post-socialist cities, in Tsenkova, S. and Nedovic-Budic, Z. (eds), *The Urban Mosaic of Post-Socialist Europe.* Heidelberg: Physica, 21–50.

Tsenkova, S. 2009. *Housing Policy Reforms in Post Socialist Europe. Lost in Transition.* Heidelberg: Springer.

Tsenkova, S. and Nedovic-Budic, Z. (eds) 2006. *The Urban Mosaic of Post-Socialist Europe.* Heidelberg: Physica.

Tuček, M., Červenka, J., Duffková, J., Kostelecký, T., Kuchař, P., Kuchařová, V., Machonin, P. Müller, K. and Šafr, J. 2003. *Dynamika české společnosti a osudy lidí na přelomu tisíciletí.* Praha: SLON.

Turok, I. and Mykhnenko, V. 2006. *Resurgent European Cities?* Glasgow: CPPR.

Turok, I. and Mykhnenko, V. 2007. The trajectories of European cities, 1960–2005. *Cities*, 24(3), 165–82.

Turowski, J. 1979. *Środowisko mieszkalne w świadomości ludności miejskiej.* Wrocław: Ossolineum.

Uchman, R. and Adamski, J. 2003. How to meet the market rules and social goals for housing? Local government and housing in Poland, in Lux, M. (ed.), *Housing Policy: An End or a New Beginning?* Budapest: Local Government and Public Service Reform Initiative, Open Society Institute, 117–81.

Uherek, Z. 2003. Cizinecké komunity a městský prostor v České republice. *Sociologický časopis*, 39, 193–216.

Uherek, Z. 2009. *Foreigners in Brno and Ostrava*, unpublished manuscript. Prague.

Uhlenberg, P. 2005. Demography of aging, in Poston, D.L. and Micklin, M. (eds), *The Handbook of Population.* New York et al.: Kluwer Plenum, 143–67.

UML [Urząd miasta Łodzi] 2008. *Łódź w liczbach 2007.* Łódź.

UN [Population Division, Department of Economic and Social Affairs United Nations Secretariat] 2001. *Replacement Migration: Is It a Solution to Declining and Ageing Populations?* New York: United Nations Publications.

UN ECE [United Nations Economic Commission for Europe] 2006. *Conference of European Statisticians. Recommendations for the 2010 Censuses of Population and Housing.* Geneva, New York.

USG [Urząd Statystyczny w Gdańsku] 1999. *Informator o Sytuacji Społeczno-Gospodarczej Gdańska za 1998 rok.* Gdańsk.

USG [Urząd Statystyczny w Gdańsku] 2009. *Województwo pomorskie. Podregiony, powiaty, gminy, 2009.* Gdańsk.

USG [Urząd Statystyczny w Gdańsku] and UMB [Urząd Miejski w Gdańsku] 2009. *Informator o Sytuacji Społeczno-Gospodarczej Gdańska za 2008 rok.* Gdańsk.

Vaishar, A. and Zapletalová, J. 2003. Problems of European inner cities and their residential environments. *Moravian Geographical Reports*, 11(2), 24–35.

Vaishar, A., Klusáček, P., Krejčí, T., Martinát, S., Pavelčíková, N., Pospíšilová, J. and Zapletalová, J. 2009. *Současný vývoj vnitřních částí Brna a Ostravy.* Brno, Ostrava: Ústav geoniky AV ČR, v.v.i. (Studia Geographica, 100).

van Criekingen, M. 2010. 'Gentrifying the urbanisation debate', not vice versa: the uneven socio-spatial implications of changing transitions to adulthood in Brussels. *Population, Space and Place*, 16(5), 381–94.

van de Kaa, D.J. 1987. Europe's second demographic transition. *Population Bulletin*, 42, 1–57.

van de Kaa, D.J. 1994. The second demographic transition revisited: theories and expectations, in Beets, G.C.N., van den Brinkel, J.C., Cliquet, R.L., Dooghe, G. and de Jong, J. (eds), *Population and Family in the Low Countries 1993: Late Fertility and Other Current Issues.* Amsterdam: Swets and Zeitlinger, 81–126.

van de Kaa, D.J. 2004. Is the second demographic transition a useful research concept. Questions and answers. *Vienna Yearbook of Population Research 2004.* Vienna: Vienna Institute of Demography, Austrian Academy of Sciences, 4–10.

van den Berg, L., Drewett, R., Klaassen, L.H., Rossi, A. and Vijverberg, C.H.T. 1982. *Urban Europe: A Study of Growth and Decline*, vol. 1. Oxford et al.: Pergamon Press.

van Engelsdorp Gastelaars, R. and Vijgen, J. 1990. Residential differentiation in the Netherlands: the rise of new urban households, in Deben, L., Heinemeijer, W. and van der Vaart, D. (eds), *Residential Differentiation*. Amsterdam: Centruum voor Grootstedelijk Onderzoek, Universiteit van Amsterdam, 136–63.

van Gils, W. and Kraaykamp, G. 2008. The emergence of dual-earner couples. A longitudinal study of the Netherlands. *International Sociology*, 23, 345–66.

van Imhoff, E., Kuijsten, A., Hooimeijer, P. and van Wissen, L. (eds) 1995. *Household Demography and Household Modelling*. New York, London: Plenum Press (The Plenum Series on Demographic Methods and Population Analysis).

van Kempen, R., Dekker, K., Hall, S. and Tosic, I. (eds) 2005a. *Restructuring Large Housing Estates in Europe*. Bristol: The Policy Press.

van Kempen, R., Vermeulen, M. and Baan, A. 2005b. *Urban Issues and Urban Policies in the New EU Countries*. Aldershot: Ashgate.

Vašečka, I. 2002. Utváranie sa problémových rómských zoskupení v mestách ČR, in Sírovátka, T. (ed.), *Menšiny a marginalizované skupiny v České republice*. Brno: Fakulta sociálních studií MU – nakladatelství Georgetown, 245–62.

VCRR MU [Výzkumné centrum regionálního rozvoje Masarykovy univerzity] 2005. *Vývojové tendence ve městě Brně: obyvatelstvo, trh práce, průmysl*. Brno.

Verdery, K. 2002. Whither postsocialism? in Hann, C. (ed.), *Postsocialism: Ideas, Ideologies and Practices in Eurasia*. London, New York: Routledge, 15–21.

Víšek, P. 1999. Problem integrace – řešení problematiky romských obyvatel v období 1970 až 1989, in *Romové v České republice (1945-1998)*. Praha: Socioklub, 184–218.

Vojtěchovská, P. 2000. Population development in Poland, in Kučera, T., Kučerová, O.V., Opara, O.B., Schaich, E. (eds), *New Demographic Faces of Europe. The Changing Population Dynamics in Countries of Central and Eastern Europe*, Berlin: Springer, 247–66.

Völker, B. and Flap, H. 1997. The comrades' belief: intended and unintended consequences of communism for neighbourhood relations in the former GDR. *European Sociological Review*, 13(3), 241–65.

Wagner, M. and Franzmann, G. 2000. Die Pluralisierung der Lebensformen. *Zeitschrift für Bevölkerungswissenschaft*, 25, 151–73.

Waite, L.J. 2005. Marriage and family, in Poston, D.L. and Micklin, M. (eds), *The Handbook of Population*. New York et al.: Kluwer Plenum, 87–107.

Walker, A.R. 1993. Lodz: the problems associated with restructuring the urban economy of Poland's textile metropolis in the 1990s. *Urban Studies*, 30, 1065–80.

Wallis, A. 1977. *Miasto i Przestrzeń*. Warszawa: Państwowe Wydawnictwo Naukowe.

Wacquant, L. 2008. *Urban Outcast: A Comparative Sociology of Advanced Marginality*. Cambridge: Polity Press.

Warzywoda-Kruszyńska, W. and Grotowska-Leder, J. 1997. Wybrane aspekty koncentracji biedy w Łodzi. *Przegląd Socjologiczny*, 46, 125–46.

Węcławowicz G. 1988. *Struktury społeczno-przestrzenne w miastach Polski.* Wrocław: Instytut Geografii i Przestrzennego Zagospodarowania PAN, Ossolineum.

Węcławowicz, G. 1993. *Die sozialräumliche Struktur Warschaus – Ausgangslage und postkommunistische Umgestaltung.* Unter Mitarbeit von Josef Kohlbacher. Wien: Institut für Stadt- und Regionalforschung (ISR-Forschungsberichte 8).

Węcławowicz, G. 1996. *Contemporary Poland. Space and Society.* London: UCL Press (Changing Eastern Europe, 4).

Węcławowicz, G. 2000. Kształtowanie się nowego modelu zróżnicowań społeczno-przestrzennych miasta w Europie Środkowo – Wybrane elementy przejścia od miasta socjalistycznego do miasta postsocjalistycznego, in Jażdżewska, J. (ed.), *Miasto socjalistyczne. Organizacja przestrzeni miejskiej i jej przemiany.* Łódź: UL, Komisja geografii osadnictwa i ludności PTU, Polskie Towarzystwo Naukowe (XIII Konwersatorium Wiedzy o Mieście), 25–30.

Węcławowicz, G. 2001. Transformation of large cities and urban centres in Poland – selected issues, in Kitowski, J. (ed.), *Spatial Dimension of Socio-Economic Transformation Processes in Central and Eastern Europe on the Turn of the 20th Century.* Rzeszów: Institute of Geography and Spatial Organization of the Polish Academy of Sciences, Faculty of Economics Maria Curie-Skłodowska University, Commision of Geography of Communication Polish Geographical Society (Rozprawy i Monografie Wydziału Ekonomicznego Filii UMCS w Rzeszowie, 22), 337–64.

Węcławowicz, G. 2002a. From egalitarian cities in theory to non-egalitarian cities in practice: the changing social and spatial patterns in Polish cities, in Marcuse, P. and van Kempen, R. (eds), *Of States and Cities. The Partitioning of Urban Space.* Oxford: Oxford University Press, 183–99.

Węcławowicz, G. 2002b. *Przestrzeń i społeczeństwo współczesnej Polski.* Warszawa: Wydawnictwo Naukowe PWN.

Węcławowicz, G. 2003. *Geografia społeczna miast.* Warszawa: Wydawnictwo Naukowe PWN.

Węcławowicz, G. 2004. Where the grass is greener in Poland: regional and intra-urban inequalities, in Lee, R. and Smith, D.M. (ed.), *Geographies and Moralities. International Perspectives on Development, Justice and Place.* Malden M.A.: Blackwell, 62–77.

Węcławowicz, G. 2005. Poland, in van Kempen, R., Vermeulen, M. and Baan, A. (eds), *Urban issues and urban polices in the new EU countries.* Aldershot: Ashgate, 61–77.

Węcławowicz, G., Kozłowski, R. and Bajek, R. 2003. *Large Housing Estates in Poland. Overview of Developments and Problems in Warsaw.* Utrecht, Urban and Regional Research Centre, (RESTATE report 2f).

Weichhart, P. 2009. Multilokalität – Konzepte, Theoriebezüge und Forschungsfragen. *Informationen zur Raumentwicklung*, 1/2, 1–14.

Więcław, J. 1997. Zmiany funkcji dzielnicy Kazimierz w Krakowie w świetle współczesnych przekształceń społeczno-gospodarczych. *Folia Geografica*, XXIX-XXX, 125–47.

Wießner, R. 1997. Sozialräumliche Polarisierung der inneren Stadt in Budapest, in Kovacs, Z. and Wießner, R. (eds), *Prozesse und Perspektiven der Stadtentwicklung in Ostmitteleuropa*. Passau: L.I.S. Verlag, 189–202.

Williams, K., Burton, E. and Jenks, M. (eds) 2000. *Achieving Sustainable Urban Form*. London, New York: Spon Press.

Wolaniuk, A. 1997. Spatial and functional changes in the city centre of Łódź, in Liszewski, S. and Young, C. (eds), *A Comparative Study of Łódź and Manchester. Geographies of European Cities in Transition*. Łódź: Łódź University Press, 137–58.

Zborowski, A. (ed.) 2009. *Demograficzne i społeczne uwarunkowania rewitalizacji miast w Polsce*. Kraków: Instytut Rozwoju Miast.

Internet sources

ČSÚ [Český statistický úřad]. *Obyvatelstvo – roční časové řady*. [Online]. Available at: http://www.czso.cz/csu/redakce.nsf/i/obyvatelstvo_hu [accessed: 12 October 2009].

ČSÚ [Český statistický úřad]. *Veřejná databáze*. [Online]. Available at: http://vdb. czso.cz/vdb/ [accessed: 12 October 2009].

EC [European Commission] 2006. *The Demographic Future of Europe – from Challenge to Opportunity*. Luxembourg: Office for Official Publications of the European Communities. [Online]. Available at: http://ec.europa. eu/employment_social/publications/2007/ke7606057_en.pdf [accessed: 26 August 2008].

GUS [Główny Urząd Statystyczny]. *Regional Databank*. [Online]. Available at: http://www.stat.gov.pl/bdren_n/app/strona.indeks [accessed: 12 October 2009].

GUS [Główny Urząd Statystyczny] 2003: *Gospodarstwa domowe i rodziny*. [Online]. Available at: http://www.stat.gov.pl/gus/5840_757_PLK_HTML. htm [accessed: 23 March 2010].

GUS [Główny Urząd Statystyczny] 2004b. *Prognoza ludności na lata 2003–2030. Warszawa*. [Online]. Available at: http://www.stat.gov.pl/gus/5840_648_PLK_ HTML.htm [accessed: 22 March 2010].

ISEO 2008 [Online]. Available at: http://www.mvcr.cz/clanek/statistiky-pocty-obyvatel-v-obcich.aspx?q=Y2hudW09MQ%3d%3d [accessed: 15 July 2010].

Platform for student housing of the Masaryk University, Brno. [Online]. Available at: http://www.bydleni.muni.cz/ [accessed: 21 July 2010].

Stancja. Ogłoszenia wynajmu. Gdańsk. [Online]. Available at: http://www. mieszkania-studenckie.pl/gdansk/wszystkie30.php [accessed: 20 April 2009].

Urban Audit 2001 [Online]. Available at: http://www.urbanaudit.org/ [accessed: 12 October 2009].

Zákon o hospodaření s byty 1964. in *Sbírka zákonů Československé socialistické republiky*. Praha: Ministerstvo spravedlnosti. [Online]. Available at: http:// aplikace.mvcr.cz/archiv2008/sbirka/1964/sb20-64.pdf [accessed: 21 July 2010].

Index